D0843172

COMPREHENSIVE BIOCHEMISTRY

ELSEVIER SCIENCE PUBLISHERS

1 Molenwerf, P.O. Box 211, Amsterdam

ELSEVIER SCIENCE PUBLISHING Co. INC.

52, Vanderbilt Avenue, New York, N.Y. 10017

With 10 plates, 10 figures and 13 tables

Library of Congress Cataloging in Publication Data
Main entry under title:

Molecular correlates of biological concepts

(Comprehensive biochemistry; v. 34 A. Section VI,
A history of biochemistry)
Includes biographies and index.
1. Biological chemistry – History. 2. Molecular
biology – History. I. Laszlo, Pierre. II. Series:
Comprehensive biochemistry; 34A.
QD415.F54 vol. 34A 574.19'2 s [574.19'2'09] 86-8869
[QP511]
ISBN 0-444-80776-4 (U.S.)

PRINTED IN THE NETHERLANDS

COMPREHENSIVE BIOCHEMISTRY

COMPREHENSIVE BIOCHEMISTRY

COMPREHENSIVE BIOCHEMISTRY

ALBERT NEUBERGER

*Chairman of Governing Body, The Lister Institute
of Preventive Medicine, University of London,
London (Great Britain)*

LAURENS L.M. VAN DEENEN

*Professor of Biochemistry, Biochemical Laboratory,
Utrecht (The Netherlands)*

VOLUME 34A

MOLECULAR CORRELATES OF
BIOLOGICAL CONCEPTS

by

PIERRE LASZLO

*Institut de Chimie, Université de Liège,
Sart-Tilman par 4000 Liège (Belgium)*

ELSEVIER SCIENCE PUBLISHERS

AMSTERDAM · OXFORD · NEW YORK

1986

GENERAL PREFACE

The Editors are keenly aware that the literature of Biochemistry is already very large, in fact so widespread that it is increasingly difficult to assemble the most pertinent material in a given area. Beyond the ordinary textbook the subject matter of the rapidly expanding knowledge of biochemistry is spread among innumerable journals, monographs, and series of reviews. The Editors believe that there is a real place for an advanced treatise in biochemistry which assembles the principal areas of the subject in a single set of books.

It would be ideal if an individual or a small group of biochemists could produce such an advanced treatise, and within the time to keep reasonably abreast of rapid advances, but this is at least difficult if not impossible. Instead, the Editors with the advice of the Advisory Board, have assembled what they consider the best possible sequence of chapters written by competent authors; they must take the responsibility for inevitable gaps of subject matter and duplication which may result from this procedure.

Most evident to the modern biochemist, apart from the body of knowledge of the chemistry and metabolism of biological substances, is the extent to which we must draw from recent concepts of physical and organic chemistry, and in turn project into the vast field of biology. Thus in the organization of Comprehensive Biochemistry, sections II, III and IV, Chemistry of Biological Compounds, Biochemical Reaction Mechanisms, and Metabolism may be considered classical biochemistry, while the first and fifth sections provide selected material on the origins and projections of the subject.

It is hoped that sub-division of the sections into bound volumes will not only be convenient, but will find favour among students concerned with specialized areas, and will permit easier future revisions of the individual volumes. Towards the latter end particularly, the Editors will welcome all comments in their effort to produce a useful and efficient source of biochemical knowledge.

M. Florkin†

Liège/Rochester

E.H. Stotz

PREFACE TO VOLUME 34A

The title of this volume 34A, *Molecular Correlates of Biological Concepts*, expands upon the name 'Biochemistry'. It repays in explicitness what it loses in conciseness. Marcel Florkin coined it for this particular volume. Unfortunately, except for a few scattered jottings, this is all he left for Part VI of the *History of Biochemistry* he did not live to write. To replace Florkin was impossible; I have only accepted to be a substitute for him. Hence, the numerous differences in tone and in outlook from the previous volumes in this series.*

In spite of these slight differences I nevertheless hope to have conveyed – in writing this volume – something of the intense enjoyment of all things intellectual which Marcel Florkin relished. The chemist I am could not hope to match his knowledge of biochemistry and of biochemists, which he acquired during a particularly productive period of this discipline. I have only tried to fulfill his intention to chronicle the emergence of molecular concepts in biochemical science.

This historical transition phase is not over, which compounds the difficulty to write about it. It makes the writing a little easier as well: we do not suffer, yet, from a plethora of topics to choose from. A quick, back-of-the-envelope list of biological functions which have found their molecular correlates is shown in Table I.

Two aspects are noteworthy in this Table, at least to this writer. First, by and large, the normal is much better understood (in molecular terms) than the pathological. This will run as one of the minor themes in the book. Second, the successes so far of the *biochemical* approach proper constitute a rather heterogeneous mix. To report all these histories would demand enormous space, at the expense perhaps of a clear notion of *how* and *why* molecular concepts were fruitful for physiology, and, more generally, for biology.

---

*As regards typography this volume deviates from the hitherto used style in two points: (a) while the chapters in Volumes 30–33 are numbered consecutively, the present volume starts anew with Chapter 1. (b) The references follow the Harvard System, instead of being numbered. [The Publishers.]

TABLE I

The present state of biochemistry (1985): a highly personal view from Sirius

Function	Molecular correlate
Pathological	
Nerve gas action	Acetylcholinesterase
Diabetes	Insulin
Sickle-cell anemia, thalassemia, etc.	Abnormal hemoglobins
Depression	?
Schizophrenia	?
Myasthenias, muscular dystrophies	?
Cancers	?
Hypertension	?
Parkinson's Disease	?
Alzheimer's Disease	?
(...)	
Normal	
Respiration	Hemoglobin
Digestion, metabolism	Enzymes
Genetics	Nucleic acids
Hydrocarbon solubilization, drug detoxification	Cytochrome P-450
Muscular action	Actin, myosin
Nervous impulse	Acetylcholine
Vison	Retinal, opsin
Smell	?
Memory	?
Analgesia	Brain peptides
Ion transport	Ionophores, ATPases
Secondary sexual traits	Steroidal hormones
Social communication	Pheromones
(...)	

A choice had to be made, and I made it jointly with Professor
Ernest Schoffeniels: we latched upon three topics as representative
of the protracted, drawn-out emergence of biochemistry as a molecular science: respiratory pigments, enzymes, and nucleic acids. The
latter corresponds to what is now universally referred to as 'molecular biology': the double helix has captured to itself the name for the
whole shelf on which it stands!

x

The present volume is devoted mostly to the first two examples, hemoglobin and enzymes, about which far less has been written than about nucleic acids.

I wish to thank for their trust the general editors of this series, Professors Neuberger and Van Deenen; and also my colleague and friend, Professor Schoffeniels, for his esteem and for his daring in suggesting that I might be up to the task. My gratitude goes especially to some of the main characters in the story, who have been kind enough to entrust me with their recollections and to comment upon first drafts of the various sections: Professors Dorothy Crowfoot-Hodgkin, John T. Edsall, Aaron Klug, Robert B. Merrifield, Albert Neuberger, Linus Pauling, Max F. Perutz and Frederick Sanger.

Madame Nicole Dumont typed this whole manuscript, in its successive forms, with unrelenting patience and with the exacting precision demanded by such a text.

Institut de Chimie   Pierre Laszlo
Université de Liège
Liège, 1985

Section VI

A HISTORY OF BIOCHEMISTRY

xii

LIST OF PLATES

(Photographs reproduced with permission of authors, publishers, and/or owners)

CONTENTS

VOLUME 34A

A HISTORY OF BIOCHEMISTRY

Molecular correlates of biological concepts

Part I. The Long, Long Way to Molecular Order

Chapter 1. Molecules and macromolecules

Chapter 2. From culture broth to primordial soup

Chapter 3. Morphology of molecules

Chapter 4. The analogy between biomolecular structure and architecture

Chapter 5. The animate and the inanimate. Crystals and life

Chapter 6. Complementarities between molecules

Chapter 7. Philosophical undercurrents of the resistance to the notion of biomolecules as ordered polymers

Chapter 8. *Schizoid tale of two schools*

Part II. Proteins as Molecules

Chapter 9. *The name 'protein'*

Chapter 10. *The conceptualization of albumins as a class*

Chapter 11. *Crystallization of proteins and the molecular weight of hemoglobin*

Chapter 12. *Polypeptidic nature of proteins*

Chapter 13. Dorothy Wrinch: the mystique of cyclol theory, or the story of a mistaken scientific theory

Chapter 14. A note on two wise men

Chapter 15. Molecular order in proteins: the contributions from Linus Pauling

Chapter 16. Sequence analysis

Chapter 17. The Merrifield automated supported peptide synthesis

Part III. The Hemoglobin Story

Chapter 18. Historical notes about hemoglobin and its various names

Chapter 19. Cooperativity of dioxygen binding: the Bohr effect

Chapter 20. Pauling and the magnetic properties of hemoglobins

Chapter 21. Hemoglobin and the beginnings of x-ray work on proteins

Chapter 22. Determination of phases in single crystals

Chapter 30. A disquisition on an unfavorable Zeitgeist

Chapter 31. Relation of microorganisms to fermentation and putrefaction

Chapter 32. Ferments as catalysts

Chapter 33. The Buchner experiment

Chapter 34. The aftermath of the Buchner experiment. Crystallization of enzymes

Chapter 35. Enzymes as receptors complementary to their substrate

Chapter 36. Enzymatic kinetics

Chapter 37. X-ray studies on enzymes

Part V. Hemoglobin and Enzymes

Chapter 38. Tentative sketch of a comparative history

PART I

The Long, Long Way to Molecular Order

(Chapters 1–8)

Chapter 1

Molecules and Macromolecules

The history of the awakening of biochemistry (of biochemists!) to the realization that it would be sufficient to posit entities embodying order and organization at the microscopic level is also that of the transfer of the corresponding concept, that of *molecule*, from chemistry. Hence this chapter will examine the emergence of the notions of molecule and of macromolecule within chemistry. It will emphasize those aspects which would turn out to be of crucial importance for establishment of a molecular biochemistry.

1. The organic molecules of Buffon

During the eighteenth century, the prevailing meaning of *molecule* is that of an extremely small – or even fictitious – material particle.

This is the meaning which Buffon attaches to this term in his descriptions (or rather *Gedankenexperiment*) of the origins of various minerals. He writes for instance of " . . . this stone, from which they have flushed out the major part of the liquid and solid water *molecules* it contained beforehand" (Buffon, 1835a); or elsewhere, he explains the genesis of crystals and gems:

"These rain waters, and even their humid vapors, by acting upon the surface or by penetrating the substance of glassy and calcareous materials, have detached from them stony particles with which they have become loaded and which have formed new stony bodies. These *molecules* detached by water have met, and their aggregation has produced . . . " (Buffon, 1835b).

When Buffon purports to explain 'petrifications and fossils', in his *Histoire Naturelle des Minéraux* (1783–88), he clearly uses *molecule* with the meaning of an 'extremely small material particle':

"When water loaded with these calcareous, vitrous, or metallic particles, has not reduced them into *molecules* tenuous enough to penetrate in the inside of organized bodies, they are only capable to attach themselves to their surface, and to coat them with an incrustation of variable thickness" (Buffon, 1835c).

The term *molecule* appears also frequently under Buffon's pen in conjunction with the adjective *organic*. For Buffon, such *organic molecules*, endowed with motion (Brownian motion, in fact; see Ch. 2), bridge the mineral world with the animal and vegetal kingdoms. And they are responsible, by their aggregation according to a morphological plan which is a characteristic of each species, for the constitution of each living being. For Buffon, the build-up of organisms from these *organic molecules* resembles the growth of a crystal from its constituting *molecules*. A few quotations of relevant fragments from *Histoire des Animaux* will give us the gist of his conjectures. Buffon starts with the notion that:

"in order to well understand the manner of this (animal) reproduction, it is sufficient to conceive that in the food which these organized beings draw, there are *organic molecules* of various species; (. . .) that each part of the organized body, each inner mold, admits only those *organic molecules* which fit it *(qui lui sont propres)*(. . .)" (Buffon, 1835d).

Further on, he reconstructs the assembly of these *organic molecules* into spermatic particles:

"A second kind of beings of the same species (as eggs) are the organized bodies which one finds in the seed of all animals and which, such as those in the milt of the calamar, are rather natural machines than animals. These beings are actually *(sont proprement)* the first assembly resulting from the *organic molecules* which I have so much written about; they are even probably the organic parts which constitute the organized bodies of animals" (Buffon, 1835e).

Elsewhere, he writes:

"One can look at these organic bodies which move, these spermatic animals, as the
first assembly of these *organic molecules* which come from all parts of the body: when
a sufficient quantity of them gathers, they form a body which moves (. . .)" (Buffon,
1835f).

The contemporary reactions to these *organic molecules* of Buffon
were of two types; denial, with Jean-Jacques Rousseau who wrote in
L'Emile (1762):

"My mind refuses any agreement with the idea of non organized matter moving by
itself or producing any action . . . I have made every effort to conceive a living molecule, without having been able to achieve (such a goal)" (Rousseau, 1762);

and as an argument for the autonomy of the science of organized
bodies from physics, by the chemist Venel, a follower of Rouelle, who
wrote the extremely interesting and well-argued entry on chemistry,
in 1753, for the *Encyclopédie*:

"That *organic molecules* and organized bodies are submitted to laws which differ
essentially (at least in the present state of knowledge) from those which rule the
movements of purely mobile and quiescent (*quiescible*), or inert, matter; this is an
assertion in support of which we have the discoveries of M. de Buffon and also the
demonstrated errors of physicians who have tried to explain animal economy with
the laws of mechanics. Therefore the phenomena of organization must be considered
the object of a science essentially distinct from all the other parts of Physics" (Venel,
1753).

Thus, the *organic molecules* of Buffon provide the historian of biochemistry with a grand opening, presenting various strands which
will repeatedly recur in this volume: the distinction between organic
and inorganic matter; the distinction between mechanism and organism; the tension between materialists and vitalists; and the accompanying notion that biochemistry, or physiological chemistry,
should not be reducible to the laws of Newtonian physics.

The 'organic molecules' of Buffon are devoid of a crucial characteristic, with respect to the modern chemical notion of molecule,
that of being an *assembly of atoms*. Hence, they should be viewed as
a pre-scientific rather than as a scientific concept. The contempo-

rary French historian of science Georges Canguilhem sees them as an example of a *scientific ideology*. He defines these as explanatory systems whose object (the reproduction of living beings, in this case) goes beyond its own norm of what constitutes a scientific concept, as it has been imported from another, already instituted and prestigious science, whose style the *scientific ideology* seeks to emulate (Canguilhem, 1981).

In the case at hand, the transfer is from Newtonian science: the *organic molecules* of Buffon associate with one another in a like manner as point masses drawn towards one another by gravitational forces. One of the key differences is that, whereas Newtonian mechanics is capable of quantitative prediction and tests, the Buffon system remains at a qualitative and descriptive level.

Also, a central tenet of the animate *organic molecules*, is that their self-assembly should lead to macroscopic spermatic particles endowed with motion. The key counter-experiment to the numerous experiments by Needham and by Buffon, purporting to show spontaneous generation of such germs from the association of 'formative particles', was performed by Spallanzani. He showed that boiling the vials, in which Needham found spontaneous generation to arise from aqueous suspensions of decomposing substances, for one hour was sufficient to prevent any such spontaneous generation.

However, even though *organic molecules* are a part of a scientific ideology according to the definition by Canguilhem, it is useful to the historian of science not to discard abruptly such blind alleys off the main avenues of scientific knowledge, as it were. Indeed, *organic molecules* touch upon both the history of chemistry and of biochemistry, in being conceptual objects cognate with the molecule and with the biomolecule, respectively.

And they are of even greater interest to the historian of ideas. Clearly, the *organic molecules* of Buffon belong to the same *episteme* (Foucault, 1972) as the integral molecules (*molécules intégrantes*) of Haüy, who published in 1784 a theory of crystals as constituted from repeating three-dimensional units (Haüy, 1784); and as the idea of coagulation, that of a swarm of bees for instance, in Diderot's *Le Rêve de d'Alembert* (1769). A final quotation from Buffon will show the cogency of this 'generalized atomism':

·A grain of sea salt is a cube composed of an infinity of other cubes which one can recognize distinctly with a microscope; these little cubes are themselves composed of other cubes which one can see with a better microscope, and it is difficult to doubt that the primitive and constitutive parts of this salt would be also cubes so tiny that they will always escape from our sight, and even from our imagination. Animals and plants which can multiply and reproduce themselves in all of their parts are organized bodies composed of other similar organic bodies, whose primitive and constitutive parts are also organic and similar (to one another), and of which our eye can only distinguish an accumulated amount, since we can only hint at the primitive parts by reasoning and by the analogy which we have just presented.

This leads us to believe that there exists in Nature an infinity of actual organic living parts, whose substance is the same as in organized beings, in the same way as there is an infinity of inorganic (*brutes*) particles similar to inorganic (*bruts*) bodies which we know, and in the same way as maybe millions of small salt cubes have to accumulate to make a single detectable grain of sea salt, millions of organic parts similar to the whole have to (congregate) in order to form a single germ contained in an individual elm or polyp (. . .)" (Buffon, 1835).

Let us now return to the *molécules intégrantes* of Haüy. René-Just Haüy (1743–1822) was a priest, who got interested in the study of minerals. After developing in 1784 the theory of crystal structure as the three-dimensional repeat of an elementary geometry, he went on to build wooden models of crystals, together with a technical assistant, a workman by the name of Claude Pleuvin.

Haüy also wrote several important textbooks to teach the science of mineralogy, which he thus greatly advanced: his *Traité de Minéralogie* was circulated in 1801 by the Conseil des Mines to each central school – we would say high school, nowadays – in France, together with crystal models made of porcelain. His *Traité de Physique* (first edition in 1803; second edition in 1806) was also widely circulated, and enjoyed much influence. Among its progeny was a seminal paper which Ampère published in 1814.

2. The representative shapes of molecules: Ampère (1814)

André-Marie Ampère sent to *Annales de Chimie* a letter 'On the determination of the proportions in which bodies combine according to the number and respective disposition of the molecules composing their integral particles' (Ampère, 1814; since Ampère's time, terminology has evolved: whenever the reader will find the term 'particle' he should read 'molecule' instead; and when he sees the term 'molecule', he should understand 'atom').

Atomic theory had been devised by Dalton in 1803-04, and exposed systematically by Thomas Thomson's *System of Chemistry* (1807) and by Dalton's *New System of Chemical Philosophy* (Part 1, 1808; Part 2, 1810; Part 3, 1827); there is no doubt that Ampère, in 1814, was familiar with the idea. Dalton referred to atoms as 'particles', and he termed 'compound atoms' what we call molecules.

The 1814 paper of Ampère is a classic in the history of chemistry: just a few years before, in 1811, another physicist, Amedeo Avogadro, had interpreted Gay-Lussac's law of combining volumes - volumes of combining gases are in the ratio of small integers - to mean that, in our modern terminology equal volumes of any gases, under the same temperature and pressure, will contain the same number of molecules. Ampère had rediscovered the same concept, independently from Avogadro (see Laszlo, 1980), and it is referred to nowadays as the Avogadro-Ampère law. The very success of this law - or rather hypothesis, as both Avogadro and Ampère named it - has put into the shadow, somewhat, the rest of Ampère's 1814 paper.

This is very unfortunate, because the letter to *Annales de Chimie* reports also a conceptual breakthrough of prime importance: molecular geometry is defined in terms of simple polyhedra, in which atoms occupy the corners. To quote Ampère, and with the above proviso (i.e. changing 'molecule' into 'atom', and 'particle' into 'molecule'):

"One should consider a molecule as the assembly of a given number of atoms in a given situation, enclosing between them a space incomparably greater than the volume of the atoms, (. . .) this space will be a polyhedron in which each atom will occupy a corner, and it will be sufficient to name this polyhedron in order to express the respective position of the atoms composing a molecule. I shall name this polyhedron *the representative shape of the molecule*" (Ampère, 1814).

Ampère then proceeds, by analogy with crystal structure, to define five elementary geometries:

"If we now consider the primitive shapes of crystals recognized by mineralogists, and if we regard them as representative shapes for the simplest molecule, and if we let in each of these molecules as many atoms as the corresponding shapes have corners, we find that there are five: the tetrahedron, the octahedron, the parallelepiped, the hexagonal prism, and the rhomboedral dodecahedron. The molecules corresponding to these representative shapes are composed of 4, 6, 8, 12 and 14 atoms" (Ampère, 1814).

The key notion, which Ampère formulates with utmost precision and clarity, as befits the mathematician of the first rank he also was, is that of *representative shape of a molecule*. The drawings he provides at the end of his publication can be claimed to be among the first molecular models to have been designed (Laszlo, 1980).

3. Molecular geometries for organic compounds: Kekulé (1858)

A young German chemist was appointed in 1858 a professor at the University of Ghent, in Belgium. Kekulé, who was only 29 and had got the job upon the recommendation of Jean Servais Stas (who wanted to inject new blood into the teaching of chemistry in Belgium), then made a dazzling demonstration of his scientific eminence: the same year 1858, he showed - simultaneously with the Scot Couper - that two postulates suffice for a description of the structure of organic molecules: tetravalence of the carbon atom, which he had discovered himself the preceding year, and the connection of these carbon atoms into chains by way of bonds between them.

The first major scientific contribution of Kekulé capped the new notion of *valence*, at which, besides Kekulé, Couper, and the Russian Boutlerov, Frankland, Williamson, Olding and Wurtz had been working actively since the beginning of the 1850s.

1858 indeed marks an achievement: on the basis of Kekulé's structural theory, it is at last possible to rationalize the coexistence of several carbon atoms within a single organic molecule, an analytical fact which had greatly puzzled the chemical community.

Furthermore, Kekulé was a model-builder: he constructed representative shapes of atoms and molecules, and made great use of such models in his teaching. For instance, he would project these molecular models onto a planar surface: from the shadows cast by atoms upon the blackboard or upon a piece of paper, Kekulé would derive what he called 'graphic formulas'. One should note, however, that when Kekulé thus proposed his 'rational formulas' for organic molecules, what he had in mind was not 'to represent the arrangement of groups of atoms in existing compounds'. Rather, he wanted his formulas to serve as the 'expression of certain relationships during the metamorphoses [i.e. during chemical reactions]'.

All the biographers of Kekulé agree that there may have been a link between his thwarted vocation – he had started to study architecture before he became a chemist, from the influence of Justus von Liebig – and his active interest for structural theories in chemistry. Likewise, he could apply his graphical talent, his gift for spatial organization, into the making of molecular models.

The molecular structure for which Kekulé is best known is that of benzene, a molecule which Faraday had discovered in 1825. Kekulé worked at it during the years 1862–65, accumulating experiments on aromatic compounds. He finally announced his benzene theory at a meeting of the Société Chimique, in Paris, chaired by Louis Pasteur, on January 27, 1865 (Kekulé, 1865; 1866). The benzene formula published in 1865 is a *cyclic, planar* formula. Several decades later, Kekulé would claim that he had the vision of the benzene formula in a dream-like trance, when he contemplated a chain of carbon atoms folding back on itself, like a snake biting its own tail. There is reason to believe that, when Kekulé told this story (which keeps cropping up in accounts and theories of scientific creativity), in the 1890s, he was influenced by pictures of the Ouroboros, one of the central alchemical symbols, which he may have seen in some of the contemporary writings of Berthelot on alchemy (Laszlo, 1979).

The 1865 benzene formula has localized double bonds, in the manner which Kekulé was drawing, in 1866, already, as:

In 1872, Kekulé published a new theory (Kekulé, 1872) in order to explain the existence of a single constitutional isomer when two adjacent carbons on the benzene ring bear different substituents. He proposed that the three double bonds in benzene, rather than being fixed, oscillate between two adjoining positions. Thus, two formulas rather than one must be used to characterize the benzene structure:

Thus, in the 1860s already, the theory of the structure of organic molecules was fairly complete, and chemists could start to account for their observations in terms of molecules, with definite shapes, made up of carbon, oxygen, nitrogen and hydrogen atoms, with their characteristic valences. Unsaturation had been explained by Kekulé in 1862. And, in the 1870s, Le Bel and Van 't Hoff discovered the key notion of asymmetric carbon, as the explanation for the optical activity of numerous organic molecules. As for the concept of planar aromatic molecules, it would in due time be expanded to include heterocyclic systems (such as the aromatic nitrogen bases, the purines and pyrimidines, the building stones of nucleic acids).

4. The notion of macromolecules (Staudinger, 1922)

Quite a few studies have been devoted to the definition by Staudinger of a macromolecule. Besides Hermann Staudinger's scientific autobiography, dating back to 1961 for the German edition (Staudinger, 1970), we have the well-documented and thoughtful book by Claus Priesner (1980) and a number of useful articles, including those by Herman Mark (1981), Robert Olby (1970), Pierre Laszlo (1974), L.M. Pritykin (1981) and Yasu Furukawa (1982).

First, we shall borrow from Pritykin (1981) his periodization which appears to be secure: (i) initial stage (1833–1862); (ii) colloidal-chemical stage (1862–1929); (iii) macromolecular stage (1929–1942); (iv) molecular stage (from 1942).

The initial stage was marked by the Berzelius definition of polymers (Berzelius, 1833), as aggregates formed by 'molecules (particles) of the same type or of a very similar nature' (Pritykin, 1981). Then, in 1861, Thomas Graham based an hypothetical microscopic explanation upon a macroscopic phenomenological distinction, that between *colloids* and *crystalloids* (Graham, 1861). The latter are substances easy to crystallize and which diffuse rapidly through a membrane. Conversely, the former do not crystallize, and they diffuse very slowly through a membrane. Graham (1862) proposed that colloids are aggregates of crystalloids: in his words 'the question suggests itself whether the molecule of a colloid substance may not be formed by the combining of a number of small crystal molecules'.

During the colloidal-chemical stage, which Pritykin places between the years 1862 and 1929, a number of substances were found with a very high molecular weight. The cryoscopic method gave a value of 32 400 for 'soluble starch' (Brown and Morris, 1889) and values between 6500 and 'extremely high' for rubber (Gladstone and Hibbert, 1889). It is useful to reproduce here the table of the molecular weights of proteins, as obtained during this period 1862–1929, which Olby (1970) has compiled (Table II). At the turn of the cen-

TABLE II

Molecular weights of proteins (Olby, 1970)

Date	Authors	Substance	Method[a]	Mol. wt.
1886	Zinoffsky	hemoglobin	QA	16 700
1891	Sabanjeff and Alexander	egg albumin	DP	14 000
1908	Herzog	egg albumin	D	17 000
		pepsin	D	13 000
		emulsin	D	45 000
1910	Herzog	egg albumin	D	73 000
1919	Sørensen	egg albumin	OP	34 000
1920	Dakin	gelatin	AQ	10 000
1922-23	Cohn and Hendry	casein	QA	12 800
1925	Adair	hemoglobin	OP	66 700
1925	Sørensen	egg albumin	OP	43 000
1926	Svedberg and Fåhraeus	hemoglobin	ES	68 000
1926	Svedberg and Nichols	egg albumin	ES	45 000
1926	Svedberg and Sjögren	serum albumin	ES	67 000
1928	Svedberg and Chirnogar	hemocyanin	ES	5 000 000

[a]D, diffusion; DP, depression of freezing point; ES, equilibrium sedimentation; OP, osmotic pressure; QA, quantitative analysis.

tury, when Emil Fischer was confronted with estimates of molecular weights around 15 000 for natural proteins (Table II), he rejected them on the basis of probable heterogeneity (Fischer, 1907):

"In my opinion these numbers are based on very uncertain assumptions, since we do not have any guarantee that the natural proteins are homogeneous substances".

He went on to establish firmly the peptide theory of protein structure. With some 30 different amino acids as building blocks, the

"large number of possible isomers, he thought, would suffice to explain the wide variation in the property of proteins. Hence, he concluded that there is no need to assume the existence of giant natural polypeptides. On the other hand, he admitted a molecular weight of 4021 for a starch derivative that he and Karl Freudenburg synthesized. In 1913 he claimed that this value of 4021 was the highest molecular weight found for 'all hitherto obtained synthetic products of well-defined individuality and known structure' (Fischer, 1913a) and 'natural proteins' (Fischer, 1913b). Fischer's authority made the claim that compounds of a molecular weight greater than 5000 do not exist very influential" (Furukawa, 1982).

During the same period 1862-1929, the concept of the micelle, which had been introduced in 1858 by the Swiss botanist Carl Nägeli, gained a wide acceptance. This was a revival of Graham's ideas: crystalloid molecules could pack into an aggregate comprising many molecules, and with colloidal properties. Chemists viewed organic substances such as gelatin, egg albumin, starch or cellulose, as consisting of such Nägeli micelles. Sometimes, they pictured these micelles, in the manner of Kekulé's theory of the benzene ring, as consisting of cyclic polymers held up by secondary valences or forces.

This last idea, of secondary valences or forces, intervening in addition to the normal Kekulé valences, had been given a new impetus in the 1890s through the works of Alfred Werner and Johannes F.K. Thiele. The former needed secondary valences – a residual affinity retained in the atom after formation of the primary valence bonds – to explain the presence of neutral molecules (H_2O, NH_3) as component particles in the complexes he had discovered and synthesized (Werner, 1902). The latter coined likewise the idea of a residual affinity, of what he also called a 'partial valence' to explain the addition potential of double bonds in unsaturated molecules (Thiele, 1899). The proponents of the aggregate theory of colloidal structure were quick to seize upon the analogy, in order to explain micellar cohesion.

Another potent driving force in the conceptualization of biopolymers as having a micellar structure, was the institutionalization of colloidal chemistry as a respectable sub-field of chemistry. Wolfgang Ostwald, the eldest son of the reputed physical chemist, Wilhelm Ostwald, made it his special province. He founded the *Kolloid Gesellschaft* and edited the two leading German periodicals in the field, *Zeitschrift für Chemie und Industrie der Kolloide*, and *Kolloidchemische Beihefte* (Furukawa, 1982; see also Ch. 8). Thus,

Wolfgang Ostwald had a vested interest in the creation of a physical chemistry of colloids.

The way he wrote about it in 1915 is extremely interesting (Ostwald, 1917):

"Physics has until recently busied itself chiefly with the properties of matter in mass; chemistry, on the other hand, has dealt chiefly with the smallest particles of matter such as atoms and molecules. Relatively speaking, we know much of the properties of large masses and we talk much, also, of the properties of molecules and atoms. It is because of this that we have been led to regard everything about us from the standpoint of physical theory or from that of molecular or atomic theory. We have entirely overlooked the fact that between matter in mass and matter in molecular form there exists a realm in which a whole world of remarkable phenomena occur, governed neither by the laws controlling the behavior of matter in mass nor yet those which govern materials possessed of molecular dimensions ... We have only recently come to learn that every structure assumes special properties and a special behavior when its particles are so small that they can no longer be recognized microscopically while they are still too large to be called molecules. Only now has this true significance of this region of the colloid dimensions - THE WORLD OF NEGLECTED DIMENSIONS - become manifest to us".

Ostwald then proceeded to define colloids as dispersed systems consisting of molecular aggregates ranging in size between 10^{-7} and 10^{-9} m (10–1000 Å).

The Ostwald program was rapidly fulfilled, and a number of biopolymers were pushed into the procrustean bed of colloidal chemistry:

"Between 1900 and the 1920s the aggregate structure was proposed by Carl Harries and Rudolf Pummerer for rubber, by Kurt Hess and Paul Karrer for cellulose, by Karrer and Max Bergmann for starch, and by Bergmann for proteins. (...) Harries claimed that colloid particles in a rubber solution are the aggregates of eight-membered cyclic molecules (dimethyl cyclooctadiene) held together by partial valences which are derived from the carbon–carbon double bonds in the unit molecules. (...) The aggregate theory gained further support from X-ray crystallography in the 1920s when X-ray diffraction was employed to examine the structure of polymers. (...) X-ray analysis of fibrous polymers such as cellulose indicated that the unit cells (...) are as small as the size of ordinary molecules. During this period, many crystallographers held that the molecule could not be larger than the unit cell. From this, some scientists, including Herzog (the director of physical chemistry at the Kaiser-Wilhelm-Institut für Faserstoffchemie), concluded that the molecular size of these polymers is likewise small" (Furukawa, 1982).

Enters Staudinger. This is where the fascinating part of the story starts since, contrary to a heroic conception of the history of science, it is not sufficient for a reconceptualization to be right for the rival theories to evaporate in thin air right away.

Having presented the tenets of colloid chemistry, viz. the creed in large micellar aggregates held together by secondary valences, we shall now give voice to the opposite macromolecular viewpoint as uttered by Staudinger. We shall relate briefly the clash between the two opposed analyses and research programs, which occurred in 1926, some six years after Staudinger's first publications on the subject. We shall try to identify also the complex reasons which led to the formulation by Staudinger of this novel concept of macromolecules, and to his rejection of colloids as an explanation for the physicochemical properties of biopolymers in solution.

As we already stated, the dominant paradigm in the 1920s carried with it the not unconsiderable weight of Emil Fischer's name and prestige. It was a firm belief in the impossibility for organic molecules to have molecular weights greater than ca. 5000. Polymers were thus necessarily molecules with a normal size. The peculiar characteristics ('the colloidal character') of their solutions in many solvents were ascribed to the existence of these secondary bonds, or intermolecular forces as we would term them nowadays, responsible for the assembly of these small molecules into large aggregates.

Staudinger challenged this paradigm in his 1920 article (Staudinger, 1970):

"And therefore I believe that with the available experimental data it is not necessary to make such an assumption (of molecular compounds which are held together by secondary valences) in order to explain the nature of the different polymerization products, and that normal valence formulas are an adequate representation; in organic chemistry, especially, one will try as far as possible to describe the properties of such compounds by formulas with normal valence bonds".

This statement marked the beginning of the polemic. For instance, H. Wieland made the following recommendation to Staudinger (Staudinger, 1970a):

"Dear Colleague, drop the idea of large molecules: organic molecules with a molecular weight higher than 5000 do not exist. Purify your products, for example your rubber, then it will crystallize and prove to be a low molecular weight compound".

Such an intuition of an homogeneity of all polymers was indeed rapidly strengthened by the discovery of the crystallization of certain biopolymers (see Part II, Chapter 11). Staudinger could only counter it with his own intuitive feeling that synthetic polymers were heterogeneous.

One could multiply such examples of incommunicability. It is striking for the intellectual historian to find that the opponents do not argue about the same things, and that they struggle about questions which they fancy as identical but which hindsight proves to be complementary. Neither can they agree about the crucial experiments. Such key experiments go by pairs!

For instance, Staudinger was rapidly confronted with what his opponents held as the ultimate proof: x-ray studies had shown that the unit cell of a crystalline biopolymer such as cellulose is small. A molecule must necessarily, they argued, be smaller than the volume element in which it is enclosed. Ergo, the cellulose molecules had to be small molecules.

Staudinger does not answer or counter this argument. It is only during the dramatic encounter of 1926 in Düsseldorf that this powerful intuition was shown by Herman Mark not to be necessarily true (see below). For Staudinger, the key experiment is the following: the colloidal character of solutions of polymers does not have to be interpreted in terms of micellar aggregates, as soaps are prone to form. It should be ascribed to the enormous size of the molecules themselves, each of which forms a globular particle. Hence, polymers give rise to genuine colloidal solutions. They are *eucolloids*, according to the Staudinger terminology; whereas other substances can become colloidal only by accident.

Staudinger demonstrates this through hydrogenation of natural rubber (which of course would remove the 'secondary valences', due to unsaturation). The consensus at that time was that it had a low molecular weight. R. Pummerer was convinced that its hydrogenation would lead to a volatile hydrocarbon. On the contrary, hydrogenation of rubber produced a substance whose solutions had *a very similar colloidal character* to solutions of rubber itself. From this experiment, Staudinger and Fritschi (1922) inferred that biopolymers such as rubber, or hydrogenated rubber, existed in the form of

Plate 1. Professor Hermann Staudinger, ca. 1930 (Courtesy of Mrs. Magda Staudinger.)

giant molecules which they termed, for the first time, *Makromole-küle*.

This act of denomination was a very important step. And Staudinger insisted that in so doing he was sticking firmly to the concepts of normal, classical organic chemistry (Staudinger, 1924):

"For those colloidal particles in which the molecule is identical with the primary particle and in which the individual atoms of the colloidal molecule are linked by normal valences, we proposed the term *macromolecules*".

In his autobiography, Staudinger insists upon his adherence to the principles of traditional organic chemistry (1970b):

"The essential proof for the existence of macromolecules was adduced by classical organic chemical methods via polymer analogous reactions (. . .). Wöhler and Liebig in their research on the radicals ethyl and benzoyl in 1832 (. . .) were able to convert organic compounds into derivatives with other properties, whereas a large part of the molecule – the radical – remained unchanged in size. (. . .) In the same manner it can be demonstrated that under suitable conditions macromolecules leave their 'macroradicals' unchanged with respect to size when they are converted into their derivatives".

To quote again from the fine article of Furukawa (1982), 'Staudinger's arguments in favor of the macromolecular theory were based on the principles and methods of organic chemistry, with few references to X-ray diffraction'. Whereas opponents of the macromolecular concept held as their main card the x-ray evidence: 'leading crystallographers like P. Niggli and P. Groth insisted that an organic molecule cannot be larger than the crystallographic elementary cell' (Mark, 1981). I shall now quote verbatim from Herman Mark's account (1981) of the famous clash at the annual meeting of the Society of German Naturalists and Physicians in Düsseldorf:

"On September 23, 1926, R. Willstaeter opened the famous meeting and called as first speaker, Professor Max Bergmann who presented several arguments for the aggregation structure of inulin and certain proteins. These arguments, together with a small elementary cell determination for silk were used by him to refute Staudinger's macromolecular postulate. Professor H. Pringsheim proceeded in a similar vein, referring to inulin and a few other polysaccharides. Both scientists quoted P. Karrer, K. Hess, and R. Plummerer and referred to the well-known Werner complex compounds and to the high viscosity of many colloidal solutions. Staudinger presented extensive material on rubber and some synthetic polymers and based his contentions essentially on the high viscosity of polymer solutions and on some hydrogenation experiments with rubber, polystyrene, and polyindene. He had no data on cellulose and protein but suggested macromolecular structure for them, too.

The principal objection to Staudinger's proposal was the small elementary cell and that was just the topic about which I had to speak. I pointed out that there are cases, clearly discussed in several articles by A. Reis and K. Weissenberg, where the molecule can be larger than the elementary cell. This is always the case when true chemical *main valences* penetrate through the main crystal. This could happen only in one direction, as in the case of cellulose and silk; in two directions, as in the case of graphite; and also in all three directions as in the diamond structure. Generally this means that small elementary cells do not exclude the presence of large molecules, but it also did not prove them either. It is a pity that neither the discussions nor the concluding remarks of Willstaeter have been published. As far as I can remember, Willstaeter thanked all lecturers and discussion speakers in friendly words and said: 'For me, as an organic chemist, the concept that a molecule can have a molecular weight of 100 000 is somewhat terrifying, but, on the basis of what we have heard today, it seems that I shall have to slowly adjust to this thought' ".

After this epochal meeting, indeed, the concept of macromolecules gained gradual acceptance, grudgingly at first in the 1930s, from the chemical community. 'His [Staudinger's] theory was now supported by molecular weight measurements with the ultracentrifuge which Theodor Svedberg introduced [see Table II], and which estimated the molecular weight of some proteins to be several millions. The American organic chemist, Wallace Hume Carothers was also establishing the macromolecular view through a series of investigations on the mechanism of polymerization' (Furukawa, 1982).

It remains to relate Staudinger's stand in the controversy to his psychological make-up, and to his science-sociological status. As often is the case with a scientific pioneer - think for instance of Alfred Wegener, a meteorologist, who revolutionized geology with his hypothesis of continental drift - Hermann Staudinger was or felt to be an outsider. He had prepared himself for a career in botany. However, his father prevailed upon him to nevertheless start studying chemistry and he continued in chemistry. Again, when he embarked upon the study of polymers, he had the feeling of leaving the fold of organic chemistry for a terra incognita (Staudinger, 1970c).

"When I took over the directorship of the chemical laboratories at the University of Freiburg in 1926 I discontinued further work in this field [of small molecules]. My colleagues were very skeptical about this change, and those who knew my publications in the field of low molecular chemistry asked me why I was neglecting this interesting field and instead was working on very unpleasant and poorly defined compounds, like rubber and synthetic polymers. At that time the chemistry of these compounds often was designated, in view of their properties, as *Schmierenchemie* ('grease chemistry')".

It is perhaps thus not surprising that Staudinger had to compensate for his *centrifugal* tendencies, which we have just referred to, by numerous countervailing *centripetal* forces. This is how one can explain his repeated insistence, his obstinacy to stress a continuity, in proclaiming the operational value of traditional concepts from organic chemistry as applied to the class of new molecular objects he had recognized, the macromolecules. In his book (1970), he returns again and again to the normal, usual character of organic macromolecules – these are compounds of carbon, hydrogen, oxygen and nitrogen in which bonds are made solely from primary valences, in conformity with the classical formulation of Kekulé. Likewise, solvation of these macromolecules obeys similar rules as solvation of normal organic molecules. In a like manner, Staudinger complements his conceptual orthodoxy with technical orthodoxy: he insists upon using classical techniques of chemistry, microanalyses in particular, in the new field he has opened up, rather than relying, as his opponents did, on the new technique of x-ray crystallography. Hence, Staudinger can be characterized in jest as a sort of 'retrograde pioneer'!

REFERENCES

Ampère, A.M. (1814) Ann. Chim., XC, 43–86.

Berzelius, J. (1833) Jahresbericht über die Fortschritte der Chemie, 12, 1–70 (p. 63).

Brown, H.T. and Morris, G.H. (1889) J. Chem. Soc., 55, 462–474 (p. 473).

Buffon (1835) Oeuvres complètes, avec les Suppléments, Augmentées de la Classification de G. Cuvier, 1. Duménil, Paris; (a) vol. 1, p. 335; (b) vol. 3, p. 223; (c) vol. 3, p. 165; (d) vol. 3, p. 394; (e) vol. 3, p. 462; (f) vol. 3, p. 473; (g) vol. 3, pp. 382–383 'De la reproduction en général'.

Canguilhem, G. (1981) Idéologie et Rationalité dans l'Histoire des Sciences de la Vie, G. Vrin, Paris, 2nd ed., pp. 33-45.

Fischer, E.H. (1907) Ber. D. Chem. Ges., 40, 1754-1767 (pp. 1757-1758).

Fischer, E.H. (1913a) Ber. D. Chem. Ges., 46, 1116-1138 (pp. 1119-1120).

Fischer, E.H. (1913b) Ber. D. Chem. Ges., 46, 3253-3289 (p. 3288).

Foucault, M. (1972) L'Archéologie du Savoir, Gallimard, Paris.

Furukawa, Y. (1982) Hist. Scientiarum, 22, 1-18.

Gladstone, J.H. and Hibbert, W. (1889) Phil. Mag., 28, 38-42 (pp. 39, 42).

Graham, T. (1861) Phil. Trans. Roy. Soc. (London), 151, 183-224 (p. 183).

Graham, T. (1862) Liebigs Ann. Chem., 121, 1-71 (pp. 68, 71).

Haüy, V. (1784) Essai d'une théorie sur la structure des cristaux appliquée à plusieurs genres de substances cristallisées.

Kekulé, A. (1865) Bull. Soc. Chim. France, (2) 3, 98; (1866) Liebigs Ann. Chem., 137, 129; (1872) Liebigs Ann. Chem., 162, 77.

Laszlo, P. (1974) Critique, 237-328, 782-789.

Laszlo, P. (1979) La Recherche, 10, 1019-1020.

Laszlo, P. (1980) Nouv. J. Chim., 4, 699-701.

Mark, H. (1981) J. Chem. Educ., 58, 529-539.

Olby, R. (1970) J. Chem. Educ., 47, 168-174.

Ostwald, An Introduction to Theoretical and Applied Colloid Chemistry: The World of Neglected Dimensions, (Fische, M.H., transl.), Wiley, New York, 1917, p. 76. Quoted by Furukawa (1982).

Priesner, C. (1980) H. Staudinger, H. Mark und K.H. Meyer: Thesen zür Grösse und Struktur der Makromoleküle: Ursachen und Hintergründe eines akademischen Disputes, Verlag Chemie, Weinheim.

Pritykin, L.M. (1981) ISIS, 72, 446-456.

Rousseau, J-J. (1762) L'Emile, IV, p. 329, note (quoted by Robert, Dictionnaire de la langue française, entry on 'molecule').

Staudinger, H. (1920) Ber. D. Chem. Ges., 53, 1073-1085 (p. 1073).

Staudinger, H. (1924) Ber. D. Chem. Ges., 57, 1203-1208 (p. 1206).

Staudinger, H. (1970) From Organic Chemistry to Macromolecules. A Scientific Autobiography Based on my Original Papers, Wiley-Interscience, New York. (a) p. 79; (b) pp. 104-105; (c). p. 77.

Staudinger, H. and Fritschi, J. (1922) Helv. Chim. Acta, 5, 785-806 (p. 788).

Thiele, J. (1899) Justus Liebig's Ann. Chem., 306, 87-142 (p. 89).

Venel, G-F., in D. Diderot et al. (1751-1780) Encyclopédie, ou Dictionnaire Raisonné des Sciences, des Arts et des Métiers (35 vols.), Neuchâtel, article Chimie, vol. 3, pp. 407-437.

Werner, A. (1902) Justus Liebig's Ann. Chem., 322, 261-296 (pp. 268 et seq.).

Chapter 2

From Culture Broth to Primordial Soup

1. Introduction

We get a tantalizing glance at the clash of paradigms - colloidal chemistry vs. macromolecular chemistry - by latching onto the pivotal year 1926, that of the conference in Düsseldorf in which the notion of a macromolecule grudgingly gained the beginning of its acceptance (see Chapter 1, section 4). The year 1926 also saw the award of the Nobel prize for chemistry to Theodor (The) Svedberg (1884-1971) who had just determined the molecular weight of hemoglobin and shown that its particles were all of the same size (isodisperse).

As luck has it, the same speaker, Professor H.G. Söderbaum, Secretary of the Royal Swedish Academy of Sciences, delivered the Presentation Speech for the award of the 1925 Nobel chemistry prize to Richard Adolf Zsigmondy, and for the award of the 1926 Nobel chemistry prize to Theodor Svedberg. Zsigmondy earned his prize for work utterly within the concepts and methodology of colloidal chemistry. He was a *paradigmatic* colloid chemist, and the citation for his Nobel prize commended him 'for his demonstration of the heterogeneous nature of colloid solutions and for the methods he used, which have since become fundamental in modern colloid chemistry'. In his presentation speech for Zsigmondy, delivered on December 11, 1926, Professor Söderbaum mentioned explicitly proteins, in the exact following words: 'Other examples of colloids well known to everyone are proteins (. . .)' (Söderbaum, 1926). Less than five months later, the same Professor Söderbaum was introducing The Svedberg, who had just demonstrated the homogeneity of solutions of the protein hemoglobin; yet, Dr. Söderbaum had

failed apparently to realize the contradiction between Svedberg's
results and the concept of proteins as colloids; and he concluded his
presentation speech with a romantic 'ode to colloids', in the following words:

"Inorganic chemistry has revealed more and more cases where only a colloid-chemical approach was able to clarify the observed phenomena.
    For physical chemistry colloids form a rich and rewarding field of research.
    In organic chemistry we meet perhaps most important colloids, the proteins and
the polymeric carbohydrates, which cannot be studied without the aid of colloid
research.
    As living matter is built up largely from organic colloids, the importance of colloid
research for physiology and the medical sciences is obvious" (Söderbaum, 1927).

In this section, first and in order to lay the background for the contribution by The Svedberg, we shall provide a short history of
Brownian motion from its discovery by Brown to its theory by Einstein: diffusion of a particle in a fluid would form the basis for the
experimental determinations by Svedberg. We shall also allude
briefly to the colloidal view of matter, which was dominant during
the whole period 1870-1930, to such an extent that Professor Söderbaum could still affirm, somewhat naively, this dominance in his
presentation speeches, not only in 1926, but still in 1927: as if even
the Nobel Committee had been caught unawares by some of Svedberg's most recent results, and had failed to grasp their full significance.

2. The Brownian motion

Exactly 100 years earlier, in 1827, the British botanist Robert
Brown (1773-1853) was observing under his microscope the fertilization of a pistil with pollen. His instrument could enlarge about
500-fold because of the recent introduction of achromatic lenses.
What was he trying to see? Brown was chasing the *organic molecules* of Buffon (see Chapter 1), these living particles which the biological theory of the period believed to be concentrated in the semen
of animals and plants. One of the chief expected characteristics of
these essential principles of life would be motion.

Since Brown indeed discovered particles in rapid movement, this appeared as a handsome confirmation of biological theory, of the ideas of Buffon and of Needham. Brown saw tiny bodies undergoing fast unceasing motion; the smallest particles were the most rapid and motion appeared to be inherent to the granules. Brown proceeded to perform control experiments – and this is where he proved his worth: he tried not only pollen from long-dead plants, but he also experimented with particles of soot, or from fossilized wood, with pieces of broken glass, with mineral powders. All these samples gave rise, when suspended in a liquid, to the same erratic zig-zag motions.

When Brown published his observations, he found that the Frenchman Brongniart had also made similar findings on grains of pollen, reported in a paper read on December 26, 1826, to the Académie des Sciences. The paper by Brown was rapidly translated into French and German and it triggered a flurry of investigations; they all confirmed his key finding: no external physical cause appeared to be responsible for the observed motion.

Subsequently, Brownian motion became quickly forgotten: microscopists knew about the phenomenon, and that was all; till 1877, when Father Delsaux perceived the pronounced analogy with the kinetic theory of gases.

In a gas, the position of a molecule is ever-changing. Likewise, Delsaux explained Brownian motion by collisions of the molecules from the liquid upon the suspended solid particle: if the particle is small enough, collisions on its various sides will become unequal in number and will be distributed at random in time and space. Hence, the observed haphazard motion.

Let us interject a reflexion at this point. From the very beginning, when Brown made his observations in 1827, the ancient linkage of life and motion, which had been the joint belief of Heraclitus, Aristotle, Epicurus, and Lucretius, had been loosened. This is the concept behind the word animal, as an animated being. And this concept had been hit, since inert minerals were also capable of motion. As soon as Brownian motion had been explained convincingly (and somewhat ironically, by a clergyman) in terms of the permanent thermal agitation of molecules in a fluid, vitalist beliefs had suffered a defeat. The dualistic division of reality between the organic and the inorganic worlds, based on motion of the former but not of the latter, was no longer tenable. It had to be replaced with another dichotomy, that between the macroscopic and the microscopic realms; with the change of scale, in a subtle way phenomena changed their nature.

This was precisely the contention of Wolfgang Ostwald, whom we have quoted in Chapter 1, section 4 to the effect that the range of dimensions 10^{-9}–10^{-7} m in between the macroscopic masses physicists deal with, and the microscopic masses of atoms and molecules which constitute the province of the chemist, corresponded for dispersed systems to molecular aggregates in a new state of matter, the colloidal state of matter.

The colloidal state, when first defined by Graham (1861), had been endowed by him with motion. In his words, the colloidal state was 'a dynamical state of matter, the crystalloidal being the static condition. The colloid possesses ENERGIA. It may be looked upon as the probable primary source of the force appearing in the phenomena of vitality'. Thus, with Graham one has gone full circle: it is as if intellectual history had reverted to the pre-1827 period, back to the original impetus for the experiments of Brown, with the notion of colloids as consisting of animated particles or molecules of life!

3. Measuring dimensions of colloidal particles

To R.A. Zsigmondy (1865-1929), from the benefit of hindsight, fell the useful role of laying down the groundwork for comparisons later to be made by others, led by Svedberg, between proteins and colloidal systems. His work belonged squarely to the colloidal school or paradigm. Zsigmondy's first contribution was the preparation of colloidal gold, by the reduction of gold chloride with formaldehyde in weakly alkaline solution. Subsequently, he found that Michael Faraday had also achieved extremely fine divisions of gold, using phosphorus as the reducing agent. Combining both techniques, Zsigmondy achieved extremely fine divisions of gold. The resulting solutions were stable and homogeneous. They behaved in dialysis like colloidal solutions (or sols): the gold particles did not pass through a parchment membrane. Zsigmondy proved also the importance of the negative electrical charge borne by the gold particles in such colloidal solutions; addition of salts, by boosting the ionic strength, reduces Coulombic repulsion between the suspended particles, colloidal gold changes color from red to blue, and coagulates.

A second major contribution of Richard A. Zsigmondy was his invention of the ultramicroscope, jointly with the optician Siedentopf from the Zeiss company in Jena, at the turn of the century. In the ultramicroscope, the sample is viewed perpendicularly to the direction of the incident light beam. With immersion ultramicroscopes, Zsigmondy could achieve spatial resolutions of ca. 5 nm (50 Å). In this manner, Zsigmondy showed that gold colloids have particle sizes ranging widely 'from colloids the particles of which are invisible even in the ultramicroscope, up to ones whose particles lie at the limit of visibility in the microscope' (Söderbaum, 1926). He determined quantitatively the size of the particles in colloidal solutions of gold or silver; and these measurements proved the *heterogeneous nature of colloidal solutions*: 'the particles of colloidal solutions are larger than the molecules, and must frequently be regarded as aggregates of molecules; they may be of different size even when we are dealing with solutions of the same substance' (Zsigmondy, 1926). This was a major breakthrough: in the words of Edsall (1979) 'particles, formerly invisible, could now be seen, and their motion could be followed'. Furthermore, this technical breakthrough, the invention of the ultramicroscope, coincided with the devising by Einstein of his theory of the Brownian movement.

The very year 1905 which saw the appearance of the immensely influential book by Zsigmondy *Zur Erkenntnis de Kolloide* (Edsall, 1979), Albert Einstein published his first paper on Brownian motion entitled 'On the movement of small particles suspended in a stationary liquid demanded by the molecular-kinetic theory of heat' (Einstein, 1905). Soon afterwards, when he had become completely convinced of the identity between his conclusions and the interpretation of Brownian motion as resulting from the microdynamics of molecules that could not be seen in the microscope (Gouy, 1888), Einstein amplified his treatment to account not only for translations, but also for rotations of spherical particles undergoing Brownian motion (Einstein, 1906a). Einstein could calculate the mean free path of a Brownian particle. He showed also how Brownian motion could be exploited for the determination of molecular dimensions (Einstein, 1906b); Einstein applied his theory to the determination of the hydrodynamic radius of sugar (sucrose) molecules in water, for which he found a value of 4.9 Å.

The theoretical work of Einstein, together with that of Smoluchowski (1906), inflamed the enthusiasm of the colloid chemists: not only, using the ultramicroscope, could they see colloids which previously were invisible, but right away - by taking advantage of the theory of Brownian motion - even much smaller particles, in the 1–10 Å range, could have their translational and rotational diffusions investigated and thus could be sized up indirectly.

One of these enthusiastic colloid chemists was the young The Svedberg, born in 1884, who was then starting his own research. He has himself (Svedberg, 1946) noted the enormous influence on his research program of the appearance in 1915 of Zsigmondy's book *Zur Erkenntnis de Kolloide*. Svedberg, using the Zsigmondy–Siedentopf ultramicroscope, set out investigating inorganic colloids, such as colloidal gold or sulfur. He characterized their heterogeneity using distribution functions, i.e. making use of a statistical approach. The interplay between experiment and theory was such that Albert Einstein started both his *Theoretical Observations on the Brownian Motion* and then his popularization *The Elementary Theory of the Brownian Motion* (Einstein 1907; 1908) by referring to the experimental work of Svedberg which had been published in the same journal, *Zeitschrift für Elektrochemie* (Svedberg, 1906a, b).

4. Sedimentation of hemoglobin

Indeed Theodor (The) Svedberg, just like Richard A. Zsigmondy, performed his early work within the mainstream of colloidal physical chemistry: in his doctoral thesis (1908) *Studien zur Lehre von den Kolloiden Lösungen*, he provided confirming evidence upon the validity of the Smoluchowski–Einstein theoretical description of Brownian motion. The same year (1908), Jean Perrin made use of the Stokes' Law for determining the size of a particle.

This was the departure point for Svedberg's efforts to determine the distribution of particle sizes in coarse-grained disperse systems. In 1916–17, Odén showed that this could be done from sedimentation data in the gravitational field. In the early 1920s, Svedberg started using a centrifuge to study particle-size distribution and molecular weights. At equilibrium, after a rather long centrifugation time, the molecular weight M_r, can be determined (see Chapter 5) from the following equation:

$$M_r = \frac{2\,RT \ln (c_2/c_1)}{(1 - V\rho)\omega^2 \, (x_2^2 - x_1^2)}$$

If x_1 and x_2 denote two distances from the axis of rotation, and c_1 and c_2 the corresponding concentrations at these points, V being the partial specific volume of the substance and ρ the solvent density, ω being the angular velocity, M_r is known from determining the concentrations c_1 and c_2 at a given value of ω. In order to determine the difference in concentration between points only a few millimeters apart, or even fractions thereof, in the sample, Svedberg resorted to a microphotometric method: the cell has two parallel quartz windows through which a light beam is directed unto a photographic plate. Absorption of ultraviolet light by the sample thus serves to measure concentrations. The angular velocities ω, which in the original experiments of Svedberg corresponded to centrifugation speeds of the order of 10 000 rev./min, are measured with a stroboscope.

Svedberg performed the key experiments in 1924–25, immediately after his return from Madison, Wisconsin, where he had built the first primitive 'ultracentrifuge'. The first attempts to sediment a protein were made (with J.B. Nichols, a co-worker from the Madison period) on ovalbumin which according to the literature should have a molecular weight of about 34 000; but the results were not encouraging. On the suggestion of Robin Fåhraeus, hemoglobin was tried instead. Svedberg was reluctant because this protein, according to its content of iron, should have a still smaller molecular weight (17 000)(see Table II, p. 13).

"Svedberg has told us that he was awakened in the middle of the night by a telephone call from Fåhraeus who was watching the run and shouted: 'The, I see a dawn'. Svedberg rushed to the institute, and there was indeed a marked lightening of the color at the top of the cell. The hemoglobin was sedimenting" (Tiselius and Claesson, 1967).

The data which Fåhraeus and Svedberg proceeded then to gather, at approx. 10 000 rev./min for 39 h in a centrifugal field about 5×10^3 that of gravity, were of the following type for hemoglobin (Svedberg, 1927):

TABLE III

Molecular weights (M_r) determined by Svedberg and Fåhraeus for hemoglobin

x_2, cm	x_1, cm	c_2, %	c_1, %	M_r
4.61	4.56	1.220	1.061	71 300
4.56	4.51	1.061	0.930	67 670
4.51	4.46	0.930	0.832	58 330
4.46	4.41	0.832	0.732	67 220
4.41	4.36	0.732	0.639	72 950
4.36	4.31	0.639	0.564	60 990
4.31	4.26	0.564	0.496	76 570
4.26	4.21	0.496	0.437	69 420
4.21	4.16	0.437	0.308	66 400

From these data, based upon the sedimentation equilibrium, the molecular weight had a mean value $\bar{M}_r = 67\ 870 \pm 5700$. Another and independent method, based upon the sedimentation speed, provided a value $\bar{M}_r = 68\ 350$, i.e. very close to the preceding one. Furthermore, there was no spread in the size of the particles: 'all the hemoglobin molecules (or particles) were of the same size' (Svedberg, 1927). By contrast, the suspensions of colloidal gold which Svedberg had been studying, jointly with H. Rinde, and which he had prepared according to the method of Zsigmondy (reduction from chloroauric acid and phosphorus) were *polydisperse*. They displayed a distribution in sizes between 0.7 and 2.2 nm, with a maximum at about 1.5 nm (Svedberg, 1927). Thus, the clear-cut distinction was made for the first time between a *polydisperse* system (gold) and an *isodisperse* system (the protein).

Svedberg was almost immediately awarded the 1926 Nobel prize for chemistry for his discoveries. In his Nobel lecture, on May 19, 1927, The Svedberg could announce the molecular weights of a number of proteins, besides hemoglobin: ovalbumin (M_r = 34 500), phycocyanin (M_r = 105 900) and phycoerythrin (M_r = 207 700). A side result of this work was the clear-cut demonstration that a number of proteins consisted of sub-units.

"Thus, hemocyanin from *Helix pomatia*, which according to its copper content should have a molecular weight of 15 000 to 17 000, proved to consist of particles of a particle weight of several millions, and all of equal size (work by Svedberg and Chirnoga). (. . .) Changes in the acidity of the medium produced sub-units by (reversible) dissociation, and these were again homogeneous as regards particle (or molecular) size" (Tiselius and Claesson, 1967).

This finding was corroborated by Sørensen when he found, in 1930, that the solubility of various proteins increases as the quantity of solid phase increases: a fact correctly interpreted, on the basis of Gibbs' phase rule, as indicating coexistence of molecules and sub-units (see Part II, Chapter 11 on this point).

5. Conclusions

At the beginning of this chapter, we have quoted Professor Söderbaum as an example of a chemist from the colloidal school or point of view and called attention to the irony of the situation, when he had to present The Svedberg as the 1926 Nobel Laureate. A further irony is apparent in the scientific career of The Svedberg himself, who underwent a drastic switch from colloid chemist in his youth to molecular biophysicist in his mature age! Such a change, from one paradigm to another and opposed paradigm, undergone by a single person, must be an infrequent occurrence.

At the beginning of his career, The Svedberg probably fancied that the ultramicroscope would enable him to see protein molecules, which he must have conceived then as ill-defined glue-like colloidal aggregates. During Svedberg's career, it became obvious that, in the case of proteins, even the aided eye, the eye assisted by the microscope, could no longer see the objects of study.

However, also during Svedberg's career, another and entirely different type of instrument had become available: with the advent of x-ray crystallography, the study of the shape of proteins as objects, but at the molecular level, became feasible. Let us draw the reader's attention both on the continuity, x-rays appearing to extend visible rays to make 'visible', as it were, objects too small to be seen with a normal microscope; and to the epistemological discontinuity, since molecular models built from x-ray diffraction patterns are predicated upon a theory of diffraction, based itself upon various assumptions. This was an epochal change, especially to naive positivists trained to rely exclusively upon evidence from sensory data.

Let us recall again the conceptual background before the establishment of the monodispersity of protein molecules. During the nineteenth century, vitalists conceived biology as the realm of diversity. Using their microscopes, they saw infusoria, paramecia, which displayed their protoplasm as a viscous fluid, as a jelly in continuous movement. Furthermore, the mobility of these micro-organisms was felt to express their full vital force, their *élan vital* to borrow an appropriate term from the French philosopher Henri Bergson. Quoting Bergson (1859–1941) may appear a little anachronistic, even though he was profoundly influenced by the biological concept of the nineteenth century. To illustrate the prevalence of the vital force in this line of thought, we can refer for instance to what Justus von Liebig wrote in the early 1840s:

"A remarkable activity, a force in the resting state is to be found in the ovum, in animals, and in the seed, in plants (. . .) this force is named the *vital force*. (. . .) The vital force manifests itself as a force of attraction to the extent that new combinations, resulting from changes in the shape and in the nature of the foodstuff, and having the same composition as the mobile, become integral parts of this mobile (Mr. Liebig, notes the French translator, names *mobiles of vitality* all parts which are both organized and living)" (Liebig, 1842).

Two decades later, when Graham made the distinction between crystalloids and colloids, it was conceptually a distinction between static and dynamic particles, at the same time as it was operationally a distinction between diffusion and non-diffusion through a membrane.

Colloids, the particles of which undergo unceasing Brownian motion, were picked as convenient models for living matter ('The colloid possesses ENERGIA'; Graham, 1861). Hence, Brownian motion connected biology with physics and chemistry as a statistical phenomenon: this was the contribution of Delsaux, in 1877, and of Gouy, in 1888. (The whole period 1850-1914 is indeed one strongly dominated by statistical considerations, both in the natural sciences, the statistical thermodynamics of Boltzmann, Mendelian genetics, and in the social sciences (Quételet and Galton).)

Because of this emphasis on statistical distributions, which The Svedberg determined in his doctoral thesis for all kinds of colloidal systems, because of the widespread belief in the unattainable diversity and complexity of any and all biological entities, scientists (such as Svedberg himself) were totally unprepared for the revelation of biomolecules as well-defined systems, in the late 1920s. They could not suspect that biomolecules would turn out to be well-defined and firm, instead of fuzzy objects.

The science of colloids had started, with Graham, in the 1860s, at a time when the controversy about spontaneous generation raged between Pasteur and Pouchet. The title of this chapter refers obliquely to this period, which could be termed 'The Age of the Culture Broth'. In the 1930s, after proteins had had their molecular nature demonstrated, vitalists were put on the defensive, and materialists could extend their claims to include the origin of life. This was the time when the ideas of Oparin about a primordial soup, from which the molecules of life could self-generate spontaneously, started being taken seriously, especially by Marxist biochemists such as Haldane and Bernal. This period could be termed 'The Age of the Primordial Soup'. Hence, the title of this section, referring to the period 1860-1930, which saw the triumph and the demise of colloid science.

REFERENCES

Edsall, J.T. (1979) Ann. N.Y. Acad. Sci., 325, 53–73.
Einstein, A. (1905) Ann. Phys., 17, 549.
Einstein, A. (1906a) Ann. Phys. (4), 19, 371–381.
Einstein, A. (1906b) Ann. Phys. (4), 19, 289–306.
Einstein, A. (1907) Z. Elektrochem., 13, 41–42.
Einstein, A. (1908) Z. Elektrochem., 14, 235–239.
Gouy, M. (1888) J. Phys. (2), 7, 561.
Graham, T. (1861) Phil. Trans. R. Soc. London, 151, 184.
Liebig, J. von (1842) Chimie Organique Appliquée à la Physiologie Animale et à la
Pathologie, Fortin, Masson, Paris, pp. 1, 2, 203.
Smoluchowski, M. (1906) Ann. Phys. (Leipzig), 21, 756.
Söderbaum, H.G. (1926) Presentation Speech for R.A. Zsigmondy, Nobel Laureate in
Chemistry, Nobel Lectures. Chemistry, 1922–1941, Elsevier, Amsterdam.
Söderbaum, H.G. (1927), Presentation Speech for T. Svedberg, Nobel Laureate in
Chemistry, see Nobel Lectures. Chemistry, 1922–1941, Elsevier, Amsterdam.
Svedberg, Th. (1906a) Z. Elektrochem., 12, 853–860.
Svedberg, Th. (1906b) Z. Elektrochem., 12, 909–910.
Svedberg, T. (1927) Nobel Lecture, May 19, in Nobel Lectures. Chemistry,
1922–1941, Elsevier, Amsterdam, p. 66.
Svedberg, T. (1946) Forty years of colloid chemistry, in Harold Nordensen 60 år,
Stockholm, pp. 340–349 (quoted by Edsall, 1979).
Tiselius, A. and Claesson, S. (1967) Annu. Rev. Phys. Chem., 18, 1–8.
Zsigmondy, R.A. (1926) Nobel Lecture, December 11, in Nobel Lectures. Chemistry,
1922–1941, Elsevier, Amsterdam, p. 45.

Chapter 3

Morphology of Molecules

Some among the Greek philosophers, Democritus for instance, pro-pounded the notion that atoms composing the bodies in Nature were to be characterized themselves, not only by a given mass and size, but also by a well-defined *shape*. As with very many scientific con-cepts, an eloquent early expression of this key notion of a geometry for microscopic particles can be found in *De natura rerum*, the splendid scientific and didactic poem by Lucretius (1st century BC). In it, Lucretius explains for instance the sour taste of a substance as originating in *needle-shaped* atoms.

These concepts remained essentially invariant throughout the centuries, till the time of Lavoisier and the revolution he was so instrumental in bringing to chemistry. For instance, many authors during both the 17th and the 18th century rationalized the comple-mentariness and the properties of acids and bases by postulating that the corresponding particles were pointed, needle-shaped in the case of acids – thus explaining their sour taste – and that, conversely, bases were made of soft, spongy, porous particles into which the acidic needles could insert.

During the latter half of the 18th century, Diderot and D'Alem-bert issued the *Encyclopédie*. They entrusted the entry on CHEM-ISTRY to Gabriel-François Venel, a follower of Rouelle. In it, Venel makes an extremely interesting attempt to dissociate chemistry from Newtonian physics. Therefore, Venel argues, it should be the task of chemists to explain macroscopic properties (color, texture, density, etc.) on the basis of hypothetical modes of interaction between the composing microscopic particles, with well-defined shapes and sizes. Hence, even in this pre-Daltonian era, and dating back to Buffon's 'molécules organiques', chemistry had set for itself a research program of explaining the properties of substances on the basis of the geometry of the particles composing them.

A decisive move forward was made at the time of the French Revolution by Haüy, whom we have already referred to. Haüy (1743-1822) developed the theory of crystal structure as the three-dimensional repeat of an elementary module in 1784.

To give but one example of Haüy's penetrating intuitions and reasoning, based upon the congruence of the macroscopic form and the shape of the microscopic component bodies, he explained the shape of snow-flakes (a problem brought to the fore since the Renaissance, and on which Kepler (1611) himself worked) as 'implying necessarily a bent geometry for the water molecules', to continue using a modern terminology.

Haüy became an influential scientist, through the writing of several textbooks to teach the science mineralogy, which he had greatly advanced.

A prime example of this influence is the already referred to 1814 paper by André-Marie Ampère (1775-1836), a giant of science and a genuine polymath. This paper of Ampère is very important to our topic, since it presents in extremely relevant manner what I believe to be the first definition of *molecular shape* - this decades before Van 't Hoff, Le Bel, and Kekulé (Ampère, 1814).

The very success of the Avogadro-Ampère law - or rather, and this is an important point to which I shall return, *hypothesis*, as both Avogadro and Ampère named it - has put in the shadow, somewhat, the rest of Ampère's 1814 paper. This is quite unfortunate, for this letter reports also a conceptual breakthrough of the first magnitude: molecular geometry is defined in terms of simple polyhedra, in which atoms are placed at the corners.

Since Ampère's time, chemical terminology has evolved (some!). All of my reading of Ampère's work is based on the apparently completely legitimate decision to equate Ampère's word 'particle' with what we call a 'molecule' and to hold Ampère's word 'molecule' to mean what we call an 'atom'. Hence, in the quotations I shall give, and in order to make reading easier, I shall make the corresponding changes. Atomic theory was invented by Dalton in 1803-1804, and exposed systematically in Thomas Thomson's *System of Chemistry*

(1807) and in Dalton's own *New System of Chemical Philosophy* (Part 1, 1808; Part 2, 1810; Part 3, 1827). Dalton also referred to atoms as 'particles', and he would term 'compound atoms' what we now call molecules.

The starting point for the Ampère foray into stereochemistry is a consideration of chemical bonding:

"Consequences deduced from the theory of universal attraction, seen as the cause of cohesion, together with consideration of the ease with which light goes through transparent bodies, have led physicists to think that the ultimate atoms of bodies are held by the attractive and repulsive forces which are their own, at distances infinitely large relative to the dimensions of these atoms" (Ampère, 1814).

Therefore, since

"one should consider a molecule as the assembly of a given number of atoms in a given situation, enclosing between them a space incomparably greater than the volume of the atoms, (. . .) this space will be a polyhedron in which each atom will occupy a corner, and it will be sufficient to name this polyhedron in order to express the respective position of the atom composing a molecule. I shall name this polyhedron *the representative shape* of the molecule" (Ampère, 1814).

Ampère then proceeds, by analogy with crystal structure, to define five elementary or basic geometries (see Chapter 2).

It should be noted that Ampère selects for the 8-atoms case the parallelepiped (with rectangular faces) rather than the dodecahedron (with triangular faces), and, for the 12-atoms system, he opts in favor of the hexagonal prism (again, with rectangular faces) instead of the icosahedron (with triangular faces). As spelled out by Ampère, he is building on the foundations led by crystallographers, Haüy first and foremost. I see the Ampère work as an attempt to transpose to isolated molecules, in the gaseous state, the geometrical knowledge gained from consideration of the apparent shapes of macroscopic crystals.

Ampère proposes very detailed schemes to account for chemical combination in terms of juxtaposition of the simpler polyhedra into more complex assemblies. These occur always face-to-face, so that bonding of atoms takes place in a symmetrical manner leading to compact globular shapes which can be circumscribed by a sphere.

Here is, for example, how Ampère accounts for the observed composition of hydrochloric acid:

"Chlorine combines with hydrogen at equal volumes, and the resulting muriatic acid occupies a volume equal to the sum of the volumes of its two components. One could rationalize such a mode of combination, by assuming that the representative shapes of the chlorine molecules are isolated tetrahedra such as those for oxygen, nitrogen, and hydrogen; that for the molecules of muriatic acid would then be a tetrahedron; but it is also possible to explain it by considering each molecule of chlorine as formed by reunion of two tetrahedra into a parallelepiped, and as containing therefore eight atoms. This last hypothesis is the only one to agree with proportions observed in other combinations of chlorine, with the phenomena they display and with their characteristic properties. By making this assumption, one finds that each molecule of muriatic acid, containing half of a hydrogen molecule, half of a chlorine molecule, has for representative shape an octahedron composed of two atoms of hydrogen and four atoms of chlorine. When muriatic acid combines with ammonia gas, each of its octahedra combines with a cubic molecule of this gas; hence it must absorb an equal volume of it, as observed, and the molecules of the resulting salt must have for their representative shape a rhomboedral dodecahedron" (Ampère, 1814).

(The underlying notion is, evidently, that of valence, which was set in the 1850s, following the hunch of Edward Frankland, by Williamson, Olding, Wurtz, Couper, and Kekulé.)

The key notion, which Ampère formulates with utmost precision and clarity, as befits the good mathematician he was, is that of *representative shape of a molecule*. The drawings he provides at the end of his publication can be claimed to be among the first molecular models to have been designed.

Why was such a crucial contribution overlooked? There is one major reason: in Pauling's words, the

"value of (Avogadro's) law remained unrecognized by chemists from 1811 until 1858. In this year Stanislao Cannizzaro (1826-1910), an Italian chemist working in Geneva, showed how to apply the law systematically, and immediately the uncertainty regarding the correct atomic weights of the elements and the correct formulas of compounds disappeared. Before 1858, many chemists used the formula OH for water and accepted 8 as the atomic weight of oxygen; since that year, H_2O has been accepted as the formula for water by everyone. (. . .) The failure of chemists to accept Avogadro's law during the period from 1811 to 1858 seems to have been due to a feeling that molecules were too 'theoretical' to deserve serious consideration" (Pauling, 1956).

Every word of this assessment applies to reception of the Ampère (1814) paper equally well. And one can even pinpoint the cause of the neglect of both the Avogadro and the Ampère contributions to their use of the word 'hypothesis'. What we now know as the Avogadro–Ampère law was only an *hypothesis* in the eye of its begetters.

And 'hypothesis' was not a well-accepted word at the time. For Romantics, it 'seems to have implied mechanical and atomistic explanations, mistaken attempts to penetrate behind the phenomena.' (Knight, 1970).

At the time of Avogadro and Ampère (later on Van 't Hoff would encounter similar trouble), imagination in science was frowned upon. Naive realism got into the way of acceptance of all-important concepts, such as atomic theory. And it is an interesting paradox that molecular geometry – which we now go to the opposite extreme of holding as a concrete fact, forgetting that we are dealing only, in Ampère's words, with *representative shapes*, with figments of our imagination (Mislow, 1977) – was held for at least two generations as a complete fiction, as a flight of fancy, which serious scientists should not bother with!

I have adapted this chapter from a recent article (Laszlo, 1980). The title I have given it stems from two developments. The advent of the microscope, first in the 18th century due to Leeuwenhoek, and then after the 1830s when achromatic lenses came into general use, made all the natural scientists much keener to observe the microscopic aspects of reality, and to give detailed descriptions of their aspects and shapes. Microscopic morphology became an important part of natural science, as well as of the general imagination (*Gulliver's Travels* are a prime example, as Marjorie Hope Nicolson has beautifully reminded us).

Second, during the first decades of the 19th century, when Ampère published his paper (which, and this is tantalizing to the historian, might have become a seminal contribution), a powerful, a dominant *episteme* in the sense of Michel Foucault was that of morphology. This was the time when pathology was related to anatomy by the leaders in the medical field. This was the time when Cuvier pioneered morphological criteria in paleontology and comparative anatomy. This was the time when Goethe, who also engaged in paleontological, geological, and botanical studies, was convinced that he had evolved a general science of morphology. Ampère's gallant and elegant thrust for the shape of molecules should be seen against this background of a general tendency to seek explanations and descriptions of a morphological nature.

REFERENCES

Ampère, A.M. (1814) Ann. Chim., XC, 43 (my translations).

Kepler, J. (1611) De Nive sexangula. In Opera (Fritsch, ed.), VII, 723.

Knight, D.M. (1970) Hist. Sci., 9, 54.

Laszlo, P. (1980) Nouv. J. Chim., 4, 699.

Mislow, K. (1977) Bull. Soc. Chim. Belg., 86, 595.

Pauling, L. (1956) General Chemistry, Freeman, San Francisco, 2nd ed., pp. 298-299.

Chapter 4

The Analogy between Biomolecular Structure and Architecture

An important epistemological step was reached when even with the aided eye it became impossible to see the object of study; moreover, the microscope was limited to sizes down to 10 000 Å or so. Other tools, such as x-ray diffraction and electron microscopy, were thus devised to get beyond this limit and to probe into the atomic and molecular scales. It then became necessary to evolve constructs at an intermediate level, both conceptually and physically, to serve as models and to be looked at, examined and scrutinized. Model building has been a key feature of many an important discovery in modern biochemistry, such as the presence of α helices and β sheets in proteins (Pauling and Corey, 1951), the DNA double helix (Watson and Crick, 1961), or the structure of viruses (Caspar and Klug, 1962).

1. Linus Pauling as the pioneer molecular architect

In 1946, Linus Pauling, who at the time was chairman of the Division of Chemistry and Chemical Engineering at Cal Tech, published a lucid and seminal article in *Chemical and Engineering News* entitled 'Molecular Architecture and Biological Reactions' (Pauling, 1946).

Pauling had been a model maker from early on, because of his training and early work in crystallography. He is quite explicit on this point and answers the question: 'if you can recall them, what were the main factors behind your model-building approach to the structure of proteins?' thus (Pauling, 1983):

"My model-building approach to the structure of proteins was based on my similar approach to the study of minerals and other inorganic crystals. My first paper, published with my mentor Roscoe Gilkey Dickinson in 1923 (Dickinson and Pauling, 1923), was on the crystal structure of molybdenite, MoS_2. Dickinson had some frameworks of metal rods and wires on which he placed wads of plasticene to indicate the positions of the atoms in a crystal. I became accustomed to making similar models of crystal structures, by means of tetrahedra, octahedra, and other coordination groups, which were attached together by corners, edges, and faces to give a complete crystal structure. I think that this method was included in a paper by me in 1928, and especially in a paper by my student J.H. Sturdivant and me, dealing with the forms of titanium dioxide (Pauling and Sturdivant, 1928).

I had become interested in interatomic distances and atomic radii already in 1922, just after I began my graduate work and research in the field of x-ray crystallography. I began work on organic molecules in 1930, mainly by the electron-diffraction method for determining the structure of gas molecules. The first organic molecule for which bond lengths and bond angles were determined was hexamethylenetetramine, the structure of which was reported by R.G. Dickinson and A.L. Raymond in 1923. This work also was done in the California Institute of Technology. During the years following 1930 my students and I, and also other investigators, reported bond lengths and bond angles for many organic compounds, and I began formulating a systematic theory of the values. By 1937 I had reached some conclusions about the probable values of bond lengths and bond angles in polypeptide chains. In that year Dr. Robert B. Corey and I decided to start a program of determination of the structure of amino acids and simple peptides, in order to get experimental information bearing on the problem of the structure of proteins.

The ball-and-stick models of carbon compounds had come into use. I do not know how early they became available. They did not attempt to indicate bond lengths and bond angles with any accuracy, but essentially were based on the tetrahedral arrangement of the four bonds of the carbon atom and the assumption that other bond angles probably were close to the tetrahedral value. Dr. Corey and I began to build molecular models that were accurate with respect to bond lengths and bond angles. I think that this work started about 1937 or 1938. The accurate models were built in the machine shop of the Division of Chemistry and Chemical Engineering in the California Institute of Technology, under the supervision of Dr. J.H. Sturdivant, Dr. Robert B. Corey, and me".

Not only was Linus Pauling familiar with the building of accurate molecular models from his professional experience, but, perhaps more important, his thinking has always been rooted in a mental image or idea. As he expresses it in another section of the same letter (Pauling, 1983):

"I do not remember what I was thinking about when I wrote my 1946 article in *Chemical and Engineering News* (Pauling, 1946). I realized at that time that my original ideas in the field of science usually came or perhaps always came from my having a picture or model relating to some phenomenon. Often then I would attempt to make quantum mechanical calculations in order to support the idea, but the idea came first, and the calculations next. In this attack I differed from, for example, J.H. Van Vleck, who told me that all of his contributions to physics had resulted from his manipulating the equations, and none of them from just having a picture or an idea. W. Pauli also told me that he had done only one thing in his life that was not mathematical, and that he had received the Nobel Prize for his one non-mathematical idea – the idea that constitutes the Pauli exclusion principle".

Hence, Pauling's approach to chemistry, and likewise to biochemistry, is to complement intuitions or arguments based upon thermochemical considerations – evaluating by experiment or estimating from approximate expressions quantities such as bond energies – by considering a detailed three-dimensional geometry, obtained either from the x-ray structure, or from a rational reconstruction of what the structure should be, or could look like (i.e., the building of a model).

The history of biochemistry has repeated the history of its parent and predecessor, organic chemistry, in that the era of isolation and characterization of substances has been followed by a period dominated by structural determination using physical methods. We have become so used to structure as an essential component of molecular personality, as it were, that we tend to forget how relatively recent, in biochemistry, are structural considerations. In 1946, when Pauling published his article in *Chemical and Engineering News*, there were simply not enough structural data to build on. Pauling (1946) formulated his far-from-trite statement by commenting upon a quote from the physicist Eddington:

"Eddington has said that the study of the physical world is a search for structure rather than a search for substance. If we ignore the philosophical implications of the words, we may say that the chemist and the biologist in their study of living organisms must carry on both a search for structure and a search for substance, and that

44 STRUCTURE AND ARCHITECTURE

the second of these must precede the first. Investigators have had great success in isolating chemical substances from living organisms, and in determining the chemical composition of the simpler of these substances. (. . .) We may consider this work of isolation and identification of active chemical substances as the search for substance in biology".

Pauling (1946) then proceeds in his article to provide his general reader with orders of magnitude for the dimensions accessible (then) through various instruments. A visual microscope could get down to 10 000 Å. X-rays could probe particles with dimensions in the range of 10–10^{-3} Å. The electron microscope could reveal details down to 100 Å. Hence, there remained, at the time of writing (1946), a gap covering the range, crucial for a better understanding of protein structure and function, of 10–100 Å:

"There are many ways," continued Pauling (1946) "of investigating this region – by X-rays, ultracentrifuges, light-scattering techniques, the study of chemical equilibria, the techniques of degradation, isolation, identification, and synthesis used by the organic chemist, serological methods, chemical genetics, the use of both radioactive and nonradioactive tracers, the use of electron microscopes of improved resolving power – but no one method is good enough to solve the problem, and all these methods must be applied as effectively as possible if the problem is to be solved".

The problem to be solved is that,

"at the present time (1946) (. . .) we do not know in detail how the amino acids are combined to form proteins. We do not know, except very roughly, even the shapes of such important molecules as serum proteins, enzymes, genes, the substances which make up the protoplasm (. . .)".

In order to give his readers a sense of what the problem was and entailed, Linus Pauling resorted to the classic metaphor of the observer from Sirius, changing only the scale to make it commensurate with normal laboratory practice (Pauling, 1946):

"Let us imagine ourselves increased in size by the linear factor 250 000 000 – the commonly used factor in molecular models, which makes 1 Å, 10^{-8} cm, become approximately 1 inch (. . .) with this magnification we would become about equal in height to the distance from the earth to the moon".

Pauling has then his gigantic observer looking down at Manhattan: if he uses a microscope, the resolution is about 1000 feet (Central Park, Rockefeller Center, etc.). If he resorts to x-rays, then he will be able to gain 'complete information about the structure of objects smaller than about 1 foot in diameter, such as a storage battery, a small electric motor, a piece of cable, a small gear wheel, a bolt or rivet' (Pauling, 1946). By using an electron microscope, with a resolving power of about 10 feet, much additional information would be gained, for instance, about the overall shape of the Empire State Building. Yet, because of the unavailability 'of a method of exploring objects in the range 1 to 10 feet' (Pauling, 1946), human beings would be invisible, cars would appear as barely discernible and shapeless particles, and traffic through the 'veins and arteries' (Pauling, 1946) of New York City could not be followed.

By his neat little parabola, Pauling emphasizes the desirability of information about the molecular architecture of proteins:

"Until this problem is solved all discussions of the exact basis of biological reactions remain in some degree speculative. The polypeptide-chain structure of proteins proposed by Fischer is now generally accepted, and there is little doubt that the picture of folded chains held by hydrogen bonds, Van der Waal's forces, and related weak interactions in more or less well-defined configurations, as discussed eleven years ago by Mirsky and me (Mirsky and Pauling, 1936) is essentially correct. But this whole picture remains very vague (. . .)" (Pauling, 1946).

The main epistemological interest of this 1946 article is the analogy it vividly and forcefully propounds between the structure of a protein and the architecture of a building such as the Empire State: while Pauling refrains understandably of pushing the analogy too far and avoids making such detailed comparisons, it is clear that, to him, bonds between atoms are much like 'the steel girders of which the building is constructed'. Hence, the principles of structural chemistry (the existing knowledge about bond lengths and angles, atomic radii), as obtained on small organic molecules, will provide the blueprints, as it were, for the construction of the huge biological macromolecules. This is indeed what Pauling proceeded to perform in his work of the late 40s and early 50s on protein structure (Pauling and Corey, 1951).

And, on a more personal note, Pauling sees clearly that the study of biomolecular structure will allow him to bring together the two strands in his research (Pauling, 1946):

"There are two subjects that I am deeply interested in – structure, the detailed nature of molecules, crystals and cells, described in terms of their constituent atoms, with interatomic distances determined to within 0.01 Å, an interest that began in my youth and has received most of my attention until recent years, and the basis of the physiological activity of substances, an interest that is more recent but just as keen".

We shall turn now to a second example or case study documenting also the same analogy between macroscopic architecture of buildings and microscopic biomolecular structure. It is no accident if the analogy was perceived again by an x-ray crystallographer who, like Pauling, was also influential in making biochemists become more receptive to concepts of structural chemistry.

2. Aaron Klug, and the input from the geodesic domes of Buckminster Fuller*

Dr. Aaron Klug was trained in x-ray crystallography, and was a co-worker of Rosalind Franklin. He took over the leadership of the group after her untimely death in 1958. Dr. Klug received the Nobel Prize for chemistry in 1982, for devising special crystallographic methods for processing data from the optical diffraction of electron micrographs in order to 'see' the two-dimensional arrays which make up the protein coats of viruses, among other contributions.

One of his discoveries we shall be concerned with here is the concept of quasi-equivalence, as applied to isometric particles such as viruses. The original idea for the structural build-up of viruses (Crick and Watson, 1956) is the regular arrangement of identical sub-units on a surface. The biological advantage of using identical

---

*This section borrows heavily from a published article (Laszlo, 1983), and has greatly benefited as well from an exchange of letters with Dr. A. Klug.

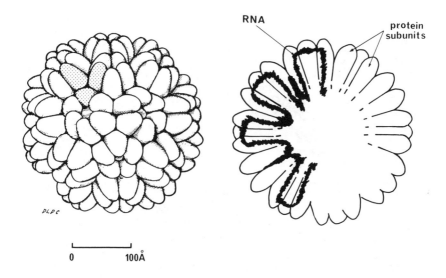

Fig. 3. (Left) A drawing of the turnip yellow mosaic virus as revealed by negative staining. It consists of 180 structure units. (Right) Schematic drawing to indicate the relation of the gross RNA distribution in the virus to the arrangement of protein subunits. (Courtesy of Prof. Aaron Klug.)

sub-units is that hence a minimum amount of information has to be encoded in the viral nucleic acid (Crick and Watson, 1956). Crick and Watson suggested that rod-shaped viruses are built by helical arrangements, and spherical viruses by cubic arrangements of protein sub-units. Let us focus on the second class, the spherical viruses. They 'would have to adopt the symmetry of one of the classical cubic point groups, namely tetrahedral, octahedral, or icosahedral' (Klug, 1979).

Let us remind the reader, in passing, that the icosahedron is one of the five Platonic solids, having 20 faces in the shape of equilateral triangles, 12 corners, and 30 edges.

Because of these three types of cubic symmetry, tetrahedral (point group 2:3), octahedral (4:3:2), and icosahedral (5:3:2), the maximum numbers of equivalent sub-units are 12, 24, and 60, respectively.

Plate 2. Professor Aaron Klug at the time of the discovery of quasi-equivalence. (Courtesy of Prof. Aaron Klug.)

"No regular arrangement of more than 60 is possible. The first experimental evidence from X-ray studies (. . .) by Finch and myself on turnip yellow mosaic virus (Klug and Finch, 1960) (. . .) pointed to icosahedral symmetry, a gratifying result, yet suggesting that there was somehow another special principle at work. Moreover, the first electron microscope and chemical data did not apparently agree with the X-ray ones: turnip yellow mosaic virus had 32 morphological units on its outside and contained about 150 protein units, whereas the X-ray results required 60, or a multiple of 60" (Klug, 1979).

The rescue or hint came from quite another sector! The visionary architect and engineer R. Buckminster Fuller had sent to Aaron Klug, 'when he heard about the earlier work by Finch and myself on poliovirus with its conclusion about icosahedral symmetry' (Klug, 1980), a book about his geodesic structures (Marks, 1960). Klug sent Caspar this book, and from the 'many pictures of triangular geodesic domes (. . .) Caspar made the connection; [Klug] came to the idea in a rather different way' (Klug, 1980). As he writes (Klug, 1980): 'Indeed, it was this that gave me the notion of quasi-equivalence'.

Quasi-equivalence is the recognition that, in forming a stable polyhedral shell, protein sub-units will tend to form as many bonds as possible with one another, and that 'the sub-units need not be in exactly equivalent environments to minimize the free energy of the system' (Makowski, 1980). And the greatest degree of quasi-equivalence is associated with the icosahedron (Caspar and Klug, 1962), which is also advantageous over the alternate cubic symmetries because it allows the packing of more sub-units: 60, instead of 12 or 24, as indicated above.

The discovery of quasi-equivalence came to Klug (1980), in his word from:

"one happy instance of misreading or 'misprision' [see below the explanation of this term] on page 196 of the above mentioned book (Marks, 1960). There is a picture of identical rods connected by strings which were built up into a large object with 270 struts (not a multiple of 60, but really composed of 4×60+60 halves, since dyad axes pass through 30 of the struts). I thought that the lack of equivalence was compensated by little wheels on which the strings ran and realised that the lengths of the strings could be adjusted by the wheels running along. To translate this into molecular language, this would mean that the building units are identical but the 'bonds' between them adjust themselves to build a larger structure with a multiple of 60. The irony of this is that some years later I received further details of this construction from Buckminster Fuller and it turned out that what I thought were wheels were not so at all, but were rather turn-buckles placed at fixed positions on the string and that Fuller had calculated the position for each class of turn-buckle!" (Klug, 1980).

This first-hand account of the discovery of quasi-equivalence is noteworthy in several respects.

(*i*) The first person to see the connection, i.e. to be alert to the possibility of his ideas extending into another field, was the remarkable polymath R. Buckminster Fuller; the definition he gave (Buckminster Fuller, 1970) of his geodesic domes came remarkably close to the principle responsible for the ordering of protein sub-units into a sheath around a virus:

"Three or more sets of angularly independent circularly continued push-pull paths must inherently triangulate by push-pull stabilization of opposite angles. Triangulation means self-stabilization, which creates omni-directional symmetry, which makes an inherent three-way spherical symmetry grid, which is the geodesic structure" (Buckminster Fuller, 1970).

(*ii*) We are thus provided with another key example of interdisciplinary cross-fertilization, again (as with Pauling) from architecture (geodesic domes, in this case) to biochemistry (icosahedral viruses).

(*iii*) This handsome story of the not-to-be-expected input from Buckminster Fuller is consistent with the cumulative character of the scientific enterprise (or, equivalently, of intellectual history as we reconstruct it), but with the added twist that every single step or contribution builds on its predecessor only after having slightly displaced, misunderstood or misinterpreted it.

(*iv*) The key word here is that of 'misprision', which Klug (1980) uses knowingly. This technical term is borrowed from literary criticism*; it was devised by Harold Bloom and presented most extensively by him in his book *A Map of Misreading* (Bloom, 1975). As indicated by this book title, this is the notion that each major writer

---

*When Klug (1980) explains to his correspondent the borrowing from architecture, he himself borrows a notion from another field (literary theory) in order to apply it to history of science.

is predominantly influenced by an earlier writer whom *he miscon-strues and thus reinterprets* in his own texts. Thus when Klug makes use of this sophisticated word in his letter to Dr. Matthews (Klug, 1980), he is referring to what could be called the generalized problem of translation: transfer of ideas from one sphere to another occurs necessarily with a slant or a deformation of some sort.

To sum up this chapter, a significant factor in the gradual awakening of a molecular biology has been the input from architectural ideas, through the agency of x-ray crystallographers, Linus Pauling first and foremost. Because of their professional training and traditions, in which model-building was natural, one might say part of the routine, model-building became installed, from early on, as an essential component for the understanding, first of protein function, later of nucleic acids and of other classes of biological macromolecules.

REFERENCES

Bloom, H. (1975) A Map of Misreading, Oxford University Press, Oxford.

Buckminster Fuller, R. (1970) I Seem to Be a Verb, Bantam Books, New York.

Caspar, D.L.D. and Klug, A. (1962) Cold Spring Harbor Symp. Quant. Biol., 27, 1-24.

Crick, F.H.C. and Watson, J.D. (1956) Nature, 177, 473.

Dickinson, R.G. and Pauling, L. (1923) J. Am. Chem. Soc., 45, 1466.

Klug, A. (1979) Heineken Prize lecture to the Royal Netherlands Academy of Sciences (courtesy of Dr. A. Klug).

Klug, A. (1980) Letter to Dr. R.E.F. Matthews, University of Auckland, New Zealand, of October 23 (courtesy of Dr. A. Klug).

Klug, A. and Finch, J.T. (1960) J. Mol. Biol., 2, 201.

Laszlo, P. (1983) Nouveau J. Chimie, 7, 145.

Makowski, L. (1980) In Biological Recognition and Assembly, Liss, New York, pp. 233-258.

Marks, R.W. (1960) The Dymaxion World of Buckminster Fuller, Reinhold, New York.

Mirsky, A.E. and Pauling, L. (1936) Proc. Natl. Acad. Sci. USA, 22, 439.

Pauling, L. (1946) Chem. Eng. News, 24, 1375.

Pauling, L. (1983) Letter to P. Laszlo of June 13.

Pauling, L. and Corey, R.B. (1951) Proc. Natl. Acad. Sci. USA, 37, 235.

Pauling, L. and Corey, R.B. (1951) Proc. Natl. Acad. Sci. USA, 37, 241.

Pauling, L. and Corey, R.B. (1951) Proc. Natl. Acad. Sci. USA, 37, 251.

Pauling, L. and Corey, R.B. (1951) Proc. Natl. Acad. Sci. USA, 37, 256.

Pauling, L. and Corey, R.B. (1951) Proc. Natl. Acad. Sci. USA, 37, 261.

Pauling, L. and Corey, R.B. (1951) Proc. Natl. Acad. Sci. USA, 37, 272.

Pauling, L. and Corey, R.B. (1951) Proc. Natl. Acad. Sci. USA, 37, 282.

Pauling, L. and Corey, R.B. (1951) Nature, 168, 550.

Pauling, L. and Corey, R.B. (1951) Proc. Natl. Acad. Sci. USA, 37, 729.

Pauling, L. and Sturdivant, J.H. (1928) Z. Kristallog., 68, 239.

Pauling, L., Corey, R.B. and Branson, H.R. (1951) Proc. Natl. Acad. Sci. USA, 37, 205.

Chapter 5

The Animate and the Inanimate. Crystals and Life

Our present-day familiarity with molecular models and, thus, with the architectural metaphor they embody has tended to obscure the basic, deep-rooted opposition between the symmetrical and highly geometric shapes associated with the inorganic world of minerals, and the baroque exuberance of living forms with their convoluted shapes. As beautifully expressed by D'Arcy Wentworth Thompson, the difference stems from the modes of construction themselves:

"(...) older naturalists called attention to a new distinction or contrast of form between organic and inorganic objects, in that the contours of the former tended to roundness and curvature, and those of the latter to be bounded by straight lines, planes and sharp angles, we see that this contrast was not a new and different one, but only another aspect of their former statement, and an immediate consequence of the difference between the processes of agglutination and intussusception" (D'Arcy Wentworth Thompson, 1942a).

Almost immediately, the British seminal thinker proceeds to dispel the false notion of symmetry as uniquely characteristic of the inorganic, while giving at the same time an explanation for its widespread occurrence among organic species:

"In all cases where the principle of maxima and minima comes into play, as it conspicuously does in films at rest under surface-tension the configurations so produced are characterised by obvious and remarkable *symmetry*. Such symmetry is highly characteristic of organic forms, and is rarely absent in living things – save in such few cases as *Amoeba*, where the rest and equilibrium on which symmetry depends are likewise lacking. And if we ask what physical equilibrium has to do with formal symmetry and structural regularity, the reason is not far to seek, nor can it be better put than in these words of Mach's: 'In every symmetrical system every deformation that tends to

[53]

destroy the symmetry is complemented by an equal and opposite deformation that tends to restore it. In each deformation, positive and negative work is done. One condition, therefore, though not an absolutely sufficient one, that a maximum or minimum of work corresponds to the form of equilibrium, is thus supplied by symmetry. Regularity is successive symmetry; there is no reason, therefore, to be astonished that the forms of equilibrium are often symmetrical and regular' (Mach, 1902).

A crystal is the perfection of symmetry and regularity; symmetry is displayed in its external form, and regularity revealed in its internal lattices. Complex and obscure as the attractions, rotations, vibrations and what not within the crystal may be, we rest assured that the configuration, repeated again and again, of the component atoms is precisely that for which the energy is a minimum; and recognize that this minimal distribution is of itself tantamount to symmetry and to stability" (D'Arcy Wentworth Thompson, 1942b).

This second quotation makes it plain that rest and equilibrium are prerequisites for the high symmetry of the crystal. Nowadays, with the advent of Prigogine's ideas about life structuring itself, far from thermodynamic equilibrium, when it acquires patterns of order which he terms 'dissipative structures', we have the reassuring feeling of treading again on firm ground, with such a restored clear-cut distinction between the animate and the inanimate.

For centuries, however, there had been a deep ambivalence about this very point: to many, the similarities between living forms and crystals seemed almost as important as their obvious differences. The oxymoron of 'the living crystal' has been one of the most extraordinarily powerful and productive metaphors. For a while, it did not occupy the scientific central stage. But literary writers kept it alive. And this metaphor is, I submit, an important factor in the truly predominant role x-ray crystallography and x-ray crystallographers have played, during the first half of the 20th century, in the emergence of biochemistry.

Thus, I have seen fit to devote a single chapter to this topic. Because, dating back to the alchemists, this image of 'the living crystal' has been shared by science and literature, it is impossible to avoid quoting some literary texts as well. I shall do so for the insight they give into some scientific concepts absolutely central to the history chronicled in this book.

1. Living crystallizations

As a start, let me immediately give an example – related to the subject matter in part III of this book, which deals with the hemoglobin story – of the impact of this metaphor in chemical physiology. It will illustrate also the resistance to this notion. I can do no better for this purpose than quote from the recent book by Debru:

"(. . .) in 1852 and 1853, Funke, Lehmann, Teichmann and others as well had succeeded in crystallizing the proteic content of the red cells from the arterial blood of numerous vertebrates. Henri Milne-Edwards declares in 1857 that the study of this hematocrystallin 'will probably shed a new light upon the intimate nature of the proteic substances, whose chemical composition has been represented to this day by arbitrary formulas only' (Milne-Edwards, 1857). Crystallization of hemoglobin, however, was challenged as to its reality. Robin and Verdeil, who were theoreticians of anatomical chemistry with a positivistic slant, severely criticized the idea of these crystals being formed indeed from the proteic substance of the red cells. These organized and anatomical objects, proteins, certainly cannot crystallize in the manner of chemical compounds, they argued. According to them, these crystals are to be identified with sodium phosphate from the serum which carried into its crystallization some proteic substances together with the pigment from the red cells (Robin and Verdeil, 1853a).

This unfair criticism based itself also on the extremely poor extant knowledge, at the time, about the role of hematin in respiration, and about the functional nature in vivo of organic compounds, such as the combination of iron with hematin. (. . .) As Lehmann underlines with respect to hematin, the intimate relationship between chemistry and physiology still had to be established (Lehmann, 1853). Indeed Lehmann believes that crystallization of the coloring matter of blood will have important consequences for founding a protein chemistry. It will provide also a more precise idea of the elemental analysis and of the constitution of the coloring matter. But the coloring matter corresponds to a class of objects, it crystallizes in systems differing from one animal species to another; it does not constitute an identical object, but rather a manifold of homologous substances whose comparative analysis should be made. The crystal, Lehmann says, shows an homology (Lehmann, 1853b)" (Debru, 1983).

Robin and Verdeil, as positivists, needed an opposition between the organized structures of the living state and the crystallinity of the inorganic realm. Such an opposition is relatively recent.

Traditional thought, especially alchemical thought, merged the inorganic with the organic; or more precisely, modelled the former on the latter. For the alchemists, metals germinate in the womb of the earth from seeds sown as emanations from the corresponding celestial objects, planets (Mars for iron, Venus for copper, etc.) or a star (the Sun for gold) (Eliade, 1978).

This is the notion of inorganic matter undergoing a growth (however slow a process it is) akin to development of an embryo. The reverse idea has also been extremely widespread and singularly productive, viz. that crystallization provides a model for the living state:

"The forces responsible for such arborescent growths of silver amalgam (*arbor Dianae*) and *arbre de Mars* (iron silicate and potassium carbonate) were often compared with those responsible for the formation of living structures (Florkin and Stotz, 1977), and such analogies can be found in the writings of Bacon, Hooke, and Coxe. Nehemiah Grew (1674) wrote that the processes of growth were the same as those responsible for the crystallization of salt and that plants were composed of crystalline formations (Ritterbusch, 1968). French scientists Beaumont (1676), Maupertuis (1744), and Buffon likewise employed crystals as useful analogues to living organisms; Guéneau de Montbéliard presented spontaneous generation as a crystalline process, and de la Métherie (1811) saw in crystallization the mechanism of organic growth and development: 'The seeds of the male and of the female, being mixed, act as two salts would and the result is the crystallization of the fetus' (...)" (Gilbert, 1982).

The contemporary French philosopher and literary critic, Gaston Bachelard, who pioneered a psychoanalysis of poetry based upon deep-seated images, in particular those of the four elements, devotes a whole chapter of his book *La Terre et les Rêveries de la Volonté* to 'crystalline dreams'. I shall mention briefly here what pertains to the present topic. Bachelard roots 'in the deep and rare beauty of an isolated crystal' such crystalline *rêverie*. As he writes:

"Once taken this lesson from the crystal, the imagination will transport such dreaming everywhere. A chemist from the seventeenth century, Guillaume Davisson, extends the principle of crystallization, not only to salts and to mineral substances, but also to honeycomb cells and to certain parts of plants, such as leaves and flower petals (Hoefer, ca. 1872)" (Bachelard, 1948).

Bachelard goes on to quote the identical fragment from de la Métherie and Gilbert, above; he has also this perceptive note:

"For every dreamer, the crystal is an active *center*, it draws to itself the crystalline matter. One often says that a crystal feeds from its mother-liquor. An author has inversed this image and claims that digestion is a crystallization" (Bachelard, 1948).

Sometimes, there is a more intimate and disturbing link between the crystal and an organic substance or fluid, as in this folk tale, quoted by a 17th century author to which Bachelard in turn refers:

"Diamond which resists to the strongest forces in the universe, to iron and to fire, concedes defeat, according to Pliny, and gives when presented with blood from the he-goat, if it comes straight from the animal, and still warm" (François, 1657, quoted by Bachelard, 1948).

All such fantasies about crystals, rather than vanishing into thin air, came to renewed vigor when the microscope led the enthralled viewers into the marvels and intricacies of a new world, resplendent with Pythagorean harmony, which many saw as a microcosm sheltering within its fold, cosmos-like, the grand design of God:

"What must we then say, when we behold every Species working as it were on a different Plan, producing Cubes, Rhombs, Pyramids, Pentagons, Hexagons, Octagons, or some other curious Figures peculiar to itself; or composing a variety of Ramifications, Lines and Angles, with a greater Mathematical Exactness than the most skillful Hand could draw them? (. . .) When therefore these Particles of Salts are seen to move in Rank and File, obedient to unalterable Laws, and compose regular and determined Figures, we must recur to that Almighty Wisdom and Power, which planned all the System of Nature, directs the Courses of the Heavens, and governs the whole Universe" (Baker, 1753).

Marjorie Nicolson has documented in detail the effect on the literary imagination of the 17th and 18th centuries of the invention of the microscope; for instance, *Gulliver's Travels* bears its unmistakable mark (Nicolson, 1976).

The contemporary literary imagination continues to thrive on images of perfection and of beauty drawn from the sight of minerals. Such imagery pervades the writings of the German Ernst Jünger

who, probably more than anyone, save the late French poet Saint-
John Perse, constantly draws upon his extensive knowledge both of
mineralogy and of alchemy. An often repeated figure of his echoes
the above quotation from Henry Baker: the microscopic world of
crystals is a model for the macrocosm. One is at a loss to select an
illustrative fragment. They are so numerous. And many would be
worth quoting, either for their total relevance to the topics under
consideration, or for the felicitous wording, or for both!

For instance, here is a page from Jünger's World War II diary as a
German officer in occupied France:

"Before breakfast a chance discovery moved me greatly. I had gone down into the
park and, with the idea of taking some token of remembrance from my stay, I
explored the white craters on the lawn looking for bomb splinters. The first cone gave
me the surprise of a much rarer find: the admirable petrification of a shell torn by the
explosion out of its chalk berth! I weighed it in my hand like a gift one has just
received. This sea-shell had a conical form, was almost of the length of the forearm,
shaped as a bulged spiral and it was cracked at the opening. Accordingly, the spindle
and the inner circonvolution, whose nacreous lustre was well preserved, were visible.
I, who had been escorted shadow-like by destruction during all these last days understood immediately the meaning of this discovery: Mount Mirail had been, in remote
times, a reef in this sea of chalk, in which lies, indestructible, the wonderful element of
which this castle itself, with its garden, is but an emanation, an ephemerary parabola,
in the manner of this shell" (Jünger, 1979).

Jünger goes on to say that such wonders remind him of the metamorphoses enabled by the Philosopher's Stone, turning everything
into gold. Before returning to our chemical theme proper, and by
way of introduction to the concept of crystals as germs for living
organisms, I can do no better than resort to another (and last) literary reference. It will be to a short story, *La Chambre Rouge*, by the
contemporary French writer André Pieyre de Mandiargues. He is
the prose equivalent of a Surrealist painter, such as Max Ernst; his
stories are memorable for the dream-like precision of the details, for
the cool description of the weirdest happenings which, more often
than not, are highly eroticized.

A young woman goes skiing together with her *fiancé* and with a
few other friends. At the end of the day, when she goes up to her

hotel room, it has a strange appearance, dominated by the color red:

"I am going to sleep in the heart of a carbuncle, had thought the young lady, or rather inside the cave-like core of a ruby not yet cut and refined, because the room had an irregular shape according to the laws of geometry and because the dominant color had less brilliance than it had a lustreless sumptuousness. On her finger, the ring which (her *fiancé*) had given her, and which held a pink beryl cut into a pear shape, glowed with a more active fire, as if the stone had lit from the furnace around it.

She undressed and got under the sheets, below the feather eiderdown. Before switching off the light, she removed her ring and had a look at it. It seemed to her, for the first time, that the pink stone and a diminutive diamond set below it assumed the curious shape of a question mark" (Pieyre de Mandiargues, 1963).

The rest of the story then unfolds: in a dream-like stance, as if he had arisen from the red crystal on her ring, in pitch darkness, and in total silence, a man – whom, at first, she believes to be her *fiancé* – rapes her.

Let us now return to the arborescent growths displayed by a number of inorganic crystals, since this analogy in shape between inanimate and animate matter served as the support for conceptualization of the living state on the model of crystallizations. I shall quote here the entry 'Végétation métallique' from the *Encyclopédie* of Diderot and d'Alembert. It was written by Gabriel-François Venel (1723–1775), a most remarkable chemical thinker, from the school of Rouelle, already mentioned and whom we shall encounter a number of times in this book:

"Even though the word 'vegetation' befits only plants properly speaking, still Chemists resort to it in order to describe some particular crystallizations, or other arrangements of any kind, whose external appearance resembles rather markedly that of plants; it is in this manner that Chemists name *Diana tree* or *philosophical tree* a silver *vegetation*, and that they name *Mars tree* another chemical *vegetation*, with some analogy with the former; this latter *vegetation* is an iron dissolution made by means of spirit of nitre (. . .)" (Venel, ca. 1753).

2. The crystallization analogy

Crystallization provided Theodor Schwann with an analogue to biological selectivity. His ideological slant in the late 1830s, intent upon banishing vital forces from biology (see Ch. 7), pushed him in this direction. He wrote thus:

"The attractive power of the cells manifests a certain degree of election in its operation; it does not attract every substance present in the cytoblastema, but only particular ones; and here a muscle cell, there a fat cell is generated from the same fluid, the blood. Yet crystals afford us an example of a precisely similar phenomenon, and one which has already been frequently adduced as analogous to assimilation. If a crystal of nitre be placed in a solution of nitre and sulphate of soda, only the nitre crystallizes; when a crystal of sulphate of soda is put in, only the sulphate of soda crystallizes. Here, therefore, there occurs just the same selection of the substances to be attracted" (Schwann, 1839; quoted by Florkin and Stotz, 1977).

Again, in a like manner as with Schwann, his ideology of a Marxist materialist may have drawn J.B.S. Haldane (1892-1964) to a similar analogy from the inanimate to the animate:

"The growth and reproduction of large molecules are not, it may be remarked, hypothetical processes. They occur, it would seem, in certain polymerizations which are familiar to organic chemists. In my opinion, the genes in the nuclei of cells double themselves in this way. The most familiar analogy to the process is crystallization. A crystal grows if placed in a supersaturated solution, but the precise arrangement of the molecules out of several possible arrangements depends on the arrangement found in the original crystal with which the solution is 'seeded'. The metaphor of seeding, used by chemists, points to an analogy with reproduction" (Haldane, 1932).

Notice in this paragraph of Haldane the back-and-forth motion: gene replication (or rather the doubling of chromosomes) is first likened to crystallization which, in turn, is likened to the reproductive process.

Rather similar metaphors would be used slightly later, in the 1940s, by Linus Pauling. He resorted to the analogy from crystallization in order to describe the complementarity between mutually adjusted parts of biomolecules, such as enzyme–substrate or antigen–antibody interactions. This metaphor of Pauling is important to our story, and I shall delve on it at some length in Chapter 6, section 4.

3. The crystal–gene analogy

In 1935, W.M. Stanley succeeded in crystallizing tobacco mosaic virus (TMV) using the standard techniques for purification of proteins, namely a combination of isoelectric precipitation and of salting out (with ammonium sulfate). This was an epochal find. That the pathological activity of a virus (which earlier could not even be seen with a microscope and whose presence could only be inferred from the symptoms once it had infected an organism) could be dormant inside a crystal, a suspension of which had a 'satin-like sheen' (Stanley, 1946), caused a sensation. Even though Stanley's isolation of crystals was quickly duplicated in many other laboratories, its announcement met with equally widespread disbelief:

"For a time there was great skepticism that the crystalline material could be tobacco mosaic, due chiefly to the old idea that viruses were living organisms" (Stanley, 1946).

But Stanley was able to show clearly that

"essentially all of the virus activity present in infectious juice could be isolated in the form of the crystalline material. (...) The same crystalline material has been obtained repeatedly from different batches of diseased plants obtained at different times of the year and under different growing conditions. The same material has been obtained from different kinds of plants, such as mosaic-diseased tomato, petunia, spinach and phlox plants" (Stanley, 1946).

Stanley sent this crystalline material to J. Desmond Bernal who, together with his co-worker Fankuchen, obtained x-ray pictures from the 280-nm-long rod-shaped particles of TMV which showed that the 'individual rods of tobacco mosaic virus have a regular inner structure of such a nature that each rod could be considered to be a crystal' (Stanley, 1946).

The crystallization by Stanley of the TMV was

"a philosophical as well as biological breakthrough, similar in its significance and interpretation to Wöhler's synthesis of urea, and it was recognized as such by biochemists. (...) Within a year, geneticist Hermann Muller was willing to claim that this crystallized virus was apparently a pure protein capable of 'autosynthesis' and that 'it represents a certain type of gene' (Olby, 1971). At the Genetical Congress in Edinburgh in 1939, no less than three crystallographers, Astbury, McKinney, and Gowen, put forth analogies between viruses and genes. Astbury said that viruses and chromosomes were the two simplest reproductive systems known. Thus, autocatalysis, which was thought to occur by crystalline duplication, was now seen to occur in viruses which could themselves be crystallized. Furthermore, since this process is also common to chromosomes, the chromosomes, too, must be crystalline in nature.

Thus, by 1940 the lines were being drawn. Biochemistry, concerned with intermediary metabolism and the energy that drives it, worked well within the tradition of flux and thermodynamics. However, the portion of the life sciences concerned with the transmission and expression of inherited characteristics rejected this view for the tradition of crystalline morphogenesis" (Gilbert, 1982).

Erwin Schrödinger published in 1944 his little book *What is Life?* in which he proposed, among other things, that genes can be thought of as aperiodic crystals. In order to explain the evolution of the ideas of the physicists about the bases of life, during the 1930s and early 1940s, as they found their expression – their paradigmatic statement, one might say – in the book by Schrödinger, I have to backtrack a little.

The departure point can be found in the ideas submitted by Niels Bohr in the early 1930s, when he reflected on the consequences and possible generalizations from the uncertainty principle; he presented these ideas in his *Light and Life* 1932 address to the International Congress of Light Therapy:

"(. . .) the existence of life must be considered as an elementary fact that cannot be explained, but must be taken as a starting point in biology, in a similar way as the quantum of action, which appears as an irrational element from the point of view of classical mechanical physics, taken together with the existence of the elementary particles, forms the foundations of atomic physics. The asserted impossibility of a physical or chemical explanation of the function peculiar to life would in this sense be analogous to the insufficiency of the mechanical analysis for the understanding of the stability of atoms" (Bohr, 1933).

These ideas of Niels Bohr were picked up by his student Max Delbrück, who gave them more substance by applying them more specifically to a sub-field of biology, genetics. In a seminal 1935 article Delbrück remarked:

"Whereas in physics all measurements must in principle be traced back to measurements of place and time, there is hardly a case in which the fundamental concept of genetics, the character difference, can be expressed meaningfully in terms of absolute units" (Timoféef-Ressovsky et al., 1935).

Accordingly, he made a strong case for the autonomy of genetics as a science, which '*must not be mixed up* with physical-chemical conceptions' (Timoféef-Ressovsky et al., 1935; emphasis added).

Delbrück notes the difference in meaning between the molecules of chemistry, large ensembles of identical molecules in an homogeneous medium, and the 'molecules' of genetics, typically a single molecule immersed in a chemically heterogeneous environment. Not only is the cellular environment irregular, it also undergoes continual change. Nevertheless, the gene as a molecule must preserve its identity:

"(. . .) when we speak of genes as molecules we are not so much thinking of their similar behavior but more generally of a well-defined union of atoms, supposing that the identity of two genes represents the same stable arrangement of the same atoms. The stability of this configuration must be especially great vis-à-vis the chemical reactions that normally proceed in the living cell; the genes can participate in general metabolism only catalytically" (Timoféef-Ressovsky et al., 1935).

Delbrück rationalized this stability of the gene molecule by having its atoms in deep potential energy wells, which they could escape from under the influx of very high energies only, an extremely rare occurrence which thus would account for the relative rarity of point mutations 'whose spontaneous frequency', Delbrück reckoned, 'could be as low as one per atom per 30 000 years if the activation energy exceeded kT by a factor of 60' (Stent, 1968).

This paper (Timoféef-Ressovsky et al., 1935), in Gilbert's words, 'was "rediscovered" for Schrödinger by the crystallographer P.P. Ewald, and it became the basis of Schrödinger's analysis of the gene as crystal' (Gilbert, 1982).

In my opinion, at least these two strands determined the maturing

64   CRYSTALS AND LIFE

of the ideas of Erwin Schrödinger: *(a)* the ideas of Niels Bohr on the
basic irreducibility of the living phenomenon, as later embodied by
the views of Delbrück on genetics, and by Delbrück's studies on the
bacteriophage; *(b)* the crystallization of the TMV by Stanley.

It is an interesting coincidence that the same year, 1935, saw the
first crystallization of a virus, appearance of the Delbrück paper, and
also the proposal by Joseph Needham for setting up a new research
institute to be devoted to the biological impact of crystal physics:
'Needham, as well as other embryologists, were aware of the recent
advances in crystallography and were impressed by Bernal's beliefs
that crystallography could help elucidate biological problems' (Gilbert, 1982).

In *What is Life?*, Schrödinger sought to explain the maintenance
of order and organization at the microscopic level. He adopted for
this purpose Delbrück's concept of the gene, which he expanded
upon by 'suggesting that the gene is an aperiodic crystal made up of a
succession of different isomeric components which carry the genetic
information: a remarkable anticipation of the Watson–Crick structure' (Hayes, 1977). Indeed, Schrödinger initiated right there the
informational school of the molecular biology that was to follow:

"Indeed the number of atoms in such a structure need not be very large, to produce an
almost unlimited number of possible arrangements. For illustration, think of the
Morse code. The two different signs of dot and dash in well ordered groups of not more
than four allow more than thirty different specifications" (Schrödinger, 1945).

This little book of Schrödinger's had a profound influence:

"The importance of this essay has been attested to by such researchers as Crick,
Watson, Wilkins, Luria, Benzer, and even Chargaff (...) Yoxen has studied the
influence of this volume and claims that it 'exerted on some scientists a powerful and
transient influence in suggesting and validating a particular line of research' (Yoxen,
1979). This line of research was predicated on the idea that crystallinity is a state
unifying all of matter and that 'the most essential part of a living cell - the chromosome fiber - may suitably be called *an aperiodic crystal*' (Schrödinger, 1945)" (Gilbert, 1982).

4. Final remarks

In rounding up this chapter, I find these to have been the prevailing and most significant statements made by the various authors cited:

(*i*) distinction of life (round, curved, organized, an energy flux) and of non-life (straight, angular, crystalline, an energetic standstill);

(*ii*) crystal growth stems from aggregation of like with like, and provides thereby THE criterion for purity of a substance;

(*iii*) the Pythagorean perfection embodied in a crystal, as well as in the Platonic solids, is a constant source of wonder;

(*iv*) the symmetry displayed by crystals is diagnostic of an equilibrium state; it is indicative of energy minimization: each atom goes to its pre-determined niche or position within the lattice; the symmetry of a crystal is synonymous with its inherent stability;

(*v*) conversely, the birth of a crystal remains highly mysterious; it provides a link with biology: it is conceptualized as akin to generation, to spontaneous generation even!

(*vi*) the arborescent growths of crystals seem also to be signs of a fundamental unity between the inorganic and the organic;

(*vii*) crystals are seen by scientists and writers alike as worlds in miniature, as replicas of the outside world; it is no surprise therefore, and in view of (*v*) and (*vi*) above – and despite point (*i*) –, that crystals provide models for more complex biological phenomena, such as the formation of tissues from cells (Schwann), replication of genes (Haldane), or biological complementarity (Pauling).

This last statement is pregnant with the point made throughout this whole chapter: rather than starting in the early 1930s with the first attempts by x-ray crystallographers at deciphering the structures of key biomolecules from diffraction patterns given out by their crystals, I submit that there has been a long and rich prehistory to this notion of 'crystals of life'. This notion proved to be extraordinarily fruitful for the emergence of a molecular biology, which became based primarily on determinations of crystal structures.

The second point I wish to make, in concluding this chapter, is the reiteration of the illuminating distinction made by Gilbert, when he contrasts the static, structural and crystal-based view of biophysics and molecular biology with the dynamic, 'whirlpool' and 'cascade'-based founding aporia of biochemistry, as the science of exchanges of metabolic energy, cf. (*i*) above (Gilbert, 1982).

REFERENCES

Bachelard, G. (1948) La Terre et les Rêveries de la Volonté, José Corti, Paris, ch. X.

Baker, H. (1753) Employment for the Microscope, Preface, iv-v; quoted by Nicolson (1976).

Bohr, N. (1933) Nature, 131, 421.

D'Arcy Wentworth Thompson (1942) On Growth and Form, 2nd ed., (a) p. 349; (b) p. 357. Cambridge University Press, Cambridge.

Debru, Cl. (1983) L'esprit des Protéines, Hermann, Paris, pp. 141-142 (my translation).

Eliade, M. (1978) The Forge and the Crucible: The Origins and Structures of Alchemy, University of Chicago Press, Chicago.

Florkin, M. and Stotz, E.M. (Eds.) (1977) History of Biochemistry, Comprehensive Biochemistry, Vol. 32, Elsevier, Amsterdam, p. 133.

François, R. (1658) Essay des Merveilles de Nature et des plus nobles Artifices, quoted by Bachelard (1948).

Gilbert, S.F. (1982) Persp. Biol. Med., 26, 151.

Haldane, J.B.S. (1932) The origin of life, in The Inequality of Man and Other Essays, Chatto and Windus, London, quoted by Gilbert (1982).

Hayes, W. (1977) Search, 8, 68.

Hoefer, F. (1872) Histoire de la Chimie, Vol. 2, p. 235, quoted by Bachelard (1948).

Jünger, E. (1979) Diary-I (1939-1940): Jardins et Routes, 1st ed., Bourgois, Paris; Plon, Paris, 1942 (my translation).

Lehmann, C.G. (1853) Lehrbuch der Physiologischen Chemie, Vol. 1, (a) p. 288; (b) p. 367, 2nd ed., Leipzig.

Mach, E. (1902) Science of Mechanics, p. 395; quoted by D'Arcy Wentworth Thompson (1942b).

Milne-Edwards, H. (1857) Leçons sur la Physiologie et l'Anatomie comparée de l'Homme et des Animaux, Vol. 1, p. 175, Paris.

Nicolson, M. (1976) Science and Imagination, Archon Books, Hamden, CT (1st ed., Cornell University Press, New York, 1956).

Olby, R. (1971) The Path to the Double Helix, Macmillan, London.

Pieyre de Mandiargues, A. (1963) Nouvelle Revue Française, 123, 393 (my translation).

Ritterbusch, P.C. (1968) The Art of Organic Forms, Smithsonian Institution Press, Washington, DC, p. 14.

Robin, C. and Verdeil, F., (1853) Traité de Chimie Anatomique et Physiologique Normale et Pathologique, Paris, vol. 2, p. 335.

Schwann, Th. (1839) quoted by Florkin and Stotz (19..).

Schrödinger, E. (1945) What is Life?, Cambridge University Press, New York.

Stanley, W.M. (1946) Nobel Lecture, December 12, in Nobel Lectures. Chemistry, 1942-1962, Elsevier, Amsterdam, pp. 137-159.

Stent, G.S. (1968) Science, 160, 390.

Timoféef-Ressovsky, N.W., Zimmer, K.G. and Delbrück, M. (1935) Nachr. Ges.
    Wiss., 6, 189; quoted by Stent (1968).
Venel, G.-F. (ca. 1753), entry 'VEGETATION METALLIQUE', Encyclopédie, ou
    Dictionnaire Raisonné des Sciences, des Arts et des Métiers, Diderot, D. et al.
    (1751-1780), 35 vols.; Neuchâtel, Vol. 16, p. 944 (my translation).
Yoxen, E.J. (1979) Hist. Sci., 17, 17.

Chapter 6

Complementarities between Molecules

Numerous molecular correlates of biological function rely upon the mutual fit of complementary parts in biomolecules. Examples of such complementarities are enzyme–substrate, drug–receptor, antigen–antibody, and nucleic base pairing interactions.

In this chapter, I shall first document the prehistory of this concept in Chinese (section 1) and in Western (section 2) alchemy. As a time probe, I have chosen the key-and-lock analogy of Emil Fischer at the end of the 19th century (section 3) (Paul Ehrlich evolved during the same period the related notion of antibody receptors). Finally, to examine a closer period yet, I shall delve into the concept of complementarity held by Linus Pauling: as we shall see, it remained a constant factor in his thinking, something like Ariadne's thread to guide him through the maze of biological complexity (section 4).

1. The affinity between chemical complementarities in Chinese thought

I shall borrow heavily in this section from the writings of Joseph Needham, the biochemist and historian of science who has written the monumental *Science and Civilisation in China* series. I shall draw in particular from an article extracted from Part 4, Volume V of that series (Needham, 1977–1979). Well before the Hellenistic proto-chemists, Chinese thought had a coherent body of doctrine about interactions and mutual resonances between objects belonging to the same categories.

"Things behaved in particular ways (. . .) because their position in the ever-moving cyclical universe was such that they were endowed with intrinsic natures which made that behaviour inevitable for them. (. . .) They were thus organic parts in existential dependence upon the whole world-organism. And they reacted upon one another not so much by mechanical impulsion or causation as by a kind of mysterious resonance (. . .) Chinese thinkers (instinctively adhered) to what was essentially a prototypal wave-theory, the reciprocally dependent rise and fall of the Yin and Yang forces (. . .)".

And, among a number of quotations he makes from various Chinese treatises, Needham gives this eloquent and amusing example from Li Shih, in the 12th century:

"When the tiger roars, the wind rises. When the dragon gives tongue the clouds gather. The lodestone attracts needles. Amber attracts bits of straw (literally: mustard seeds). After coming into contact with crabs lacquer will not concrete. Lacquer added to hemp (-seed oil) makes it bubble. (. . .) Crude salt preserves piles of eggs. The gall of the otter cracks (literally: divides) wine-cups. [All these phenomena occur because the *Chki* (pneumata) of these things are in sympathy (*chki chhi shuang chih*) and thus bring about mutual resonance (*hsiang kuan kan yeh*).]"

During the period between the 3rd and the 7th century an alchemical treatise was written entitled *Tshan Thung Chhi Wu Hsiang Lei Pi Yao*, which 'was presented to the throne with a commentary by Lu Thien-Chi (. . .) between +1111 and +1117'. Needham gives the following table of chemical categories taken from this work.

He makes this important remark upon examination of this table:

"We wish to remark only on two especially interesting entries, No. 7 and *b*, which indicate clearly there was a gradation of Yang-ness and that, while mercury might be female to sulphur, it would act as male to silver."

2. Complementarities in Western alchemy

The notion of complementarity was basic to alchemy. One might say that each step in the alchemical opus required achieving some form

TABLE IV

A Chinese table of chemical categories (after Needham, 1977–1979)

Lei	Yang	Yin	Neutral
1	cinnabar	mercury	
2	realgar	orpiment	
3	realgar	sal ammoniac	
4	honey and fritillary corns	orpiment	
5		arsenious acid	clay
6	sulphur	magnetite	
7	sulphur	mercury	
8		mercury; orpiment	
9	cinnabar	vinegar	
10	lead	mulberry ashes	
11	litharge	tin	five coloured clays
12	cinnabar	bronze coins	
13	Persian brass fragments	mercury	
14	copper carbonate	mutton fat	
15	blue copper carbonate	red haematite	
a	lead	calomel	
b	mercury	silver	
c	red salt	alum	
d	copper carbonate	silver	
e	red salt	calomel	

of completeness. I shall use first a quotation from a French alchemical treatise of the 17th century:

"For this reason they are all named imperfect metals, as if one said they were neither made nor completed, but yet had to be perfected, and thus they desire and always yearn for perfection, being on their way to acquire it; which they can do only by the Elixir or by the Stone perfect in the white or in the red, because they are dead as soon as they are detached from their matrix in the mine; the Elixir is alive, however, and animates the Mercury of all Metals, being their seed, and performs a kind of resurrection somewhat similar to that undergone by the various seeds of plants" (*Le Filet d'Ariadne*, 1695a).

This fragment is suffused with the Platonic notion of complementarity, which fertilized all of Western alchemy, starting with the Alexandrine alchemists. Here is for instance what Socrates explains in Plato's *Symposium*:

"Hence, this man as well as anyone else desires something; it is what he does not have access to he desires, it is what is not present; and what he does not own, what he is not personally, these are more or less the objects of his desire, of his love" (Plato, ca. 368 BC).

Hence, for the alchemists it was all-important to identify and make combined complementary entities, such as solvent and solute:

"One will be the dissolvent and the other will be the dissolved matter; the one is fixed and the other (is) volatile, and from the union of these two the so marvellous child of the Sun is born" (*Le Filet d'Ariadne*, 1695b),

or such as liquid and vapor:

"The union of the spirit with the humid radical is so much the stronger if the composite is freed from the excremental impurities; it is, as the Philosophers say, (like) the sky and the earth conjoined and reunited; it is the brother and the sister, the husband and the wife embracing one another very closely" (Le Breton, 1722a).

The same Le Breton writes elsewhere that 'the invisible is volatile and the visible is fixed' (Le Breton, 1722b), and that

"The secret animal is figured with a circle made of two snakes, one winged, the other wingless; which stand for the two spirits, fixed and volatile, united together" (Le Breton, 1722c).

An alchemical textbook, also French, from the latter part of the 17th century, will provide an illustration of the Platonic myth of the androgynous, as

"The male and the female united in the Philosophal Mercury; i.e. when the two sexes, the male and the female, are conjoined in the black color very black, which is perfect putrefaction: then water is converted into earth, and the former enemies are made into friends (...)" (*Dictionnaire Hermétique*, 1695).

There was a long distinguished posterity to this alchemical notion of the essential complementarity of opposites. Without having to invoke Hegelian dialectics and Karl Marx seizing upon this notion as the driving force for historical change, a couple of examples of the first rank, much closer to chemistry and to biological chemistry, will suffice. Newton, it is well known, maintained throughout his career an active interest in alchemy. From his jottings on his copy of the alchemical book *Secrets Revealed*, some of the continuity of his alchemical thinking can be reconstituted. Here is, for example, how Newton conceived of the stage following that of the black, dead *materia* just mentioned:

"The utter desolation of the dead material is soon relieved. The soul acts as an attractive 'medium' or mediator, and it begins to attract and unite the body and spirit to itself. The three essential principles of the material are thus 'reconjoined in order to a resurrection' " (Dobbs, 1979).

Another distinguished thinker, who was not only a poet and novelist and playwright, but also an amateur scientist whose influence on biochemical thinking should not be undervalued, was Goethe. During his formative period, in his late teens, he became extremely interested in alchemy, and became familiar with the treatise by Kirchweger, *Aurea Catena*. In this book, Kirchweger gave the following list of complementary principles:

Sulfur	*Salt*
Acid	Alkali
Spirit	Body
Father	Mother
Male seed	Female seed
Universal active principle	Universal passive principle
Heaven and air	Water and earth
Steel	Magnet
Hammer	Anvil

(Kirchweger, 1723).

This dualistic list was taken over by Goethe, who presented it as follows, in his *Theory of Colors*:

Plus	*Minus*
Yellow	Blue
Effectiveness	Deprivation
Light	Shadow
Bright	Dark
Power	Weakness
Warmth	Cold
Proximity	Distance
Repulsion	Attraction
Related to acids	Related to alkalis

(Goethe, 1810).

This table of Goethe is interesting, not only for its illumination of some of Goethe's literary writings, such as *Wahlverwandtschaften (Elective Affinities);* but also because it illustrates a prevailing theme in Western thought. This is the notion that qualities are determined by quantities: any observable property is accountable in terms of admixture to some pure essence, A, of variable amounts of its opposite $\bar{A}$.

A case at hand, basic to chemistry, is the notion of the strength of acids as determined by the quantity of the opposite and complementary principle, which might be termed alkalinity. This is a 'leitmotiv' already in the 18th century:

"Acids are bodies stiff, elongated, pointed, cutting and altogether apt to insinuate themselves in some sort of sheath of porous and spongy bodies named alkalis.

Bases or alkalis are spongy and porous bodies in which, like so many kinds of sheaths, home themselves stiff, long, pointed and cutting bodies named acids.

To give a perceivable idea of the ones and the others, one often compares an acid enclosed in its alkali to a sword pushed back into its scabbard" (Paulian, 1781).

Likewise, when in 1753 Venel is intent upon presenting in the best possible light chemistry and chemical thinking against Newtonian physics, at the time triumphant, in his 'Chemistry' entry in the

Encyclopédie, he gives a very similar image of dissolutions (which clearly antedates the concept of an enzyme–substrate interaction):

"(. . .) chemists imagine dissolutions with the very concrete image of a solvent (*menstrue*) armed with stiff, solid, massive, cutting parts, on one hand; and as a body pierced with an infinity of *pores proportioned to the mass and even to the shape* of the parts of the solvent, on the other hand (. . .)" (Venel, 1753).

These brief indications will suffice to point to the ancient origin of concepts of modern chemistry and biochemistry, such as complementary acids and bases as defined by Gilbert Lewis, or nucleophilic humps fitting into electrophilic hollows as proposed by Christopher Ingold to help elucidate organic reaction mechanisms. Emil Fischer had in mind something closely similar when he formulated in 1894 his lock-and-key analogy to illustrate the stereoselective action of an enzyme on a sugar molecule.

3. The concept of complementarity and the lock-and-key analogy of Emil Fischer

In the final years of the 19th century, during the period 1894–1898, Emil Fischer started studying the action of yeast upon various hexose sugars. He found a relationship between the efficiency of the enzyme and the configuration of the asymmetric carbon center undergoing the reaction (Fischer, 1894). The results were so clearcut that Fischer felt the need for an analogy: the end of his article is devoted to this simile. Fischer pointed to the resemblance with a lock and key, an image which has durably impressed itself since on the collective mind of biochemists.

If I may be forgiven a longish quotation, here is, in English translation, what Emil Fischer wrote:

"To take an image, I claim that the enzyme and the glucoside have to be put together as the lock and the key [*Schloss und Schlüssel*], in order to effect a mutual chemical action. Such a representation, in any case, has gained in plausibility and in value for such stereochemical researches, after the phenomenon itself has been brought from the biological realm to (that of) pure chemistry [*die Erscheinung selbst (ist) aus dem biologischen auf das rein chemische Gebiet verlegt*]. This phenomenon contributes to broadening the theory of asymmetry, without being a direct consequence of it; for the notion, according to which the geometric construction of molecules, by way of the [enantiomeric] forms [*Spielbildformen*] has such a great influence on the play of chemical affinities, could in my opinion be established only on the basis of new experimental observations. The experience we have so far, that the two salts built for asymmetric components can be recognized by their solubility and melting points, is assuredly not sufficient. Whether one finds such facts for complicated [*complicirte*] enzymes only, or also soon with simpler asymmetric agents, is as little doubtful as the necessity of the enzyme for the discovery of the configuration of asymmetric substances.

The notion that enzyme efficiency to such a high degree is determined by molecular geometry should be profitable also to physiological research. Yet more important in this respect, since it seems to me to prove it, are that the multiple differences found earlier between chemical activity of the living cell and the working of chemical agents do not exist really in relation to molecular asymmetry. Through this, the analogy stressed so often by Berzelius, Liebig, and others, of the 'living and inanimate ferment' [*lebenden und leblosen Fermente*] is restored to its full significance [literally: is put back again into not non-existent points]".

Assuredly, Fischer formulates the analogy with the actual turning of a key in a lock, an action which opens or closes a door, which switches on (or off) a process, with hydrolysis reactions in mind.

Among aldohexoses, only D-glucose, D-mannose, and D-galactose (the latter, to a much reduced extent); among ketohexoses, only D-fructose, are liable to the action of the ferment; and their L-enantiomers are not affected. Furthermore, Emil Fischer recognized the presence of a structure–activity correlation: carbons 3, 4 and 5 had similar configurations in the above mentioned sugars, shown below (enclosed in a box, for clarity):

D-fructose   D-glucose   D-mannose   D-galactose

If the glucoside has the adequate configuration for a particular enzyme, then it is acted upon; if it does not have the correct configuration, then it escapes the reaction unscathed. Hence, we see that the 'lock-and-key' analogy is equivalent, in logical terms, to an A+B pair of propositions of the form: 'if A (is true), then B (is also true)'. As Fischer wrote, 'the enzyme and the glucoside have to be put together as the key and the lock in order to [react together]'.

Another point stressed explicitly by Emil Fischer is the high efficiency of enzymes; he refers to 'enzyme efficiency to such a high degree'. As soon as the key has entered the lock and opened it, enzymatic catalysis unfurls and the reaction proceeds rapidly to completion.

The lock-and-key analogy is also appropriate on another count: in real life, keys and locks are constructed as rather complex combinations complementary to one another, so that a given key can only fit (ideally) a single lock. Likewise, an enzyme is a complicated entity for Emil Fischer.

However, Fischer hastens to assure his reader that, far from being a piece of magic, enzymes can be understood by reference to the basic concepts of the organic chemistry of simpler molecules: in the same way as his own work brings the phenomenon of enzymatic action 'from the biological realm to (that of) pure chemistry', likewise there is continuity between enzymes and simpler molecules. To further the same, crucial point Emil Fischer makes use of the fascinating, oxymoronic reference [the oxymoron is a figure of speech binding two opposites; the 'black sun' is an oxymoron] to 'the living and inanimate ferment': for him, the enzyme has an hybrid nature which does not restrict itself to the sphere of biology or to the sphere of chemistry. The enzyme acts as a go-between. It introduces chemicals into biological pathways. The enzyme itself ignores and transcends the distinction between life and non-life (and perhaps here, unconsciously or semi-consciously, Emil Fischer draws upon the image of a door, of a partition between two separate compartments evoked by the lock-and-key analogy).

This apparent paradox, which ran counter to the vitalist dogma, was important to Emil Fischer. He felt a little uneasy about it: this is the one point where, in the text I have quoted, he comes close to using an argument from authority and where he summons for support the names of prestigious chemists from the past, 'Berzelius, Liebig, and others'.

The attack upon vitalism is present; I am not making it up. It occurs in two stages: at the beginning of the first paragraph I have quoted, Emil Fischer shows rather delicately his hand with the statement about translating the phenomenon of enzymatic activity from the biological sphere to the chemical; and in the second paragraph, he more forcefully reduces to nil the 'multiple differences (. . .) between chemical activity of the living cell and the working of chemical agents'.

The lock-and-key analogy, in its original presentation under analysis here, is also valuable and perceptive because it is formulated with reference to chiral substrates. Fischer stresses repeatedly the relationships between the way enzymes function, and the properties of asymmetric substances. He refers obviously to the theory of the asymmetric carbon of Le Bel and Van 't Hoff.

The 'theory of asymmetry' was formulated as a *geometrical* concept. Likewise, the lock-and-key image Fischer resorts to is that of the mutual fit of two three-dimensional objects. Emil Fischer in this seminal text or definition thus sets the program of future researches on enzymes, which would unfold for a little over half-a-century after publication of his article: to elucidate in detail the geometrical complementarities between an enzyme molecule and its substrate.

To close with this cursory textual analysis, I shall submit a conjecture: maybe, an added impetus for the choice by Emil Fischer of his felicitous analogy was the assonance, in the German language, of the two nouns 'Schloss' and 'Schlüssel'. The nice euphony of these words superimposes itself on the harmony between their referents.

4. The concept of complementarity for Linus Pauling

Style in science is all-important. Not only does it distinguish between the individual contributions of two scientists who otherwise shared in the same *episteme* (Foucault) or *paradigm* (Kuhn). In this first sense, style could be taken merely as the way of presenting the results – as the icing on the cake, one might say. But there is also another, deeper-seated sense in which the style of the creative scientist is a prime component of the scientific enterprise. This second sense has to do with design of the operational procedure and with the key concepts informing (implicitly or explicitly) research; defining the field of action, identifying a research problem, imagining a line of attack - all these depend upon the mental make-up of the individual scientists, upon a few basic ideas which one might term *themata*, to borrow and extend somewhat this term from Gerard Holton. This second sense has been emphasized also by Gunther S. Stent as the essential factor ensuring the *uniqueness* of a scientific discovery (Stent, 1972).

The contribution of Linus Pauling to modern biochemistry has been of overwhelming importance. He has been responsible, perhaps (or probably) more than any other single individual, for the penetration of molecular concepts into biochemistry. Hence it is worthwhile, I submit, to go beyond the factual listing of his achievements; and to study also at least the main components of his scientific style.

Nevertheless, however important the contributions by any individual scientists - and those by Linus Pauling are a prime example - the present book cannot turn into biography, let alone hagiography, nor into parcelled-out scientific history. And, even though my personal interest (or prejudice) goes to intellectual history rather than to what I feel is the more restricted scientific history, this book emphasizes more the collective endeavor than the individual breakthroughs, however inspired the individual. Because of these limitations in the thrust of this volume and in its length, I can merely sketch here some of the most outstanding features of Pauling's scientific style as they influenced his science.

This will be illustrated by Pauling's concept of complementarity. He first wrote about complementarity in an article he published with Max Delbrück in *Science* to refute some ideas of P. Jordan about the biological importance of interactions between identical or like partners (Pauling and Delbrück, 1940). Since this is the first statement by Pauling of his views about complementarity, it is worth quoting at some length:

"It is our opinion that the processes of synthesis and folding of highly complex molecules in the living cell involve, in addition to covalent-bond formation, only the intermolecular interactions of Van der Waals attraction and repulsion, electrostatic inter-

Plate 3. Linus Pauling (photograph by Jill A. Cannefax, courtesy of Professor Pauling).

actions, hydrogen-bond formation, etc., which are now rather well understood. These interactions are such as to give stability to a system of two molecules with complementary structures in juxtaposition, rather than of two molecules with necessarily identical structures; we accordingly feel that complementariness should be given primary consideration in the discussion of the specific attraction between molecules and the enzymatic synthesis of molecules.

A general argument regarding complementariness may be given. Attractive forces between molecules vary inversely with a power of the distance, and maximum stability of a complex is achieved by bringing the molecules as close together as possible, in such a way that positively charged groups are brought near to negatively charged groups, electric dipoles are brought into suitable mutual orientations, etc. The minimum distances of approach of atoms are determined by their repulsive potentials, which may be expressed in terms of Van der Waals radii; in order to achieve the maximum stability, the two molecules must have complementary surfaces, like die and coin, and also a complementary distribution of active groups".

The die-and-coin analogy is interesting; we shall see how it will evolve later into a more chemically based and yet more fruitful image. Indeed, Pauling refers repeatedly to the same metaphor when he wants to convey the perfection, the exquisite mutual compatibility of complementary chemical groups engaged in biological recognition at the molecular level, whether in enzyme–substrate, antigen–antibody, or drug–receptor interactions. Let me quote first from his 'Molecular Architecture' article in *Chemical and Engineering News* (Pauling, 1946):

" . . . the specificity of the physiological activity of substances is determined by the size and shape of molecules, and (there is now very strong evidence) that the size and shape find expression by determining the extent to which certain surface regions of two molecules (at least one of which is usually a protein) can be brought into juxtaposition – that is, the extent to which these regions of the two molecules are complementary in structure. (. . .) The one general chemical phenomenon with high specificity is closely analogous in both its nature and its structural basis to biological specificity [sic]: this phenomenon is crystallization. There can be grown from a solution containing molecules of hundreds of different species, crystals of one substance which are essentially pure. The reason for the great specificity of the phenomenon of crystallization is that a crystal from which one molecule has been removed is very closely complementary in structure to that molecule, and molecules of other kinds cannot in general fit into the cavity in the crystal or are attracted to the cavity less strongly than a molecule of the substance itself".

I should like, before turning to another use by Pauling of the same metaphor, to list what I see as its main ingredients:

(i) recourse to an analogy from the inorganic realm, the crystallization phenomenon, to explain 'physiological activity';

(ii) the notion of complementarity as caused by incompleteness of structure: a vacancy is created in the regular structure of a crystal, and the cavity thus generated is specific only, or predominantly, to the molecule removed originally from it;

(iii) differential affinity for the cavity between various potential guests reflects their greater or smaller 'distance' from the structure of the ideal guest molecule.

Let us consider now the reiteration by Pauling of this same metaphor in his article 'Nature of forces between large molecules of biological interest' in *Nature* (Pauling, 1948) (this paper reproduces Pauling's Friday Evening Discourse, at the Royal Institution of Great Britain, London, delivered on 27 February 1948):

"There is a highly specific phenomenon, of the chemistry of simpler substances that is closely analogous in its nature and its cause to the highly specific phenomenon of serological interaction; namely the phenomenon of crystallization. Chemists are accustomed to using the process of crystallization as a method of purification: a crystal growing in a complex mixture of molecules is able to select in the mixture just the molecules of one kind, rejecting all others. Thus pure crystals of sugar may deposit from a jam in which there are molecules of thousands of different substances. The specificity of crystallization is the result of the same striving toward complementariness and the operation of the same interatomic and intermolecular forces that are responsible for the specificity of antibodies. A molecular crystal has the structure that gives it the greatest stability which would result from the maximum amount of attraction for each molecule in the crystal and the surrounding molecules. Each molecule in the crystal is then in a cavity that conforms in shape to the shape of the molecule itself. The molecule may be described as complementary in structure to the remainder of the crystal; and other molecules, with different shape and structure, would not fit into this cavity nearly as well, and in general would not be incorporated in the growing crystal. We may hence say that life has borrowed from inanimate processes the same basic mechanism used in producing those striking structures that are crystals, with their beautiful plane faces, their unfailingly constant interfacial angles, and their wonderfully complex geometrical forms".

To comment again very briefly on this quotation before proceeding to yet another:

(iv) biological specificity is the *analogon* of the self-selection of a pure substance from a complex mixture;
(v) there is a 'striving for complementariness';
(vi) the enormous molecules of biological interest share with 'inanimate' crystals this feature of mutual adaptation or fit between 'wonderfully complex geometrical forms'.

I shall quote finally from Pauling's Harrison Howe lecture, which he gave in Rochester, NY, on February 4, 1946 (Pauling, 1946a), a text which chronologically antedates the two I have just quoted from:

"The reaction shown by simple chemical substances which is analogous to that of specific combination of antigen and antibody is the formation of a crystal of a substance from solution. A crystal of a molecular substance is stable because all of the molecules pile themselves into such a configuration that each molecule is surrounded as closely as possible by other molecules – that is, if a molecule were to be removed from the interior of a crystal, the cavity that it would leave would have very nearly the shape of the molecule itself. We can say that the part of a crystal other than a given molecule is very closely complementary to that molecule. Other molecules, with different shape and structure, would not fit into this cavity nearly so well, and in consequence other molecules in general would not be incorporated in a growing crystal. This is the explanation of the astounding chemical process of purification by crystallization – from a very complicated system, such as, for example, grape jelly, containing hundreds of different kinds of molecules, crystals which are nearly chemically pure may be formed, such as crystals of cream of tartar, potassium hydrogen tartrate".

Features noteworthy here (and which of course are also present in the other quoted occurrences of the same crystallization metaphor) are:
(vii) the *Gedankenexperiment* of removing a single molecule from a crystal;
(viii) purification by crystallization is an 'astounding chemical process', because it is such a highly specific recombination of like with like;
(ix) conceptually, the driving force in this recombination of like with like, rather than being simply the attraction of a molecule by its neighbors is more subtly the attraction of a molecule by its host cavity; in other terms, when exit of a molecule creates a local vacan-

cy, there is a very strong restoring force between the empty space thus generated and the abstracted molecule;

(x) the growing crystal is like a sieve capable of extremely fine discrimination within even very complex mixtures.

After having gone through these three versions, or expressions, of a single notion formulated when Linus Pauling was in his mid-40s, we can formulate now some of the elements of his scientific 'style':

(a) The unity of chemistry and biochemistry: there is continuity between the properties of large biomolecules and the chemistry of simpler substances. Analogies can and should be found between the most wonderful phenomena of physiological activity, such as serological agglutinations, and the more familiar phenomena of elementary test-tube chemistry.

(b) Structural explanations are the most powerful; they have an unrivalled basic simplicity, hard to challenge.

(c) When studying a complex phenomenon, the search for an *analogon* or model is all-important: by turning to a better understood question, the mind will be able to identify more quickly, more clearly, those features which are the key for understanding the conceptually more remote phenomenon.

(d) The interlocking or assembly of parts, as a skilled craftsman might perform, as a child also clumsily emulates when playing with blocks, or yet better when putting together a jigsaw puzzle, is an image for the perfection and harmony in the productions of nature.

REFERENCES

Dictionnaire Hermétique (1695) Laurent d'Houry, Paris, p. 9.
Dobbs, B.J.T. (1979) Ambix, 26, 145.
Fischer, E. (1894) Ber. Dtsch. Chem. Ges., 27, 2985.
Goethe, J.W. von (1810) Farbenlehre, para. 695–696.
Gray, R.D. (1952) A Study of Alchemical Symbolism in Goethe's Literary and Scientific Works, Cambridge University Press, Cambridge, pp. 10 and 111.
Kirschweger, J. (1723) Aurea Catena Homeri, Frankfurt; Leipzig, p. 93; quoted by Gray (1952).
Le Breton (1722) Les Clefs de la Philosophie Spagyrique, Claude Jombert, Paris (my translation) (a) p. 12; (b) p. 25; (c) p. 32.
Le Filet d'Ariadne. Pour entrer avec Sureté dans le Labirinthe de la Philosophie Hermétique (1695) Laurent d'Houry, Paris (my translation) (a) p. 27; (b) p. 55.
Needham, J. (1977–1979) Ἐπετηρις του Κεντρον Ἐπιστημονιχων Ἐρευνων, IX, Λευκωσια, p. 21.
Paulian, A.H. (1781) Dictionnaire de Physique, 8th ed. (my translation).
Pauling, L. (1946a) Chem. Eng. News, 24, 1064.
Pauling, L. (1946b) Chem. Eng. News, 24, 1375.
Pauling, L. (1948) Nature, 161, 707.
Pauling, L. and Delbrück, M. (1940) Science, 92, 77.
Plato (ca. 368 BC) Symposium 200 (with reliance upon Ryle, 1966 for the date of composition of this dialogue, and in my translation based upon the French translation of Robin, 1950).
Robin, L. (1950) transl. Platon, Le Banquet, Gallimard, Paris.
Ryle, G. (1966) Plato's Progress, Cambridge University Press, Cambridge, pp. 218–219.
Stent, G.S. (1972) Sci. Am., 227, 84.
Venel, G.-F. (1753) in Diderot, D. and d'Alembert, Encyclopédie, ou Dictionnaire raisonné des Sciences, des Arts et des Métiers (35 vols., Neuchâtel, 1751–1780), entry 'CHIMIE', vol. 3, pp. 407–437 (emphasis added, my translation).

Chapter 7

Philosophical Undercurrents of the Resistance to the Notion of Biomolecules as Ordered Polymers

That the molecule of DNA is the reservoir for heredity, that the molecules of enzymes perform the reactions of metabolism – to pick but two examples out of a much longer series – have by now become well-established and quite ordinary. It is hard for us to imagine any other conceptual framework. Yet these are extremely new notions in intellectual history. One might argue, with considerable justification, that their protracted emergence has been due to deeply ingrained vitalistic prejudices among physiologists and biological chemists during the whole 19th century and the first quarter of the 20th century.

Hence this chapter will chronicle selected episodes, during the former period. These will illustrate the successful resistance by original thinkers, in France and in Germany, to the threatening takeover by Newtonian rational mechanics of their field of physiology. With the crucial help of philosophies, those of Condillac and of Kant primarily, they evolved a modernized version of vitalism, which was intellectually respectable, and which could guide their research.

As we shall see, we owe to a large extent the autonomy of biochemistry as a discipline to these philosophical discussions that occurred at the end of the 18th and at the beginning of the 19th centuries. Do the benefits from what may also be considered as a rear-guard action by the vitalists outweigh the liabilities? This is not for an objective history to adjudicate, especially when the fight between materialists and vitalists during these periods looks very much like a game having ended in a draw.

1. The mechanistic view of life processes

My starting point is the etymology of the word 'mechanism'. It is
worthwhile, for this purpose, to quote an astute dissection of this
term by a great contemporary French philosopher of science:

"Mechanism, it is known, comes from μηχανη whose meaning of *device* (*engin*) brings
together the two meanings of lure and stratagem, on one hand, and of machine, on the
other. One may question if these two meanings do not merge into one. Invention and
use of machines by men, technical activity in general, are they not what Hegel terms
the trickery of reason (*Logic of the Small Encyclopaedia*, § 209)? The trickery of
reason, this is to reach one's own ends by means of objects acting upon one another in
conformity with their own nature. A machine, in essence, offers a mediation; or, as
mechanics are wont to say, a relay. A mechanism does not create anything, and this is
what its inertia (*in-ars*) consists of, but it can be built only by art and it remains
basically a trick (*une ruse*). Mechanism, whether as scientific method or as philosophy, is thus the postulate implicit in any use of machines. (. . .) But is it not legitimate to conclude thus that the theory of the living state as a machine is a human
trickery which, were it to be taken literally, would eradicate life"? (Canguilhem,
1965a).

This unfamiliar view of mechanism is nevertheless fully consistent
with the strong original meaning of the term. A mechanism, for the
Greeks, was the product of a crafty ingenuity, of the sort of practical
and wily intelligence they called *metis*. Daedalus is the chief
mechanical engineer depicted in Greek mythology. His construc-
tions display the cunning which Ulysses, in other contexts, was capable of. The Trojan horse was such an artifact, a *daidalon*. As were
the flying machines which Daedalus and Icarus used.

   Thus, from the beginnings of Western philosophy, mechanism
served to describe any artifact which, to some extent, could – at least
seemingly – violate the laws of nature. If we now jump through the
centuries, to the 17th, we find a sudden about-turn: mechanism
indicates now a belief in the widespread applicability of certain laws
of nature. Indeed, mechanistic explanations for living processes, of
the type of interest here, came to the fore as a consequence of the
discovery of universal gravitation by Isaac Newton. The New
Science of (what we now term) classical mechanics seemed to be
capable of anything. It could account. in like manner, for sub-lunar

motions, those of a pendulum for instance, and also for the motions of remote celestial bodies. These achievements were extraordinarily impressive; and it was quite tempting to try and extend the dominion of mechanics to other fields than the physical sciences, where they seemed to be leaping from success to success.

The French seventeenth century philosopher René Descartes was also instrumental in making mechanism a favored mode of explanation:

"The great spiritualist philosophy in the seventeenth century, cartesianism, is (. . .) mechanistic. Descartes makes of thought the essence of the soul, he makes of space the essence of matter; the body of animals being material, it follows that all the operations of life should be reducible to motions in space; and thus, life will be a mere mechanism, animals will be nothing but machines" (Fonsegrive, ca. 1870).

As the same Georges Canguilhem lucidly remarks, biology as an autonomous science has a vested interest in denying mechanism, and in espousing vitalism:

"(. . .) There is little to expect from a biology so much fascinated by the prestige of the physico-chemical sciences that it is reduced or that it reduces itself to being just a satellite to these sciences. Such a reduced biology has for a corollary the vanishing of the biological object as such, viz. devalued in its specificity. Hence a biology, autonomous as to its subject, and as to its grasp of it - which does not mean a biology ignoring or despising the sciences of matter - always incurs to some degree being branded as vitalistic" (Canguilhem, 1965b).

Let us now remind the reader of a few important statements of the vitalistic creed since the Renaissance.

2. The vitalistic view of life processes

Vitalism stems from a holistic, organicist view of the world. To quote again from the same articulate discussion by Canguilhem of the fight between mechanism and vitalism:

"Plato, Aristotle, Galen, all the thinkers of the Middle Ages and to a large extent those of the Renaissance were (. . .) vitalists. They considered the Universe as an organism, that is to say as an harmonious system ordered both according to laws and according to ends. They conceived of themselves as an organized part of the Universe, as a sort of a cell within the Universe taken as the organism; all the cells were unified by an internal sympathy, in such a way that the fate of the partial organ seemed to them to be linked naturally with the emotions of the heavens" (Canguilhem, 1965c).

Jan Baptist van Helmont (1580–1644), a strange and endearing figure, an original thinker and an intellectual gadfly even in his own age, came out very strongly in favor of vitalism. To quote Canguilhem once more, van Helmont

"holds experimental science and mechanism as both jesuitic and diabolical. He refuses mechanism because it is an hypothesis, i.e. a trick intelligence performs on reality. Truth, according to him, is reality, it exists. And thought is but a reflection. Truth goes through man like lightning" (Canguilhem, 1965d).

A key figure for the resurgence in modern times of vitalistic notions is Georg Ernst Stahl (1660–1734). He was brought up on the essential unity of chemistry and medicine. His friend from student days and colleague at the newly founded University of Halle, Friedrich Hoffmann (1660–1742) pushed with great force mechanistic explanations for life processes. Because of his dissatisfaction with such explanations, Stahl evolved a system of his own, and entered into a tremendous intellectual rivalry and battle with Hoffmann: their friendship did not survive. The doctrine Stahl evolved is that of a vital force, a residual inexplicable in terms of material motions, which comes as a necessary supplement. He termed it *anima*. Stahl was the successor there, not only of van Helmont, of Robert Fludd too, who had also posed the existence of a number of vital principles, in addition to the soul, as essential components in all living creatures. Stahl published his theory in a major and lengthy work, *Theoria medica vera* (1st edition, Hall, 1707).

3. And what was the question?

Physiology, in the 19th century, especially during the first half of the century, saw a heated debate between the mechanists and the vitalists. Throughout the 19th century, chemical physiology reflected this tension:

"All these processes which are part of nutrition: digestion, absorption, secretion, etc., are they explicable by the present laws of chemistry and physics, or should one invoke a mysterious force, outside of the laws heretofore known and termed the vital force? The whole history of physiology is impelled by the fight between physicists and physico-chemists, on one hand, and the vitalists, on the other" (Langlois, ca. 1870).

To a limited extent, this debate reflected national differences, French physiologists such as Cabanis, Bichat, Magendie, being vitalists, while German physiologists, such as Du Bois-Reymond, Schwann, Helmholtz, were or became materialists (Temkin, 1946).

In the 20th century, after biochemistry had achieved the status of an independent discipline, this same debate was to flare up again, the most vocal being Driesch on the side of the vitalists and Jacques Loeb on the materialist side.

4. Kant raises the level of the debate

We come now to a turning point, to an episode in this history valuable for its dialectic content. It is a transitive tale of the transition from what might be termed the classical (and visceral) vitalism of van Helmont and Stahl to the more moderate mechanistic vitalism influential among German physiologists during the first half of the 19th century. Because Immanuel Kant influenced indirectly this whole generation (or two) of German physiologists, and because in turn Kant had formulated his notions of the proper methodology for biology in response to the ideas of the 18th century theorist of biology, Blumenbach, we have first to briefly outline the main features in Blumenbach's thinking.

Johann Friedrich Blumenbach (1752–1840) was an anthropologist and a comparative anatomist. In his dissertation, *De generis humani varietate nativa* (1776), he

"argued that the various (human) races were all degenerations of the Caucasian race, resulting from climatic variation produced by migration, changes in nutrition corresponding to the difference in the new habitat, and differences in the mode of upbringing due to cultural differences" (Lenoir, 1980).

In this doctoral work, Blumenbach espoused the preformationist views of Albrecht von Haller (the embryo is preformed in the maternal egg). He had to temper them soon afterwards, though, in the face of the evidence of fertile hybrids, both human (mulattos) and non-human (the tobacco plants *Nicotiana rustica* × *Nicotiana paniculata*). Yet, Blumenbach did not renounce a central tenet of Haller's preformationism:

"Organisation could not be accounted for in terms of physico-mechanical causes but had to be treated as primary" (Lenoir, 1980).

Blumenbach took as his central observation regeneration of the missing part of a polyp, after it has been removed: it grows back, but slightly reduced in size.

This, Blumenbach took as expressing the vital force of the organism; for him, this force was not a blind striving for growth, nor a ubiquitous and all-powerful soul, but a coherent structuring principle which he named the *Bildungstrieb*:

" . . . in all living organisms, a special inborn *Trieb* exists which is active throughout the entire lifespan of the organism, by means of which they receive a determinate shape originally, then maintain it, and when it is destroyed repair it where possible" (Blumenbach, 1781).

The *Bildungstrieb* enabled an organism to adapt its various functions to its environment, while it limited possible variations to those consistent with the original plan:

"A variation in a single part entailed (furthermore) a correlative variation in other parts of the organism" (Lenoir, 1980).

The task of the naturalist was, according to Blumenbach, genealogy: reconstruction of the *Bildungstrieb* specific to a class of organisms.

And the manner in which he conceived of this *Bildungstrieb* was as a

"teleological agent which had its antecedents ultimately in the inorganic realm but which was an emergent vital force" (Lenoir, 1980).

These views appealed to Kant. In a letter to Blumenbach of August 1790 he wrote:

" . . . your recent unification of the two principles, namely the physicomechanical and the teleological – which everyone had otherwise thought to be incompatible – has a very close relation to the ideas that currently occupy me (. . .)" (Kant, 1790).

The ideas then occupying Kant can be found especially in *Kritik der Urteilskraft (Critique of Judgment,* 1790). He addresses there the following very basic antinomy:

"On the one hand all material things must be explained by mechanical laws alone, and on the other hand there are things (namely, living organisms) that cannot be explained in this way, but demand final causes for their explanation" (Roll-Hansen, 1976).

Indeed Kant was 'completely convinced that man will never be able fully to understand or to create living organisms' (Roll-Hansen, 1976). Kant wrote, in the *Critique of Judgment,* that

"No human reason (. . .) can hope to understand the creation even of a grass-straw from mechanical laws alone" (Kant, 1790).

Kant, still in the *Critique of Judgment,* spelled out the antinomy according to these two rules:

"The first rule is reductionist: biological phenomena must be explained in mechanistic terms; otherwise there will be no 'real knowledge of nature'. The second rule is anti-reductionist. It says that some biological phenomena cannot be explained by physical laws alone. Their explanation demands a teleological principle" (Roll-Hansen, 1976).

To Kant this teleological principle (*Purposiveness of Nature,* or *Prinzip der Zweckmässigkeit der Natur* as he termed it) is a conse-

quence of the limitations of human reason. Kant names it a 'critical
principle' because he infers it from a critique of the powers of the
human mind - and here I shall place a lengthy quotation from a
pertinent recent discussion:

"Since the a priori laws of human understanding are insufficient for ordering the
manifold forms of nature, we are quite dependent on empirical concepts and laws
supplied by the reflective power of judgment. (...) Reason therefore presupposes a
principle that can secure unity of the empirical laws. (...) This principle for the
unity of nature can be no other, according to Kant, than that we conceive the empirical laws in analogy to the a priori laws. That is, we must think of them as given a
priori by a superhuman mind [which Kant calls *intellectus archetypus*] in such a way
as to fit the demands of our reason. (...) This principle of the purposiveness of
nature is valid only for the conduct of our mind and not for nature in itself. (...) [In
other terms], the concept of an inner purpose is necessary for our theory of the organism, but not for its observation" (Roll-Hansen, 1976).

Hence, for Kant a principle such as the purposiveness of nature, or
Blumenbach's *Bildungstrieb* is merely a heuristic device. It serves to
make sense of observations not readily explainable by physico-
mechanical laws. Kant warned, however, that

"even such attenuated teleological principles had a tendency to drive out mechanical
principles, and that *the purely regulative procedure very easily became constitutive.*
Purpose, in spite of its hypothetical status, tended to be used as a ground for the
explication of phenomena" (Larson, 1979; emphasis added).

As we shall see, despite this warning, purposeful explanations quickly returned through the back door, so to speak.

5. The aftermath to the *Critique of Judgment*

There were immediate consequences: Blumenbach revised his ideas.
And there were more distant consequences: Blumenbach translated
the message from the Königsberg sage into the program for a school
of mechanical vitalists.

Of the latter, two distinct but not mutually exclusive presenta-
tions have been given. In the jaded description, Kant's distinctions

progressively got blurred, and vitalism regained strength (Larson, 1979). In a more glowing appraisal, Blumenbach is seen having set a whole coherent research program, based on the Kantian ideas, and having this research program influence German physiology for a long time (Lenoir, 1981).

Blumenbach assimilated the lesson from Kant. He reworked his concepts, inclusive of the central notion in his system. He spelled it out henceforth in conformity with the clarifying distinctions offered by Kant:

" . . . the term *Bildungstrieb* (. . .) explains nothing itself, rather it is intended to designate a particular force whose constant effect is to be recognized from the phenomena of experience, but whose cause, just like the cause of all other universally recognized natural forces, *remains for us an occult quality.* That does not hinder us in any way whatsoever, however, from attempting to investigate the effects of this force through empirical observations and to bring them under general laws" (Blumenbach, 1797a; emphasis added).

As comments Lenoir:

"Fashioned in the language of the General Scholium to Newton's *Principia,* this passage revealed Blumenbach's goal of doing for organic bodies what Newton had accomplished for inert matter (. . .) it was the task of the naturalist to reconstruct the *Bildungstrieb* for each class of organism by unifying the regularities found in reproduction, generation, and nutrition under a general law" (Lenoir, 1980).

In the same 1797 edition of his *Handbuch der Naturgeschichte,* Blumenbach echoed many other Kantian pronouncements from the *Critique of Judgment.* He quoted directly Kant's definition of an organism as 'one in which each part is reciprocally ends and means' (Blumenbach, 1797b).

Like Blumenbach, Kant had also pondered the differences between the various human groups. He wrote a paper on this topic (Kant, 1785). The gist of his argument was that the empirical finding of various human races, differing very greatly in many aspects, was ascribable to differentiation from a single stem, or generative stock, the *Stamm,* endowed with seeds (*Keime*) for the different varieties – a clone theory 'avant la lettre' – and with adaptative abilities (*Anlagen*) permitting various combinations between the

Keime. Kant insisted that the *Stamm* could not be identified with any of the existing human groups.

Blumenbach had earlier held another and quite different view: the other human races were just degenerations from the Caucasian stock. But the latter Blumenbach, in the closing years of the century, rallied to – to us a more palatable position, intellectually as well as morally – the Kantian analysis. And he 'recognized that it was completely consistent with his own conception of the *Bildungstrieb*' (Lenoir, 1980).

Christoph Girtanner (1760-1800), a former student of Blumenbach (1780-1782), welded together Blumenbach's *Bildungstrieb* and Kant's *Stamm.*

"Central to Kant's model of the *Stammrasse* was that races of the same species, no matter how apparently divergent morphologically, always produce fertile hybrid offspring, and that successive crossbreeds between hybrid and parental stocks ultimately result in the extinction of the hybrid form, whereas true racial lines continue infallibly once they are established" (Lenoir, 1980).

Girtanner tested these predictions successfully by experiments on tobacco plants (Girtanner, 1796).

Carl Friedrich Kielmeyer (1765-1844) was another student of Blumenbach's in Göttingen, during the years 1786-1788. He put renewed emphasis on the genealogical method advocated by Kant for establishing a general physiology. This new discipline was meant to concern itself with development (also termed metamorphosis, at the time) of organisms. Kielmeyer specified in his program that general physiology should set itself a dual task:

" ... historically through description of changes occurring at significant moments and the determination of their temporal relations, duration, succession, and co-existence with one another and with respect to external conditions in so far as these can be determined for the organisms and their classes. Theoretically through the determination of the internal and external more or less general conditions of change in the significant individual moments; through determination of their laws; through reduction of the variations themselves to classes valid for the different classes of organisms" (Kielmeyer, 1790-93).

Kielmeyer viewed the diversity of biological species as states or stages in the unfolding of the underlying *Bildungstrieb*:

"Many species have apparently emerged from other species, just as now the butterfly emerges from the caterpillar . . . They were originally developmental stages and only later achieved the rank of independent species; they are transformed developmental stages. Others, on the other hand, are original children of the earth. Perhaps, however, all of these primitive ancestors have died out" (Kielmeyer, quoted by Lenoir, 1981).

This is a fine program of study.

In actuality, Kielmeyer set about to describe the biological sphere in economic terms, as an interchange of vital forces. Rather than studying a phenomenon in depth, he would assume some vital force responsible for it. His work bred a proliferation of vital forces, one for each physiological function (Larson, 1979).

Alexander von Humboldt (1769–1859), Heinrich Friedrich Link (1767–1851), and Godfried Reinhold Treviranus (1776–1837) were other students of Blumenbach in the same mould. Lenoir, in the article I follow rather closely here, holds that the viewpoint of Blumenbach's entire school concerning the results to be expected from paleontology is best summarized in the words of Treviranus (Lenoir, 1980):

"(. . .) Contrary to what is commonly said, the animals of the prehistoric world were not destroyed by great catastrophes; rather many of these forms have survived, but they have disappeared from nature because the species to which they belong have been transformed into other species" (Treviranus, 1805).

In physiology, the ideas of Treviranus revert actually to a naive, circular form of vitalism (Larson; 1979). The life force, posed as a hyperphysical quantity, serves to explain the maintenance of organization in the face of outside perturbations. In this role, the vital force becomes the irritability of the organism.

Karl Ernst von Baer (1792–1876) was on every count an outstanding scientist. Steeped in the material vitalism of Blumenbach and Kielmeyer, he worked in the same conceptual framework - following what Lenoir (1981) terms the teleomechanical program. He was not one to jettison materialism in favor of rampant vitalism:

"Thus in his lectures on anthropology (1824) von Baer emphasized that an understanding of physiological function must be traced ultimately to the chemical level. Citing the analyses of Berzelius, Thénard and Gay-Lussac as the basis of his opinion, von Baer adopted the view that the basic material constituents of the body: albumen, fibrin, gluten, etc. were all slightly altered chemical combinations of carbon, hydrogen, nitrogen and oxygen, the order and arrangement of these constituents being the principal determinant of the properties and ultimately the function of those substances" (Lenoir, 1981).

And von Baer was convinced he had pushed back the border between materialism and vitalism when, in 1827, he thought he had located in the center of the ovum the force responsible for the subsequent development of the embryo, a force which he named *Gestaltungskraft* (a latter-day incarnation of the *Bildungstrieb*). The important discovery he had made was of course that of the mammalian ovum.

This finding can be ascribed to von Baer's ideology; to the way his training as a material vitalist (or, as Lenoir (1981) terms it felicitously, as a developmental morphologist) made von Baer frame his inquiry. Lenoir's comment that

"each improved version of the [teleo-mechanical] research program scored spectacular empirical successes in the generation of novel facts. (. . .) von Baer's discovery of the mammalian ovum, which was directly inspired by his search for an ultimate organized structure for the localization of his developmental type" (Lenoir, 1981)

was one such resounding success.

I have been presenting in some detail Timothy Lenoir's argument because of its basic cogency. The materialistic vitalistic school started by Blumenbach deserves the choice treatment he gave it. For one thing, his presentation of the case exemplifies the beneficial influence of the Kant recommendations on German physiology.

Kant meant to protect metaphysics from total destruction by the New Science of Galileo and Newton. For this purpose, he built a wall between science and philosophy. The former, he restricted to the world of facts and to the task of making sense, of gaining knowledge out of them. The latter, he endowed with the world of values and with the task of building ethics and morals. In this way, by insulating science and metaphysics from one another, Kant hoped to save Aristotelian metaphysics from ruin.

In his renovation of metaphysics, Kant took on teleology. Before him, it tended to be an occult and immanent absolute - what might be termed hard, pre-ordained vitalism - and Kant watered it down to a figment of the imagination, to a dream of the human mind when confronted with the awesome mysteries of biology.

I cannot resist pointing here to the analogy with complementarity, as devised by Niels Bohr. Here again, the notion of intrinsic limits to human reason, of undecidability (if I may be forgiven this borrowing of a mathematical term) served to blunt philosophical pain at the sharp edge of scientific advances. As writes Sir Karl Popper:

"There seems to me a fairly obvious connection between Bohr's 'principle of complementarity' and this metaphysical view of an unknowable reality - a view that suggests the 'renunciation' (to use a favourite term of Bohr's) of our aspirations to knowledge, and the restriction of our physical studies to appearances and their interrelations. (. . .) The thing in itself is unknowable: we can only know its appearances which are to be understood (as pointed out by Kant) as resulting from the thing in itself, and from our own perceiving apparatus. Thus the appearances result from a kind of interaction between the things in themselves and ourselves. (. . .) We try to catch, as it were, the thing in itself, but we never succeed: we only find appearances in our traps" (Popper, 1968).

However, Sir Karl, as a rationalist, cannot stand black holes of this sort, because they allow metaphysics to re-enter. And he concludes thus:

"From a rational point of view, all these arguments [such as Bohr's] are inadmissible. I do not doubt that there is an interesting intuitive idea behind Bohr's principle of complementarity. But neither he nor any other member of his school has been able to explain it (. . .)" (Popper, 1968).

Indeed Popper has a point. And, just like the Kant-inspired school of materialist vitalists in Germany, at the beginning of the 19th century, tended to revert to a primitive and staunch form of vitalism (Larson, 1979), we have too often seen in recent years Bohr's complementarity degenerate into metaphysical nonsense (such as paraded about at a notorious colloquium in Cordoba, Spain, a few years ago).

To return to Kant, he makes a deep division between the world of mechanism – and therefore of technology, as he was quick to grasp – and the world of biology: never will man be able to make a leaf. Given a set of initial conditions, analytical mechanics can safely predict a trajectory for a moving body. Mechanism implies repetition, self-same events, predictability. Conversely, Kant endows biology with an eventful history. Organisms undergo changes, they diverge from their initial plan or pattern, they change their shape, they modify their size and appearance, etc. It is the job of the biologist, in Kant's view, to study the paleontology, the embryology, the metabolism, etc. (to use modern terminology), because all these disciplines are, basically, *diachronic* sciences.

Kant puts biological facts into historical perspective. Henceforth, biological species, instead of being considered in a static, collector-like manner, as Linnaeus and also Buffon to some extent had done, are set on a conveyor belt – not an escalator, this had to wait till Teilhard de Chardin. Each biological species, in the intellectual history and in biological theory from Kant onwards, becomes considered as a stage in a gradual process of unfolding, frame by frame as it were, some of the multiple potentialities all inherent in the original plan. Which is why, in the eyes of Kant, in those of Schelling and of the Nature Philosophers who became the intellectual heirs to this Kantian concept, comparative morphology is the crowning part of biology, because it enables the biologist to reconstruct the original scheme (at least in principle).

Let us note here, just in passing, that Goethe was deeply influenced by this Kantian program. His biological studies show this very clearly. His concept of the *Ur-Pflanze* in botany is an example.

Timothy Lenoir's is a revisionist goal in stressing as he does the importance of the Kant-inspired and Blumenbach-led school of materialist vitalists. He is obviously upset by the undue importance many scientific historians have given to *Naturphilosophie* as the important determinant of German biology and physiology during the crucial period from the French Revolution till 1847; an emphasis derived from the all-too-tempting analogy with literary history since Romantic writers – Hölderin, Novalis, Chamisso, E.T.A. Hoffmann, to name a few – came under the spell of Schelling's *Naturphilosophie*, if only for a relatively short time in some cases.

However, Lenoir's account smacks of the institutional history, more than it belongs to intellectual history proper. He could not resist the temptation to rigidify the flux of ideas – and God knows the German intellectual scene at the beginning of the 19th century was bubbling with new ideas – into the straightjacket of 'schools of thought'. Which is why Lenoir, who has explicitly set about showing how the materialist vitalists as a group are subsumed into a research program, as defined by Lakatos, ends up presenting them as a school of thought in the most traditional sense.

Lenoir rigidifies his history so much as to bring it to a standstill. His implicit thesis, behind the sophisticated, erudite and eloquent argument, is that of the stuttering repetition of a fixed paradigm. A master (Blumenbach or, behind him, Kant) utters the discourse of reason, and this self-same discourse is reiterated ad nauseam by the disciples.

6. Materialist vitalism in France

Just like German vital materialism arose from the reflexions of a philosopher, Kant, French vital materialism was also the child of philosophy. Its origin, beyond the *Idéologues*, goes back to the sensationalist philosophy of Condillac and Locke.

Hence the differences. The German version of vital materialism emphasized morphological development because, ultimately as we saw, this was one aspect of biology Kant felt could be sheltered from the intrusion by Newtonian mechanics. The German physiologists in the school, or in the spirit of Blumenbach exiled the vital force of Stahl in the upper, rarefied atmosphere of the general structuring or patterning of organisms according to a unified plan. Thus they pushed, they felt, vitalism outside the domain of their empirical studies of organized bodies.

In the French version of vital materialism, the physiologists were also quite careful not to invoke explicitly a single, all-powerful vital force as a *Deus ex machina*, so to speak. They saw a need to bury such a vital force deep down, beyond the tissues they studied under the microscope, in the biological fiber proper, as an intrinsic property of organic matter. Thus, they made themselves into materialists.

A brief reminder of the gist of the sensationalist philosophies of the Englishman John Locke (1632-1704) and of the Frenchman Etienne Bonnot de Condillac (1715-1780) is now in order.

John Locke conceived of the human mind as a passive receptor. He saw it as a blank slate on which sense impressions could register. This notion (like so many others) goes back to Aristotle. But Locke gave it new impetus. Here is how he put his case:

"Let us then suppose the mind to be, as we say, white paper, void of all characters, without any ideas; how comes it to be furnished? Whence comes it by that vast store, which the busy and boundless fancy of man has painted on it with an almost endless variety? . . . To this I answer in one word, from experience" (Locke, quoted by Blakemore, 1977).

Locke thus gave primacy to sensations over ideas. Sensations for him are primary. Ideas are formed from sensations: they are combined, they are ordered into series, and the brain thus moulds them into concepts.

Condillac was the main proponent and propagator in France of Locke's philosophy. But he went even further than his model. For Condillac, all human knowledge is transmuted sensation. Locke resorted to a principle of reflexion in addition to sense experiences. Condillac thought he could dispense with it. Nevertheless Condillac values the Self.

'Whether we rise to heaven, or descend to the abyss,' he wrote in 1746, 'we never get outside ourselves - it is always our own thoughts that we perceive' (Condillac, 1746). In order to convey vividly this self-perception, Condillac (who was a fine teacher) imagined the *Gedankenexperiment* - as we might now term it - of the statue. Imagine

"an animated statue which is successively endowed with each one of the senses of smell, hearing, taste, sight, and touch singly, and then in various combinations. The reader is advised to 'enter into its life, begin where it begins, have but one single sense when it has only one'. In short, he 'must fancy himself to become just what the statue is'. Essentially the statue represents a series of hypothetical individuals, some of whom are endowed with only one sense and others with various combinations of senses. (. . .) Condillac's statue thus serves the useful purpose of disentangling the learned and unlearned components in perception" (Pastore, 1971a).

Condillac gives this vivid, almost out-of-Proust example:

"If we give the statue a rose to smell, to us it is a statue smelling a rose, to itself it is smell of a rose. The statue therefore will be rose smell, pink smell, jasmine smell, according to the flower which stimulates its sense organ. In a word, in regard to itself smells are only its own modifications or modes of being. It cannot suppose itself to be anything else, since it is only capable of sensations" (Condillac, quoted by Pastore, 1971b).

This seems a blatant case of solipsism. How does Condillac skirt it? How does the mind communicate with the outer world of nature from which it ultimately must receive all of its sense experiences?

"The mind as the source of sensation is, as it were, encapsulated; how then can it reach out to the other world so as to become aware of external things? In (Condillac's) answer, access to objects is provided through touch, for this is 'the only sense which of itself can judge of externality'. It is the 'bridge of passage' between the inwardly experienced states of the mind and the external things, the gap being closed by the reach of the arm in touching the object" (Pastore, 1971c).

One might attempt a parallel between the material vitalism of Kant, systematizing biological facts (morphological, especially) into a coherent, genetically determined whole; and the sensationalist materialism of Condillac, for whom sense experiences are stored in the mind, compared, and systematized into coherent series, then projected back onto the screen of the 'outer world'.

The two books the mature Condillac wrote on 'psychological' topics, *Traité des Sensations* (1754) and *Traité des Animaux* (1755), were quite influential. They were part and parcel of a more general movement of sensationalistic materialism; La Mettrie's *L'homme-machine* (1747) and d'Holbach's *Système de la Nature* (1770) presented like ideas. Condillac, d'Holbach were very close to Diderot and to some of the other *Encyclopédistes*. And their ideas

are disseminated throughout the *Encyclopédie*. Matter in motion is
the only reality, the ideas in our mind are a mere epiphenomenon:
this was their message.

We jump now one generation, to the end of the 18th century and
to the time of the *Idéologues*:

" 'Feeling' and 'sensibility' most clearly manifested in animals became the central
phenomenon for an explanation of nature as well as the mind. Cabanis, the physiologist among the idéologues, even went so far as to suggest that gravitation and chemical affinity might be primordial forms of sensibility" (Temkin, 1946).

Pierre Jean Georges Cabanis (1757–1808), whose *Rapports du Physique et du Moral de l'Homme* (1802) is the best remembered book,
was indeed a staunch materialist. He held that thought was just the
name for 'secretions' of the brain akin to bile as a secretion from the
liver. Sensibility was for him, likewise, an inherent property of
nerves. Consciousness arose, as it did for Condillac, from sensations;
in other words it derived from material motions in the nervous
system. One could dispense altogether with the idea of a soul. This
was the lesson Cabanis had learned in his youth from mingling with
the likes of Diderot, d'Alembert, Condorcet, Condillac, and the
Baron d'Holbach. This was the lesson he passed on.

Marie François Xavier Bichat (1771–1802) belonged to a younger
generation than Cabanis. He was a great physician, specialized in
anatomy and in physiology. In his *Traité des Membranes* (1800), he
presented a theory of the origin of the various organs of the body,
differentiating from simpler units or precursors. Bichat distinguished 21 kinds of tissues whose various combinations went into
the make-up of the organs. This distinguished him from the earlier
French vitalists such as Barthez and others of the so-called Mont-
pellier School. Bichat is the first French physiologist not to resort to
an occult and undefined vital force as the sole cause of vital phenomena. He invests organized matter itself with this power, and he con-
ceives of organized matter as differing radically from inanimate
matter. In 1800, Bichat gave his classic definition:

"Life is a set of functions which resist death. Such is the mode of existence of living
bodies that everything surrounding them tends to destroy them" (Bichat, 1800).

We cannot help being reminded here of the German school of material vitalists; this statement of Bichat's is rather similar to what his
contemporary Treviranus himself was proposing.
Thus for Bichat the properties of biological tissues become the
sole causes for all vital functions. He was led to distinguish four basic
properties: *animal sensibility* from which sensations derive, and
animal contractility, responsible for voluntary movements; to which
were added *organic insensible contractility* and *organic sensibility*.
The latter were intrinsic properties of biological tissues, to which
Bichat attributed the regulated and ordered assimilation by the tissues of the various metabolites from food intake.

François Joseph Victor Broussais (1772-1838) was another very
successful French physician, active during Napoleon's reign. Following publication of his *L'examen des Doctrines Médicales* (1816),
his doctrine became the norm and the latest fashion in Paris. His
system of 'physiological medicine' derived all diseases from an irri-
tation of the gastrointestinal tract, which spread by resonance, by a
'sympathy' to neighboring organs. Whereas Cabanis had argued that
thinking was a mere secretion of the brain, Broussais held that it
was due to excitation of the brain matter.

François Magendie (1783-1855) was yet another member of this
French school of material vitalists, an experimental physiologist
who taught at the Collège de France, where he was appointed in
1831. His main claims to fame, besides having greatly influenced the
formative years of Claude Bernard, one of his students (1841-43),
are his pioneering introduction of strychnine and morphine into
medicine. He had experimented with their effects on various organs,
when he gave a shot of ovalbumin to a rabbit for the second time, and
the laboratory animal went into shock and died. His observation was
one of the first of anaphylaxis (1839).

This movement of French material vitalists also found literary
expression. In his aptly entitled series of *Philosophical Studies*
(*Etudes Philosophiques*), a part of *La Comédie Humaine*, Balzac
published in 1831 – the same year Magendie was appointed to the
Collège de France – the novel *La Peau de Chagrin*, a book title often
translated in English as *The Magic Skin*. This is the tale (or parabola) of how a desperate young man, Raphaël de Valentin, makes a

pact with a mysterious old man who owns an antiques shop. Raphaël is given a piece of parchment, the magic skin of the title, whose possession ensures fulfillment of his every wish. But the skin will keep track of the record, and will shrink every time it grants him one of his desires. When it has been reduced to the size of a dot, it will be the time for Raphaël to die.

Many readings of this book are possible and have been proposed. I have shown elsewhere some of the connections with chemistry (Laszlo, 1979). Suffice it to say here that Balzac stages in *La Peau de Chagrin* the contemporary debate between materialists and vitalists. When the hero, Raphaël, is struck with his mysterious ailment, physicians from both sides are called in consultation. Cameristus – this is a pseudonym for a real Dr. Recamier, it is believed – is a vitalist. He wants to 'seek the cause of the disease in the recesses of the mind and not in the recesses of the body' (Balzac, 1831a). Conversely, Brisset – a rather transparent pseudonym for Broussais – presents the case for the materialists. He examines and analyzes the body to find in it the 'perturbations of the vital mechanism' (Balzac, 1831b).

Balzac himself seems to have opted for a position akin to that of the French material vitalists:

"A motion of any sort is immense power, and man does not invent powers. Power is one, just like motion, the essence of power. Everything is motion. Thought is a motion. Nature is established upon motion. (. . .) and what is space? Motion is required to reveal it; without motion, it is a mere word devoid of meaning" (Balzac, 1831c).

Fortunately, the notebook in which Balzac jotted down his first idea of *La Peau de Chagrin* has been preserved: he set about explicitly to describe in his book 'human life as life and mechanism' and he chose the magic skin as an 'exact formula for the human machine' (Balzac, 1830).

7. Angry young men in Germany

In the late 1840s, four German physiologists in their prime, Ernst von Brücke, Emil Du Bois-Reymond, Hermann von Helmholtz, and Carl Ludwig forcefully challenged the pertinence of a vital force in biology; and they announced with great fanfare their intention to provide henceforth explanations for biological facts based only on elementary physical and chemical phenomena.

One of their important publications was Du Bois-Reymond's book, *Untersuchungen über Thierische Elektricität* (1848). In the preface, the young Du Bois-Reymond (he was 29) stated their general conviction that, in principle,

"an analytical mechanics of the general life process would be possible. This conviction rests on the insight, possessed even by Aristotle, that all changes in the material world within our conception reduce to motions. But again, all motions may ultimately be divided into such as result in one direction or the other along the straight line connecting two hypothetical particles. Therefore to such simple motions must even the processes within the organic state ultimately be reducible. This reduction would indeed initiate an analytical mechanics of those processes. One sees, therefore, that if the difficulty of the analysis did not exceed our ability, analytical mechanics fundamentally would extend even to the problem of personal freedom . . . " (Du Bois-Reymond, 1848).

A strong, enthusiastic statement formulated in no uncertain terms! And a statement that resounds with the same primacy of motion as my previous quotation from Balzac's *La Peau de Chagrin*.

Indeed, one may want to use *La Peau de Chagrin* as a document for scientific history. This novel presents, 16 years ahead of their outburst, all the main features of the 1847 group of Brücke, Du Bois-Reymond, Helmholtz, and Ludwig: the rebellion and despair of youth; a deep dissatisfaction with the existing social and political order; a belief in the all-powerful nature of human will; an impatience with the existing medical theories, with their lack of grip on pathologies; and likewise, a critique of the scientists from the preceding generation.

Biological systematics were purely descriptive, it was only an observational science. Physical science, conversely and whether this was physics or chemistry, was able to unleash brute power which could not be harnessed and was altogether destructive. These are key concepts in *La Peau de Chagrin*. And I venture to submit that there are rich benefits to be gained from application to the writings of the 1847 group.

The reference to analytical mechanics in the above quotation from Du Bois-Reymond's 1848 oration (it occurs repeatedly, *basso continuo*-like) is quite clear. Vitalism basically is, if not anti-rationalistic, para-rationalistic. It prides itself on the need to preserve and to respect an unexplicable residue, the vital force (or forces). By contrast, analytical mechanics stands for total rationality, encompassing the whole physical world from a speck of dust to the trajectories of huge celestial bodies.

However, analytical mechanics share with vitalism the same sort of vainglorious ambition: to account in a unified manner for the whole of nature. There was another seed for crystallization of the 1847 group. This was the instinctive urge to dispel *Naturphilosophie*:

" . . . the common pattern was for a young man to be captured by the vision of such a unified wisdom while at university and then to spend his life trying to rescue as much of it as possible by dedicated research. This sort of career is well illustrated by that of Hermann Ludwig Helmholtz, a German physiologist and physicist.

Eventually *Naturphilosophie* became an orthodoxy taught by university professors. The founders of experimental science in Germany in the 1830s and 1840s found their way blocked by them, and bitter struggles ensued. Though the scientists won, they were haunted for generations by the ghost of *Naturphilosophie*, and they reacted by curbing all speculative tendencies most severely, reinforcing the dry, inhuman style of science that the poets had found abhorrent" (Ravetz, 1979).

The last part of the preceding quotation ('Though the scientists won . . . ') oversimplifies a great deal. There were a number of eminent thinkers who refused this dichotomy of the two cultures. There were even a few who were active and creative in both. To give but an example, from the same age as Brücke, Du Bois-Reymond, Helmholtz, and Ludwig, Georg Büchner (1813–1837) - the playwright best remembered for *Danton's Death, Woyzeck,* and *Leonce and Lena* - was also an anatomist and physiologist of talent who was appointed in 1836 to a chair on the medical faculty in Zürich, shortly before he died in an epidemic.

The 1847 group announced their intention to build a reductionist biology, i.e. to erect a biophysics and a biochemistry. I shall argue, first, that this was a timely program and that they did not show any particular merit (except for the bravado) in going public and proceeding with it.

This was a timely program. The example for other sciences, which physiology could hope legitimately to emulate, was quite promising. Laplace, basing himself on Newtonian mechanics, had erected a handsome, self-consistent system in which analytical mechanics could account for the minutest details in the motions of celestial bodies. The physics of electricity and magnetism seemed also on their way to total understanding, with the studies by Ampère, Oersted, Faraday, etc.

In chemistry Wöhler's urea synthesis had been made 20 years before already (Wöhler, 1828). However,

"the synthesis of urea did not cause an immediate general abandonment of the dogma that a 'vital force' may be necessary for the synthesis of an organic compound. (. . .) The contemporary vitalists accepted the idea that simple organic compounds such as urea can be man-made, but they shifted the probable need for a vital force up the scale to the then more mysterious compounds of living matter" (Moore, 1978).

Such a counter-move by the vitalists, who furthermore often enjoyed positions of prominence, in medical faculties especially, must have been infuriating to young scientists on the rise. They were totally convinced of the merits of experimental science. They were sick and tired of the vitalist orthodoxy. They wanted to go ahead at full steam and align physiology on the hard physical sciences. As Donald Fleming puts it perceptively:

"One group of German mechanists at mid-century were a band of young experimental physiologists who simply wanted to get on with their work but found themselves intimidated by vitalists who smugly declared it was all in vain – the most fundamental processes of life would always elude a mechanistic analysis. Stung by these taunts, and perhaps to screw up their own courage, a quadrumvirate of rising physiologists – Helmholtz, Ludwig, Du Bois-Reymond, and Brücke – swore a famous mutual oath in 1845 to account for all bodily processes in physicochemical terms. This was intended as a program for research rather than the enunciation of a *Weltanschauung*. The Mechanistic Quadrumvirate and their sympathizers did not attempt to banish non-mechanistic conceptions from other spheres. Though most of them were not religious, they did not undertake to proscribe religion for others" (Fleming, 1964).

Theodor Schwann (1810–1882) was close in spirit to the '1847 Group' (Cranefield, 1957) and to their reductionist manifesto. Three members of this group, Brücke, Du Bois-Reymond, and Helmholtz had been students of Johannes Müller. Schwann had also trained with Müller. He had joined him as a student in Bonn in 1830, and then followed him to Berlin in 1833. Schwann was just a little older (by six to eight years) than the members of the '1847 Group'. There is little doubt that he not only shared their ideas, but also pioneered them to a large extent.

Before coming to this part of the story, one word about their joint master, Müller. He was a great teacher in Bonn. He must have been, to attract students of such quality. And, instead of stamping them into the mould of vitalism, he let them grow their own scientific *personas* by helping them – Socratic maïeutics at its best – to choose research problems in which they could establish the new methodology they were groping for. As writes Mendelsohn:

" . . . Müller was clearly a vitalist and the first volume of his *Handbuch* [*der Physiologie des Menschen*, 1834–1840, 2 vols.], which gave detailed expression to this point of view, was on the presses. Yet, he led his students into exploration of the very problems through which they went on to develop a radically new idea of what constituted an acceptable physiological explanation. Schwann took up the problems of muscle contraction, digestion and cellular structure, function and development; Du Bois-Reymond was led directly to his investigations of the electrical phenomena of the nervous system; and Helmholtz was encouraged to explore the nature of heat production in muscle and nerve. Each of these men examined physiological processes for which Müller had invoked special non-physical forces in his theories and hypotheses. These men emerged from their initial studies believing that the vital functions under examination could be successfully explained with recourse only to physicochemical laws" (Mendelsohn, 1965).

Theodor Schwann appears to have been the first to follow such a path. By the mid-1830s he was questioning the vitalistic dogma. At the beginning of 1835, he started toying with the idea of doing experiments on muscles, which he started carrying out at the end of the same year. In retrospect, Schwann claimed later that these were his opening shots in his campaign against vitalism (Mendelsohn, 1965). However, he had not yet converted himself to a strictly materialistic and reductionist creed when he started these experiments on muscles:

"(. . .) a letter which Schwann wrote to his brother Peter, a student of philosophy and theology, in 1835, (. . .) shows Schwann still searching and examining the status of vital forces. In (. . .) the draft of another letter to his brother, (. . .) of slightly later date, Schwann (. . .) chastise(d) his correspondent and all the others who had become used to assuming the existence of self-purposive forces which they believed necessary for the explanation of the functions of the organism" (Mendelsohn, 1965).

By the end of the 1830s, Schwann had convinced himself that vital forces should be banished from biology (Mendelsohn, 1965). He started a 'veritable campaign against the assumption of a vital force' (Temkin, 1946) and fired the first shot in 1839 when he published a book in which he attacked the viewpoints, not only of the natural philosophers but also the material vitalists, those of von Baer especially (Schwann, 1839). I borrow from Timothy Lenoir his apt distinction between the two interpretations. For von Baer, the vital force irradiated from a single point, at the center of the ovum. It resulted from the type and arrangement of the living, organized matter around it.

"Von Baer's viewpoint and that of both vital materialists and developmental morphologists generally, is best described, therefore, as 'emergent vitalism' " (Lenoir, 1981).

Schwann saw that this was an unnecessary postulate. One could get rid of this notion of an initial seed for development altogether. The properties of organic matter in themselves should suffice, in principle, to account fully for the resulting biological processes:

"The phenomena connected with life were not the results of an emergent quantum leap into a different framework of laws and physical processes. For Schwann the science of biology consisted solely in the study of order and arrangement of materials in an organic setting, acting according to normal physical laws. (. . .) Life consists *in* the order and arrangement of particles of matter *tout court*" (Lenoir, 1981).

And Schwann complemented his sweep of vital forces out of biology with an analysis of teleology which resembles the solution evolved by Immanuel Kant half-a-century earlier. And 'Schwann's solution to the problem of purpose in biology of shifting the problem from biology to the world as a whole served as a model for any materialists during the following decade' (Temkin, 1946). Schwann just assumed that

"a rational Being created matter with such forces that, by following the laws of mere mechanism, a purposive organization, in this case the solar system, would result" (Lenoir, 1981).

In this manner,

"mechanism was coextensive with scientific knowledge but not with the range of legitimate curiosity. It was a classic formulation of the contracted life-style in science – demanding enough room to do one's own research while side-stepping any cosmic pretensions" (Fleming, 1964).

Emil Du Bois-Reymond (1818–1896), who would later develop electrophysiology, followed Schwann's lead. For him, the turning point came in the early 1840s. He wrote in 1841 (Temkin, 1946; Cranefield, 1957):

"I am gradually returning to Dutrochet's view 'The more one advances in the knowledge of physiology, the more one will have reasons for ceasing to believe that the phenomena of life are essentially different from physical phenomena' " (Du Bois-Reymond, 1918a);

and he wrote in another letter the following year (1842) of having sworn with Brücke to 'make prevail the truth that in the organism no other forces are effective than the purely physical-chemical . . . ' (Du Bois-Reymond, 1918b). The same year saw publication in Rudolph Wagner's *Handwörterbuch der Physiologie* of the famous '*Lebenskraft*' article by Hermann Lotze. This consists of a careful logical analysis, which made it clear to Lotze that the 'question of the origin of biological organization lay beyond the limits of a possible scientific account' (Lenoir, 1981). Forces do not exist in nature,

Lotze pointed out, they are postulated to account for the physical changes they effect, which the human mind witnesses. Hence, a vital force was incapable of separate existence. Furthermore, Lotze dispelled the attribution of vital phenomena to operation of a vital force: consideration of the underlying physical processes sufficed. Of course, they were integrated in the organism as a whole, a whole not reducible to any of its component parts.

"The parts could not be understood without reference to the functional integration of the whole; but the systematic organization of the whole was not to be understood in terms of a force independent of the parts, imparting organization to them. (. . .) Life does not exist independently of the mechanical processes through which it is manifest; but equally important, life is not simply a mechanical process; it is inseparable from an order among mechanical processes" (Lenoir, 1981).

This article of Lotze had a big impact.

To complement this (pre)history of the program of the '1847 Group' (which also provides the substance of what it consisted of), and by way of putting into context Du Bois-Reymond's 1848 statement about material motions providing ultimately the explanation to personal freedom, in 1846 Carl Vogt made this provocative statement:

"I think that every scientist if thinking at all consistently will reach the opinion that all the faculties which we comprise under the name of psychic actions are nothing but functions of the cerebral substance, or, to express myself somewhat crudely, that thoughts have about the same relation to the brain as bile has to the liver or urine to the kidney" (Vogt, 1845-1847),

which was nothing but reiteration of the earlier statement by Cabanis, as Temkin (1946) rightly notes.

Thus one comes to the revisionist program of 1847 - not felt to be so, of course, by its originators and propagators! It is developed at length in Du Bois-Reymond's Preface to his *Untersuchungen über Thierische Elektricität* (1848). I call the program revisionist because the angry young men of 1847 were not innovating in a vacuum (however their feeling); they were restating - and borrowing from French materialists of yesteryear - an earlier *credo*. Thus, Du

Bois-Reymond interestingly uses as his epigraph the following text
by the Frenchman Vicq-d'Azyr (1748-1794), written at the turn of
the century:

"In order to discover the mechanism of the living system, one must find among its
effects those which are due to the well-established laws of chemistry or physics, and
distinguish them carefully from effects which do not have an immediate link with
these laws, or [whose link is] unknown, and whose cause is hidden from us. The latter,
Van Helmont and Stahl have placed in the dependency of an arche or of the soul,
without remarking that, since their nature had not been ascertained, what they attributed to a single agent might have been due to several. By their recourse to imaginary
causes, these great men have hidden, seemingly, their ignorance under the veil of
philosophy. They were unwilling, furthermore, to specify the point after which their
positive knowledge ended. They were probably right in saying, and we do think likewise, that certain phenomena are found only among organized bodies, and that a
particular order of motions and combinations thereof serves as the basis and determines the character [of these organized bodies].
    One was undoubtedly in error in ascribing to them hypothetical causes whose
insufficiency has been revealed at long last; however surprising they seem to us,
though, these functions are physical effects. These are more or less involved. And we
must examine their nature with all the means provided by observation and by experiment. We must not assume that [they derive from] principles on which the mind
rests itself and fancies it has completed its task when this task still begs to be done"
(Vicq-d'Azyr, 1805).

Cranefield has carefully pitted the hopes of Helmholtz, Du Bois-
Reymond, Brücke, and Ludwig, as formulated in their '1847 Program', against the actual results they and their followers have
achieved. He comes to the conclusion that, basically, they failed in
their stated aim of setting up a biochemistry and a biophysics
(Cranefield, 1957). I must agree with Cranefield. The '1847 Program' has been a failure in its own age. Its reductionist ambition was
timely and well-argued. But the limited experimental techniques
available at the time were not up to the task. They could not study in
vivo the complex machinery of living beings. They were incapable of
extracting from such complex machinery the molecular parts
responsible for the observed functions.

    And I should interject here that, not only were the available exper-
imental methods insufficient, but the conceptual framework was
also in the way. Members of the '1847 Group' were not radical

enough! Their thinking continued to be contaminated with holistic notions (see for instance the above quotation from Lenoir, which summarizes very well this aspect of their beliefs), and they would have been flabbergasted to find out that life uses such simple devices as we now know from the discoveries in molecular biology to be the case. The '1847 Group' had such a systematic view of life processes that they would not have looked for molecular explanations. Yet, Wöhler's 1828 urea synthesis was beckoning to them. Instead of following in that direction, they got sidetracked into joining the bandwagon of analytical mechanics - an attitude readily understandable at the time, given the prestige of this discipline - which ultimately can only be adjudged as detrimental to the rise of biochemistry as we know it.

Only with the emergence of biochemistry in the 20th century, when proteolytic enzymes for example became a tool, would the radical materialism dreamt by these German physiologists - which henceforth will remain very much alive, as we shall find in Chapter 8, on the example of Jacques Loeb - be at long last fully vindicated.

8. Liebig sits on the fence

Justus von Liebig (1803-1873) presents the historian with the all-too-familiar case of someone who does not practice what he preaches, of someone who talks about vital force but who really believes only in what he can weigh with a balance.

After his doctorate, which he prepared at Erlangen under Kastner's supervision (1822), Liebig went to study with Gay-Lussac (1778-1850) in his private laboratory in Paris. He learned quantitative analysis there. This experience left him with a 'strong predilection for quantitative theorizing' (Glas, 1983).

Liebig continued, throughout his career, to build upon the taste and the talent he had for quantitative analyses. As this section will make it clear, he acted like a positivistic materialist. But he maintained at the same time the high-faluting stance of a vitalist. And this dichotomy did not bother him much. For Liebig was basically a pragmatist . . .

After his Parisian stay, Liebig joined the faculty in Giessen in 1824. He was made a full professor there in 1826. During the next decade, he did mostly studies in organic chemistry. He created his own journal, *Annalen der Chemie und Pharmacie* (1831), and he set up the Giessen laboratory according to a brand-new line, that emphasized practical and analytical laboratory work in preference to the 'traditional, "encyclopedic" chemistry courses' (Glas, 1983). This Giessen laboratory organization quickly became world-famous. Students were attracted by it to Liebig from all over Europe. This can be seen as Liebig – a quick-witted and a gifted administrator – anticipating the needs of the budding German chemical industry for the skilled personnel he was training in this practical manner. Indeed, this laboratory structure devised and pioneered by Liebig quickly became a model for other German universities (Göttingen, Heidelberg, Marburg), and it became also, after the 1848 uprisings, part of a conscious political program for removing the causes of the unrest.

"The government tried to take away the economic basis of social unrest by way of an active program of industrial and scientific innovation, which consisted among other things in the establishment of research laboratories after the Giessen model" (Glas, 1983).

It is easy to summarize what Liebig's scientific style consisted of (there is some injustice in thus reducing his impressive contributions in many areas of chemistry: Liebig was active in theoretical organic chemistry, he developed chemistry of natural products, he was a pioneer in applied chemistry, especially in agricultural chemistry, he was a leader in pharmaceutical and in physiological chemistry).

He was a constant user of the 'black box'. In other words, he was not interested in the intervening details. His main concern was determining how a given input would lead to an exploitable output.

Glas writes of his theoretical view on organic chemistry that

"in 1845 he had been won over to the Dumas's view (Liebig, 1845), according to which organic substances, unlike inorganic ones, constitute one whole whose chemical properties depend first of all on its building schema or 'type' and much less on the nature of its elements" (Glas, 1983).

Liebig was led to such a holistic view because of his lack of interest in the intervening details. What counted most to him was establishing quantitative balances, comparing the initial and final states in a chemical process.

Around 1838, Liebig started investigating actively the chemical compositions of natural products, from plants and from animals. This was a natural development for him. He was a master of analytical techniques. Indeed, he had found a nice simple method for carbon and hydrogen analyses.

Among these determinations, Justus von Liebig examined the composition of albuminoids present in tissues or in bodily fluids (a practical by-product from these studies, performed during the period 1838–1842, was Liebig's beef stock extract). In 1841, he published his *Plant Chemistry*, quickly followed by *Animal Chemistry* (1842).

These two books followed *Organic Chemistry in its Application to Agriculture and Physiology* (1840), which had considerable success with agriculturalists. I see Justus von Liebig's interest in applications of organic chemistry, in its agricultural applications in particular, as stemming from the wish to emulate Humphrey Davy's trail-blazing example. Davy, it will be recalled, had examined practical problems of chemistry applied to agriculture, to geology, to designing a miner's lamp, partly because of the statutes and because of the peculiar structure of the Royal Institution, which he joined soon after its inception.

Anyway, Liebig contributed to agricultural chemistry much in the same way as he had earlier worked in organic chemistry: walking in Lavoisier's footsteps. In organic chemistry, Liebig was influenced by his stay in Paris, by the examples of Gay-Lussac and Thénard who themselves were followers of Lavoisier. In agricultural chemistry, Liebig discarded the existing theory of plants deriving their life support from the humus. He showed instead that plants lived on carbonic anhydride, water, and ammonia supplied to them from the atmosphere and from the soil. He advocated accordingly the use of mineral fertilizers to regenerate impoverished soils.

There again, both in his agricultural studies and in the physiological studies he had initiated on animals and plants, Liebig used his 'black box' approach:

"Liebig put chemical analysis in the role of furnisher of quantitative data for input/output balances, intended to reveal the gross material conditions for physiological processes in order to influence them to the benefit of man, without needing insights into the long causal chain between 'condition' and 'effect'. He described animals as 'walking stoves' regulated by input and output . . ." (Glas, 1983).

This was rather smart: in this manner, Liebig did not have to commit himself to investigating the life processes themselves. He was in the comfortable position of looking only at their end results and basing himself on rock-solid facts.

This has to be one of the main reasons why Liebig chose to clothe his physiological work in a protective layer of vitalism. This shield would give him an excuse for not attempting to elucidate the basic vital mechanisms, which he had little interest in deciphering (Glas, 1983).

Lipman (1967) has pointed at some of the resulting contradictions in Liebig's thought. This was a scientist who could declare in 1840 in his *Agricultural Chemistry* that

"the expression 'vital principle' must, in the meantime, be considered as of equal value with the terms *specific* and *dynamic* in medicine: everything is specific which we cannot explain, and dynamic is the explanation of all which we do not understand" (Liebig, 1842a).

(a most delightful poke at contemporary physicians who did not admit him into their fold on the faculty in a medical school) and who wrote in 1842 that

"everything in the organism goes under the influence of the vital force, an immaterial agency, which the chemist cannot employ at will" (Liebig, 1842b).

His first position, if taken seriously, was congruent with his early work in organic chemistry, which taught Liebig there was a continuity between organic and inorganic chemistry but an impassable difference of nature between organic chemistry and physiological chemistry. He saw organic compounds as more complex, but as not different basically from inorganic substances. Conversely,

"Liebig achieved a nonvitalistic organic chemistry because he distinguished between isolated organic compounds and organic compounds which were integrated into living tissue. Liebig believed that isolated compounds could be produced outside the organism such as in Wöhler's production of urea, but this did not mean that organized bodies could also be produced" (Lipman, 1967).

Furthermore, by the late 1830s – beginning 1840s, Liebig had come out strongly against *Naturphilosophie*. He took a firm stand, consistent with his own practice as a scientist, in favor of observation and experiment (Liebig, 1840).

In 1842 already, the full title of his *Animal Chemistry* (see Liebig, 1842b) showed that Liebig had chosen the Cartesian path from the simpler organic chemistry to the more complex physiological chemistry. As he would write in his *Familiar Letters on Chemistry*, which he started publishing in 1844,

"From organic chemistry, the laws of life, the science of physiology will be developed" (Liebig, 1859).

But the reductionist program this really entailed was strictly confined, in Liebig's practice, to his quantitative analyses of the 'black box' type. To give an example, he determined the elemental composition of 'protein' as $C_{48}H_{36}O_{14}N_6$ in close agreement with Mulder's (Glas, 1983).

Otherwise, Liebig explained away in his *Animal Chemistry* (1842b) all biological phenomena as mediated by vital forces. In this book, he asserted that the

" . . . generator of force for motion in the organism was muscular tissue. The role of the vital force in the tissue was to cause a change in its composition. As a result, a part of the muscle lost vitality and separated from the rest. In the process of compositional change, the total amount of the vital force diminished and established an equilibrium with the chemical force. (. . .) [Then] oxygen in the system combined with the separated part and completed the destruction of the living part. The change of form of the muscle tissue generated mechanical force (. . .)" (Lipman, 1967).

And Liebig, whose dogmatic and polemical bent was blatant (he hounded Schwann out of Germany, for instance), and who had achieved considerable success and obtained many distinctions for himself (he was made a baron in 1845, and he received a call from the University of Munich in 1852) could not resist denigrating materialists. He blamed them for being unable to 'show how inorganic forces completely replaced the concept of a vital force' (Lipman, 1967).

Lipman aptly concludes that Justus von Liebig occupies the

"somewhat interesting position of being reductionistic in his scientific methodology, but anti-reductionistic in his metaphysics – that is, in his explanation of living phenomena. (. . .) Since Liebig wanted some form of explanation for those areas where he thought chemistry inadequate, he took over vitalistic explanation and modified it to his own needs" (Lipman, 1967).

And it could be argued that Liebig opted for such an intellectually lazy (and politically comfortable, for the times) position, because basically he was a rather shallow thinker who did not care much about the detailed functioning of organisms.

I shall conclude this section with Liebig's concept of 'thrift in energy use', another and related factor in his vitalism. First a quotation from *Animal Chemistry*:

"Civilization is the art of saving force. Science teaches us several means for producing the greatest effects from the least possible force, it teaches us likewise how to use the means to extract the maximum force. The primitive state, the lack of civilization is thus characterized by useless or disproportionate use of force, either in agriculture, industry or science, or even in political economy" (Liebig, 1842c).

What an interesting notion! A Marxist analyst would be sure to point to Liebig's subservience to the chemical industry, interested in maximizing yields and profits for the smallest energy investment. But here I shall content myself with remarking that, for Liebig, civilization was the societal counterpart to vital forces in the organism:

"In Liebig's view the vital force worked simply through attraction and repulsion, but its hidden nature ruled out the ordinary laws of chemical combination, thereby enabling physiological transformations to take place in the most 'economical' way possible" (Glas, 1978; 1983).

I hope to analyse elsewhere this concept of 'economy' as central to Liebig's thought.

9. Claude Bernard's attempt at bridging the two viewpoints

Claude Bernard's (1813–1878) career consisted of two distinct parts. During the first, till the early 1850s, he accumulated great experimental breakthroughs, such as elucidating the nature and the role of the gastric juice; establishing the role of the pancreas in digestion; discovering the glycogenic function of the liver; his study of neuromuscular transmission, using curare as an experimental tool; and many others.

During the latter part of his career, Bernard made himself into an epistemologist. He published the book he is famous for, *Introduction to the Study of Experimental Medicine*, relatively late, at the age of 52. This book marked the beginning of his (public) discussion of the bases and of the methodology of physiology.

His starting point brings us back to John Locke (see section 6), whose philosophy permeated Claude Bernard probably indirectly, through the French positivists of the mid-19th century. Claude Bernard wrote that

"Observers then must be photographers of phenomena; their observations must accurately represent nature. We must observe without any preconceived idea; the observer's mind must be passive, that is, must hold its peace; it listens to nature and writes at nature's dictation" (Bernard, 1865).

This primacy given to observation, just as Locke and Condillac had given priority to sensation, went with the notion of the inaccessibility of things-in-themselves. Human knowledge, Bernard was convinced, could only build upon the relationships among natural objects; it could not hope to reach the essence of these objects. Hence, Bernard asserted that the primary causes of phenomena will remain outside the grasp of science. The scientific pursuit may concern itself legitimately only with secondary (or effective) causes.

122 PHILOSOPHICAL UNDERCURRENTS

Bernard thus held that

"consideration of primary causes is a task for philosophy and theology, but not for science. (Bernard) rejects both vitalism and materialism because they do not keep strictly to positive science but invoke untestable metaphysical ideas. They make the mistake of looking for primary causes" (Roll-Hansen, 1976).

And Claude Bernard thus dispels both materialism and vitalism as unacceptable doctrines.

This is the first part of his strategy, which as we shall see aims at carving a new epistemological and institutional niche for physiology. He wants his cherished and chosen science sheltered from the suffocating embraces, from the physical sciences and materialism, on one side, from descriptive biology and vitalism on the other. The specificity of physiology would vanish, physical science or traditional biology would swallow it whole. Hence, it was important to Claude Bernard to reject both materialism and vitalism, back-to-back so to speak.

The contemporary debates, nowadays, when departments of biochemistry or of molecular biology strive to erect themselves as independent entities, separate from chemistry, on one side, from biology and from medicine, on the other side, parallel the situation of physiology in Bernard's time.

To return to Claude Bernard, he argued against materialism and vitalism in that they both resorted to primary causes. The vitalists explicitly so: they held the vital force as such a primary cause. The materialists implicitly so: they purported to explain vital phenomena from material motions, or from physicochemical processes; that is, from primary causes.

"To such statements from mechanicists, Claude Bernard made a definitive answer. There is indeed an essential difference between the living and the non-living. Whereas the non-living grows from the outside, the living grows from the inside. Elements accrete around the outside of a mineral to increase its volume. Foodstuffs are introduced inside the plant or the animal. Furthermore, whereas a mineral grows along its length or along its width indifferently and indefinitely according to chance encounters with outside elements, the plant or the animal only grows along well-defined directions and under strict limits, depending upon the species it belongs to. Vital movement is thus a directed motion and a direction manifests itself in life. Finally, a mineral grows only by juxtaposition of chemically like elements, copper by juxtaposition of copper, iron by juxtaposition of iron. It cannot absorb differing elements without ceasing to exist. If sulfuric acid absorbs iron, there is no longer sulfuric acid nor iron, there is a new substance, iron sulfate.

By contrast, a living being absorbs foodstuffs that differ chemically from itself. It decomposes them and utilizes them to build its own organs. From bread, from meat it makes nerves, muscles, bones similar to those it already had or even to those its ancestors had before it did. The living being is not only a director of motions, it is also a creator of organs" (Fonsegrive, ca. 1870).

I have quoted at length from this presentation of the ideas of Claude Bernard by one of his contemporaries, because it is a fine summary. Bernard resorts to basic *themata*, isotropy versus anisotropy of growth, homogeneity of the crystal versus heterogeneity of the tissue, in order to stress the originality of the living state against the foil of inanimate matter.

Having stated this irreducible originality of life, Bernard can proceed with his strategy which, in accordance with his positivistic beliefs, was to define 'the area of validity for deterministic physical science (...) by the extent of the phenomena that can be experimentally manipulated by Man' (Roll-Hansen, 1976).

To proceed further, Claude Bernard had to avoid the mistake made by the German materialists of the 1840s (see the preceding Section 7): they meant to transfer to physiology without modifications the research methodology which had proved its value in physics and in chemistry. Yet, Bernard was suitably impressed by the success of such methods in the physical sciences. To get out of this difficulty, i.e. to justify the borrowings, while at the same time erecting physiology as an autonomous and distinct field of enquiry, Claude Bernard enlarged upon his distinction between observations and things-in-themselves. He made it into a new distinction between experimental science and 'vital morphology'.

Experimental science tackles explainable phenomena:

"A particular event is explained when it follows deductively from a set of laws plus a set of initial conditions" (Roll-Hansen, 1976).

And Bernard acknowledged that 'the general laws of physics and chemistry are valid for living systems' (Robin, 1979).

But there exist also phenomena outside the legal sphere of physicochemical conditions, phenomena that evade deterministic principles. These phenomena are, for Claude Bernard, outside the proper scope of physiology. Bernard included heredity, evolution of species, and embryology, i.e. all the great diachronic phenomena of biology, among these. Such phenomena were off-limits: they were determined by a vital force, and they belonged to the sphere of metaphysics.

Now, Bernard's stategy becomes clearer:

"By borrowing principles from both sides, Bernard constructed physiology as an autonomous discipline distinct from both of the old ones [descriptive biology, on one hand; and the experimental sciences of physics and chemistry, on the other]. This double origin of his physiology is explained in the introductory lecture of his *Leçons sur les phénomènes de la vie communs aux animaux et aux végétaux* (1878). He united the physicochemical physiology of Lavoisier and the general anatomy of Bichat into a single science, the new physiology" (Roll-Hansen, 1976).

Bernard could use his postulate of the irreducibility of morphological development to physicochemical laws in support of his claim for the autonomy of the new physiology. He could use also, conversely, the applicability of physicochemical methods to physiology in order to claim it as radically distinct from descriptive biology, or from descriptive medicine, in the old style.

To sum up – and this is the main point made by Roll-Hansen in his intelligent 1976 paper – Bernard (despite his convoluted and rhetorical prose) reaches conclusions quite similar to the conclusions of Kant arrived at earlier. And, in the same manner as Immanuel Kant had ruled out the self-ordering of inanimate matter into a leaf of grass (see Section 4), Claude Bernard ruled out manufacture by the chemist of an enzyme:

"The chemist will be able to make the products from a living being, but he won't ever succeed in making its tools, because they are the very results of organic morphology which, as we shall soon see, is outside of chemistry in the proper sense; and, seen in this manner, it is no more possible for the chemist to manufacture the simplest ferment than to manufacture a whole living being" (Bernard, 1878).

A century later, we are well on the way to proving wrong this particular prediction!

10. Concluding statement

A kindred statement of impossibility was made just a few years later by Louis Pasteur:

"You know that the most complex molecules of plant chemistry are the albumins. You know besides that these immediate principles have never been obtained in the crystalline state. May one not add that they probably cannot crystallize . . . " (Pasteur, 1886).

Pasteur had been a long-standing opponent of materialism. His demonstration, against Felix Pouchet, of the non-occurrence of spontaneous generation in a truly sterile environment, to him was a battle won against the materialistic notion of matter self-organizing into the living state. If spontaneous generation has been proven,

"what a victory for materialism," Pasteur exclaimed in 1864, "if it could be affirmed that it rests on the established fact that matter organizes itself, takes on life itself; matter which has in it already all known forces!" (Pasteur, quoted by Farley, 1978).

By the time Pasteur made this comment in an impassioned speech at the Sorbonne, the battle lines had been drawn; and they had been drawn on ideological and political grounds predominantly, as Farley points out in this particular instance of Pasteur (Farley, 1978).

Throughout this chapter I have focussed on the *philosophical* background to the controversy between materialists and vitalists. This is undoubtedly an idealized and artificial presentation. It ignores the gradually, the ever-increasing role of *ideology* and *politics* to be ascribed, among other factors, to the formation of *a public opinion* during the 19th century, moulded by the spread of literacy, by the new printing techniques which rendered possible popularization of the newly acquired scientific knowledge, by electoral reforms also.

Claude Bernard's claim of the autonomy of physiology can be discussed, and this is consonant with his own arguments, in the framework of intellectual history, as I have done here. But it belongs with institutional history and with sociology of science as well: Claude Bernard was fighting for increased support of physiology, and his epistemological writings can be interpreted also in this context (Roll-Hansen, 1976).

By the time of Claude Bernard and Louis Pasteur, the material-ist–vitalist controversy had spread outside academic politics. It had become a divisive political issue. Historically, one should probably ascribe to Napoleon I the responsibility for drawing such epistemo-logical and philosophical questions into the political arena. The action he took against the *Idéologues* at the start of the century, turning against them, pouring scorn, having them expelled from the Institut de France, even blaming French military defeats on their corrupting influence (December 1812), was decisive.

And the political lines became quickly drawn. Most of the time, and this was a recurring pattern from the beginning of the 19th century, the materialist position was assumed by young scientists with revolutionary or reformist political ideals, while older and more conservative scientists were or became vitalists. Is there a basic strain of obscurantism and anti-rationalism rampant among con-servatives, who often prefer revealed truths and institutionalized dogmas to the patient unravelling of nature's web?

The notion of a vital force, akin to the divine spark of the Scrip-tures; the notion of a reserved area science could not and should not attempt to penetrate; the notion of wonder at nature's exquisite design, as the only conceivable attitude: these were all set to appeal to conservatives, intent upon maintaining law and order. Often also, there were vested institutional interests: vitalists were well-entrenched in academic positions of power, while the younger mate-rialists did not enjoy as much security and belonged to the political left.

We could ignore here this rampant political dimension, but at the risk of de-historicizing the controversy during the period from about 1790 to the mid-19th century. Afterwards, it becomes impossible to maintain such a stance, as we shall see from the next episode, which recounts the resurgence in the early 20th century of the quarrel between materialists and vitalists, when Jacques Loeb defended the rational, materialistic viewpoint not only against Driesch but also against the colloidalists.

REFERENCES

Balzac, H. de (1830) notebook ed. by Crépet, J. (1910) Balzac, Pensées, Sujets, Fragments, Paris (my transl.).

Balzac, H. de (1831) La Peau de Chagrin, in de Sacy, S. (Ed.), Gallimard (Folio), Paris, 1974, (a) pp. 327–329; (b) p. 327; (c) p. 303 (my transl.).

Bernard, Cl. (1865) Introduction to the Study of Experimental Medicine (transl. of Introduction à la Médecine Expérimentale, Paris); Dover, New York, 1957.

Bernard, Cl. (1878) Leçons sur les Phénomènes de la Vie communs aux Animaux et aux Végétaux (2 vols.), Paris, vol. 1, p. 227.

Bichat, M.F.X. (1800) Recherches physiologiques sur la Vie et la Mort, Paris (my transl.).

Blakemore, C. (1977) Mechanics of the Mind, Cambridge University Press, Cambridge, p. 102.

Blumenbach, J.F. (1781) Über den Bildungstrieb und das Zeugungsgeschäfte, Dietrich, Göttingen; reprinted (1971) Fischer, Stuttgart, p. 12 (transl. Lenoir, 1980).

Blumenbach, J.F. (1797) Handbuch der Naturgeschichte, 5th ed., Dietrich, Göttingen, (a) p. 18; (b) p. 12 (transl. Lenoir, 1980).

Canguilhem, G. (1965) La Connaissance de la Vie, Vrin, Paris, 2nd ed., (a) p. 87; (b) p. 83; (c) p. 88; (d) p. 89.

Condillac, E. Bonnot de (1746) Essai sur l'Origine des Connaissances Humaines, Paris.

Cranefield, P.F. (1957) J. Hist. Med., 12, 407.

Du Bois-Reymond, E. (1848) Untersuchungen über Thierische Elektricität, Reimer, Berlin, XXXV (transl. Cranefield, 1957).

Du Bois-Reymond, E. (1918) Jugendbriefe von Emil du Bois-Reymond an Eduard Hallman, Reimer, Berlin, (a) p. 98; (b) p. 108 (transl. Temkin, 1946).

Farley, J. (1978) Annu. Rev. Microbiol., 32, 143.

Fleming, D. (1964) preface to Loeb, J., The Mechanistic Conception of Life, Harvard University Press, Cambridge, MA, p. viii.

Fonsegrive, G. (ca. 1870) Encyclopédie du XIXe Siècle, Paris: entry ANIMISME, Vol. 3, p. 1.

Fruton, J.S. (1976) Science, 192, 327.

Girtanner, Ch. (1797) Über das Kantische Prinzip für die Naturgeschichte, Vandenhoek and Ruprecht, Göttingen.

Glas, E. (1978) Stud. Hist. Phil. Sci., 9, 300, 302–304.

Glas, E. (1983) Stud. Hist. Phil. Sci., 14, 39.

Kant, I. (1785) Die Bestimmung des Begriffs einer Menschenrasse, in Gesammelte Schriften, Königliche Preussische Akademie der Wissenschaften, Reiner, Berlin, 1912, Vol. 8, p. 100.

Kant, I. (1790) in Gesammelte Schriften, Königliche Preussische Akademie der Wissenschaften, Reimer, Berlin, 1908, Vol. 11, p. 176.

Kant, I. (1790) Kritik der Urteilskraft, Philosophische Bibliothek, Berlin, 1926, p. 353 (transl. Roll-Hansen, 1976).

Kielmeyer, C.F. (1790-1793) Entwurf zu einer vergleichenden Zoologie, 1790-93, in Holler, H.F. (Ed.) Gesammelten Schriften, Keiper, Berlin, 1938, p. 22 (transl. Lenoir, 1980).

Langlois, J.P. (ca. 1870) Encyclopédie du XIXe Siècle, Paris: entry PHYSIOLOGIE, Vol. 26, p. 821.

Larson, J.L. (1979) ISIS, 70, 235.

Laszlo, P. (1979) Nouv. J. Chimie, 3, 209.

Lenoir, T. (1980) ISIS, 71, 77.

Lenoir, T. (1981) Stud. Hist. Phil. Sci., 12, 293.

Liebig, J. von (1842a) Chemistry in its Application to Agriculture and Physiology. (Die Chemie in ihrer Anwendung auf Agrikultur und Physiologie, 1840; transl. John Owen, p. 75) Lyon Playfair, Cambridge, MA (quoted by Lipman, 1967).

Liebig, J. von (1842b) Animal Chemistry: or Organic Chemistry in its Application to Physiology and Pathology, (Die Thier-Chemie, oder die organische Chemie in ihrer Anwendung auf Physiologie und Pathologie, 1842; transl. John Owen, p. 153) Gregory, Cambridge, MA.

Liebig, J. von (1842c) Chimie organique appliquée à la Physiologie Animale et à la Pathologie (French transl. of Die Thier-Chemie, oder die organische Chemie in ihrer Anwendung auf Physiologie und Pathologie, 1842) Fortin, Masson, Paris, pp. 85-86, my translation and emphasis.

Liebig, J. von (1845) Ann. Chem. Pharm., 53, 1.

Liebig J. von (1859) Letter I, p. 3, in Blyth, J. (Ed.), Familiar Letters on Chemistry, Walton and Maberly, London (quoted by Lipman, 1967).

Lipman, T.O. (1967) ISIS, 58, 167.

Mendelsohn, E. (1965) Br. J. Hist. Sci., 2, 208.

Pasteur, L. (1886) Conférences faites à la Société Chimique de Paris en 1883-1886, Bureau des Deux Revues, Paris, p. 36 (quoted by Fruton, 1976).

Pastore, N. (1971) Selective History of Theories of Visual Perception 1650-1950, Oxford University Press, New York, (a) p. 101; (b) p. 103; (c) p. 106.

Popper, K.R. (1968) The Logic of Scientific Discovery, Harper and Row, New York, pp. 452-456.

Ravetz, J.R. (1979) History of Science, in Encyclopedia Britannica, 15th ed., Chicago, Vol. 16, p. 366.

Robin, E.D. (1979) J. Am. Med. Assoc., 242, 1283.

Roll-Hansen, N. (1976) J. Hist. Biol., 9, 59.

Schwann, Th. (1839) Mikroskopische Untersuchungen über die Übereinstimmung in der Struktur und dem Wachstume der Thiere und Pflanzen, Sander, Berlin.

Temkin, O. (1946) Bull. Hist. Med., XX, 322.

Treviranus, G.R. (1805) Biologie: oder Philosophie der lebenden Natur, Röwer, Göttingen, (1802-1822), Vol. III, p. 163 (transl. Lenoir, 1980).

Vicq-d'Azyr, F. (1805) Oeuvres, Paris, Vol. IV, p. 14 (quoted by Du Bois-Reymond, 1848).

Vogt, C. (1845-1847) Physiologische Briefe für Gebildete aller Stände, Stuttgart and Tübingen (quoted by Temkin, 1946).

Wöhler, F. (1978) Trends Biol. Sci., 17.

Chapter 8

Schizoid Tale of Two Schools

1. An eighteenth-century natural philosopher

The location: Naples. The time: the winter of 1889-1890. The character: a young German-Jewish scientist of 30, by the name of Jacques Loeb. Loeb had had a circuitous career, before he embraced experimental biology as his life-long vocation.

A student of philosophy at first, who had been deeply moved by Schopenhauer's conception of the will as primal, he then became disgusted by what he saw as the idle speculations of metaphysical philosophy. Loeb opted instead for a scientific study of the will. 'The question he turned upon the philosophers for not resolving was the same question he put to his biological experiments: was there any such thing as freedom of the will?' (Fleming, 1964a). He had then entered medical school, taking his medical degree in 1884 in Strasbourg.

A little later, during the winter of 1886-1887, he came under the influence of Adolph Fick and Julius Sachs (1832-1897) in Würzburg, who 'initiated him into mechanistic physiology (. . .) Both put him in touch at one remove with the great Mechanistic Quadrumvirate of Helmholtz, Du Bois-Reymond, Ludwig, and Brücke' (Fleming, 1964b).

Loeb became an intimate friend of Sachs, whose research dealt with plant tropisms.

"Loeb's task in science suddenly became clear: to establish a concept of tropisms in animals by which they too could be shown to be irresistibly driven by external stimuli and impotent to interpose their wills" (Fleming, 1970-1980).

Jacques Loeb started his own work along these lines in 1888, by studying heliotropism among caterpillars.

Far from being an instinct for self-preservation, their behavior of climbing up to the tips of branches and feeding upon the buds was convincingly shown by clever experiments of Jacques Loeb to be merely a drive toward the light. These caterpillars were just photo-chemical mechanical devices enslaved to the light, as Loeb put it. After he had completed his first series of investigations, Loeb published in 1890 in Würzburg a long memoir on the heliotropism of animals and its similarity with the heliotropism of plants.

Thus we find him in Naples in 1890. He had gone there to do experiments on heteromorphosis and to study the depth migrations of animals, jointly with Theo T. Groom. He found in Naples an effervescent intellectual atmosphere dominated by Wilhelm Roux's *Entwicklungsmechanik*. But foremost and ironically, Jacques Loeb's stay in Naples was for him a discovery of America and Americans.

Loeb's own father, who had had a pronounced intellectual influence on him, had been (he died when Loeb was 16) a cultured man, a liberal, who hated Prussianism, and, by way of counterpoise, 'looked longingly toward the democratic institutions of France and of the United States' (Osterhout, 1928a).

A second factor in Jacques Loeb's budding love affair with America may have been the most kind and encouraging letter which William James (1842-1910) – the philosopher, who was keenly interested in the scientific strides taken both by physiology and psychology – wrote him after reading his maiden publication, which dealt with blindness arising from injury to the cerebral cortex (1884).

Yet another factor was the immediate friendships he formed in Naples, during this key year 1890 with young American scientists (embryologists and cytologists) he met there, such as Henry B. Ward and W.W. Norman.

At that time (1890), the United States had come into its own as a whole continent united by the newly built railroads. The American West was being developed by cattle-farmers and also by miners. They supplied the populated Eastern states with food and raw materials. America must have appealed a lot to Loeb's imagination because, not only was it a land of democratic institutions, a melting pot for immigrants of every creed and ethnic origin, it was also a country where science and technology thrived hand in hand. The single decade of 1870-1880 saw the invention of the typewriter (1873), of the telephone by Alexander Graham Bell (1876), and of the incandescent bulb by Thomas Alva Edison (1880).

To cap it, Jacques Loeb fell in love with a young American, Anne Leonard. She had earned a doctorate in philology at the University of Zürich. They quickly became engaged, traveled to America in the fall of 1891 and were married there in October. They then returned to Naples, where Loeb resumed his experiments on heteromorphosis.

Jacques Loeb, at that time, had become more and more unhappy about the Prussian militaristic, authoritarian and racist outlook. He could have continued to live, rather modestly, on the income from his inheritance. However, and even though he felt threatened by the academic life as an impingement upon his freedom (the question of freedom of the will was indeed, both personally and professionally, a central concern, something like Jacques Loeb's fundamental doubt), his new personal responsibilities convinced him to seek academic employment. Thus, it was natural that he should turn to America. This he did initially with the notion that 'he would earn his living as an oculist, devoting part of his time to practice and the rest to research' (Osterhout, 1928b). He quickly gave up this idea, though, when he realized it would prevent him from his passionate urge for full-time scientific experimentation.

Fortunately, at this stage Jacques Loeb received a call from Bryn Mawr, where he settled in November 1891. At Bryn Mawr, Loeb became an intimate friend of Thomas Hunt Morgan (1866-1945), the geneticist. Younger than Loeb by seven years, Morgan was also an iconoclast. During his formative years at Johns Hopkins, Morgan had espoused the guiding principles of his mentor, William Keith Brooks (1848-1908), who saw 'major areas of biology as problems to be solved rather than fixed and final "truths" already revealed' (Allen, 1983). This was also quite close to Loeb's own convictions.

From Bryn Mawr, Jacques Loeb soon (January 1892) moved to the University of Chicago, where he was offered better facilities. He spent 10 happy and productive years there, traveling to Woods Hole, MA, every summer to give a course in physiology and to do research.

His stay at Pacific Grove during the winter of 1898–1899 when, with his wife's help, he wrote his *Comparative Physiology of the Brain and Comparative Psychology* (Loeb, 1900) endeared California to him; he loved the climate; and the 'possibility of working on marine material all the year around' (Osterhout, 1928c) was well-nigh irresistible. Thus, when the University of California offered him a position in 1902, he accepted it readily. He stayed at Berkeley till 1910, where a laboratory of physiology was built for him (1903):

"In accepting the call to California Loeb had not realized how much he would be cut off from contact with his fellow scientists. (. . .) This isolation had much to do with his consideration of offers from Europe (especially from Budapest) and his final acceptance of a call to the Rockefeller Institute for Medical Research in 1910"(Osterhout, 1928c).

He stayed at the Rockefeller Institute till his death, in 1924.

The main elements in Loeb's scientific style were tenacity and the search for simplicity. Tenacity:

"Dropping a problem which did not seem to be leading anywhere he would nevertheless keep it always in the back of his mind so that if new facts turned up he could at once turn them to good use" (Osterhout, 1928d).

Jacques Loeb was indeed an 'indefatigable, almost frenzied worker' (Fleming, 1964c). The search for simplicity was an over-riding concern, to such an extent that

"it was sometimes said that he pictured his problem too simply and was satisfied with explanations too simple to correspond to reality" (Osterhout, 1928e).

Loeb's interaction with Fick and Sachs had nursed him with the milk of mechanistic physiology. His stay in Naples, starting in the pivotal winter of 1889-1890 had immersed him in Wilhelm Roux's (1850-1924) *Entwicklungsmechanik*: this was a mechanical conception of embryological development, seen as due solely to the operation of Newtonian forces upon the cells of the embryo. Accordingly, Jacques Loeb made it his premise (and his battle-cry, so to say) that biology was reducible to the simple laws of physics and chemistry. As his former co-worker Osterhout writes perceptively:

"His notion of biological research was simple: all the observed phenomena should be expressed in the form of equations containing no arbitrary constants" (Osterhout, 1928f).

Loeb was convinced that the organism as a whole was capable of reacting 'with that degree of machine-like precision which we find in the realm of physics and chemistry' (Loeb, 1916). For instance, when he discovered in 1899 that he could induce parthenogenesis of sterile eggs of sea urchins, simply by raising the osmotic pressure, Loeb stated that the 'importance of this discovery consisted in transferring the problem of fertilization from the domain of morphology to that of physical chemistry' (Osterhout, 1928g).

A fanatical reductionist in scientific style, Jacques Loeb was a materialistic determinist and a socialist. His philosophical studies, the reading of Schopenhauer especially, taught him the compelling necessity, the moral duty as it were, to dispel philosophical illusions, such as that of freedom of the will. Loeb and Schopenhauer

"believed, fanatically, in ripping the veil from the illusion, stripping the pretensions from the overweening automatons who fancied themselves bestriding the world. (. . . they) demanded access to the unmitigated Truth and commanded others to look upon it as steadily as themselves" (Fleming, 1964d).

As writes Fleming perceptively: 'The young Loeb found a deep repose in the contemplation of determinism' (Fleming, 1964d).

Loeb, who was a socialist, was close to Marx in many respects. Like him, he 'belonged to the class of determinists who perceive an inevitably enacted world of evils about them yet do not despair of shaping it toward some kind of millennium' (Fleming, 1964e).

Jacques Loeb, if I may be forgiven this quip, was as much a slave of mechanistic materialism as the caterpillars he had studied were the slaves of light!

He had the zeal of an evangelist in uttering his mechanistic creed. He was indeed utterly convinced that, once he had reached the truth about a living phenomenon, others had to share in his vision. In private conversation, his quicksilver mind bedazzled and overwhelmed his listeners:

"His intimate talks in the laboratory were at once the joy and despair of fellow workers. His mobile features, his expressive, eager eyes, alight with enthusiasm, were a fascinating study as he flashed from mood to mood, smiles and frown following in rapid succession. (. . .) A visitor to his laboratory was quite apt to leave in a somewhat breathless state. The rapidity with which ideas were suggested, examined, and rejected was often astonishing" (Osterhout, 1928h).

Jacques Loeb truly belonged to the 18th century. As a conversationalist. As a philosopher who looked to scientific experiment for the answer to ethical questions and to dispel metaphysical humbug. As a man who held wit as a supreme value, and who reveled in it, who was prone to cracking jokes and to cascading laughter. It is not surprising that one of his major heroes was Denis Diderot.

2. Loeb and the Arrhenius physicochemical program for biology

Jacques Loeb was an early and enthusiastic convert to the ideas of Svante Arrhenius, as already propounded in the doctoral dissertation of the Swede (1884): the properties of solutions, of ionic solutions especially, as explained by electrolytic dissociation and the attendant chemical equilibria, had universal applicability. Hence, there was all the reason to believe, Arrhenius thought and wrote already at the time of his dissertation, that all substances would conform to his theory:

" . . . perhaps one could in the future enlarge the theory given for electrolytes until it becomes, with some modifications, applicable to all substances" (Arrhenius, 1884).

And Svante Arrhenius set about to show himself right in this ambitious claim. First, and despite the initial lack of impact of his thesis, he performed numerous studies which convinced the scientific community, by the mid-1890s, of the validity and widespread applicability to chemistry of the theory of electrolytic dissociation. Then,

"the energetic Swede looked for new fields to conquer. He also wished to demonstrate that the natural sciences could be reconstituted along new lines, and he sought to fulfill the impetuous claims of his doctoral dissertation of a decade and a half earlier. In time, Arrhenius would try to interpret biology, geology, and astrophysics as specialized regions of ionic interactions" (Rubin, 1980).

Jacques Loeb came quite early under the spell of ionic theory, during these years in Würzburg when he studied under Fick and Sachs, which moulded his approach to biology. Here, I must quote at length the superb account of Fleming:

"In the same winter of 1886-1887 that Loeb experienced his initiation into mechanistic physiology at the hands of Fick and Sachs, the young Swede Svante Arrhenius was also working in Würzburg at the height of his powers. Arrhenius dated from Würzburg his famous letter of 1887 to van 't Hoff announcing his corollary to the latter's theory of solutions. Arrhenius invoked electrolytic dissociation – ionization – of salts, acids, and alkalis in solution, as the explanation for the anomalies which remained in van 't Hoff's otherwise extremely powerful conception of solutions as obeying the same laws as gases and susceptible of the same quantitative analysis. This tremendous advance in physical chemistry can hardly have been lost upon Fick. It ought not to have been lost upon Sachs, for the whole development was triggered by a long line of researches upon osmotic pressure in plants, as detailed by him in his history of botany. It is tempting to suppose that he and Fick joined in calling the attention of their protégé Loeb to the opportunities that were opening for investigation of the physiological role of ions and more generally for the assimilation of physiology to physical chemistry. It is more tempting still to suppose that Loeb and Arrhenius may actually have met in Würzburg in this year of exploding horizons for each. The incontestable facts are that Loeb became an evangelist for physical chemistry as the key to biology; he was conducting experiments from the early 1890's forward upon sea urchins immersed in salt solutions and subjected to alterations in osmotic pressure by changing the concentration; and his personal intimacy with Arrhenius in the twentieth century is well authenticated. Whenever they first met, they had more in common than an interest in physical chemistry. They participated in the same unmistakable Weltanschauung of materialism in philosophy, mechanism in science, and radicalism in politics" (Fleming, 1964f).

Accordingly, and quite early on, Jacques Loeb, with his characteristic ebullient energy, launched upon quite a number of accurate, careful studies of the role of various ions in physiology. The very titles of some of his publications are revealing:

1897 – The physiological effects of ions, I. Experiments upon muscles.
   – On the physiological effects of electrical waves.
1896 – The physiological effects of ions, II.
   – On the physiological effect of alkalis and acids in high dilution.
   – On the influence of alkalis and acids on embryonic development and growth.
1899 – On ions which are capable of calling forth rhythmical contractions in skeletal muscle.
   – On the ion-proteid compounds and their role on the mechanics of life phenomena, I. The poisonous character of a pure NaCl solution.
   – On the different effects of ions upon myogenic and neurogenic rhythmical contractions and upon embryonic and muscular tissue.
   – On artificial parthenogenesis in sea urchins.
1900 – On the significance of Ca- and K-ions for the heart activity.
etc.

Thus, Jacques Loeb is to be credited with the first application of Arrhenius' ideas to biology. He followed the research program set by the Swede and, in so doing, revolutionized physiology – showing it to be a province of physical chemistry. And the kindred scientific objectives went with personal friendship, as stated in the above quotation from Fleming.

This point deserves further elaboration: as soon as Jacques Loeb had been appointed at Berkeley, he asked Svante Arrhenius to visit. Indeed Arrhenius spent part of 1904 in Loeb's new physiological laboratory at the University of California, which had been inaugurated in October 1903, the keynote address being delivered by none other than the great physical chemist Wilhelm Ostwald, the same

Ostwald whose 'timely intercession' (Rubin, 1980) had earlier saved Arrhenius' dissertation from lapsing into obscurity. When Svante Arrhenius stayed in Jacques Loeb's laboratory at Berkeley, he came fresh from his stay in 1903 and the beginning of 1904 at Paul Ehrlich's laboratory in Frankfurt. This collaboration was thorny, it saw the start of the big public controversy about the nature of the antigen–antibody interaction between Arrhenius, bent on a chemical equilibrium explanation, and Ehrlich, who was for the formation of a covalently held compound. This controversy between Arrhenius and Ehrlich was the prototype, as it were, for the later controversy that would arise between Jacques Loeb and the colloidalists, especially between Loeb and Wilhelm Ostwald's son, Wolfgang.

3. Wolfgang Ostwald and the colloidal school

Our story has a small cast of characters!

Wilhelm Ostwald's second child, Wolfgang (1883-1943), after studying zoology at Leipzig, became Jacques Loeb's research assistant at Berkeley from 1904 till 1906 - no doubt upon his father's recommendation. There he formed a friendship with Martin H. Fischer, a physiologist and physician. There also he became interested, upon Loeb's suggestion, in the colloidal state of matter.

This notion had been introduced by Thomas Graham (1805-1869) in 1861. Graham emphasized the contrast between two states of matter, the colloidal and the crystalloidal, and he drew attention to the relevance of this opposition to the distinction between inanimate and organized matter:

"I may be allowed to advert again to the radical distinction assumed in this paper to exist between colloids and crystalloids in their intimate molecular constitution. Every physical and chemical property is characteristically modified in each class. They appear like different worlds of matter, and give occasion to a corresponding division of chemical science. The distinction between these kinds of matter is that subsisting between the material of a mineral and the material of an organized mass" (Graham, 1861).

Morphology distinguishes indeed between the mineral and the colloidal. In the beautiful description by D'Arcy Wentworth Thompson, which I shall quote here at length:

"A crystal 'grows' by deposition of new molecules, one by one and layer by layer, each one superimposed on the solid substratum already formed. Each particle would seem to be influenced only by the particles in its immediate neigbourhood, and to be in a state of freedom and independence from the influence, either direct or indirect, of its remoter neighbours. So Lavoisier was the first to say. And Kelvin and others later on explained the formation and the resulting forms of crystals, so we believe that each added particle takes up its position in relation to its immediate neighbours already arranged, in the holes and corners that their arrangement leaves, and in closest contact with the greatest number; hence we may repeat or initiate this process of arrangement, with great or apparently even with precise accuracy (in the case of the simpler crystalline systems) by piling up spherical pills or grains of shot. In so doing, we must have regard to the fact that each particle must drop into the place where it can go most easily, or where no easier place offers. In more technical language, each particle is free to take up, and does take up, its position of least potential energy relative to those already there: in other words, for each particle motion is induced until the energy of the system is so distributed that no tendency or resultant force remains to move it more. This has been shewn to lead to the production of *plane* surfaces (in all cases where, by the limitation of material, surfaces *must* occur); where we have planes, there straight edges and solid angles must obviously occur also, and, if equilibrium is to follow, must occur symmetrically. Our piling up of shot to make mimic crystals gives us visible demonstration that the result is actually to obtain, as in the natural crystal, plane surfaces and sharp angles symmetrically disposed.

But the living cell grows in a totally different way, very much as a piece of glue swells up in water, by 'imbibition', or by interpenetration into and throughout its entire substance. The semi-fluid colloid mass takes up water, partly to combine chemically with its individual molecules; partly by physical diffusion into the interstices between molecules or micellae, and partly, as it would seem, in other ways; so that the entire phenomenon is a complex and even an obscure one. But, so far as we are concerned, the net result is very simple. For the equilibrium, or tendency to equilibrium, of fluid pressure in all parts of its interior while the process of imbibition is going on, the constant rearrangement of its fluid mass, the contrast in short with the crystalline method of growth where each particle comes to rest to move (relatively to the whole) no more, lead the mass of jelly to swell up very much as a bladder into which we blow air, and so, by a *graded* and harmonious distribution of forces, to assume everywhere a rounded and more or less bubble-like external form. So, when (...) older naturalists called attention to a new distinction or contrast of form between organic and inorganic objects, in that the contours of the former tended to roundness and curvature, and those of the latter to be bounded by straight lines, planes and sharp angles, we see that this contrast was not a new and different one, but only another aspect of their former statement, and an immediate consequence of the difference between the processes of agglutination and intussusception" (D'Arcy Wentworth Thompson, 1942).

The main *operational* distinction between colloids and crystalloids was that the latter, but not the former, could cross a membrane. Biologists were quick to seize upon this timely addition to the vocabulary of their discipline. As writes percipiently Joseph Fruton, describing the ideas prevailing at the turn of the century:

"Not only were colloids retained by membranes that allowed the passage of water and salts, but like protoplasm they also imbibed water and adsorbed substances on their surface. Such adsorption phenomena had been studied qualitatively at mid-century, but the new physical chemistry provided a consistent theory to explain them. (. . .) the colloid chemistry of protoplasm offered to biologists a more satisfying guide to the molecular explanation of physiological phenomena than did the organic chemistry of Emil Fischer" (Fruton, 1976).

Many substances formed colloids: oil suspensions in water, aqueous solutions of proteins such as ovalbumin, and suspensions of small (1–100 nm in size) metallic particles such as gold. Since all these systems appeared to share very similar properties, quickly the distinctions – those of chemical nature, in the first place – became blurred and attention became focussed instead on the similarities. One major generalization was made by the colloid chemists: 'solutions' of colloids were 'in reality suspensions (. . .) hence the general laws of the physical chemistry of solutions were essentially inapplicable to them' (Edsall, 1962). The French scientist Jacques Duclaux (son of Emile Duclaux (1840–1904)) published in 1904 a book on this topic, *Recherches sur les Substances Colloïdales* (Duclaux, 1904). There is little doubt that this volume was eagerly assimilated by the young Wolfgang Ostwald, at the very time of his budding interest in colloids. In this book, Duclaux started by reiterating the distinction originally made by Graham:

"All living matters are colloids and all the colloids are, in many ways, living material" (Duclaux, 1904a).

He went on to forceful statements about the total insolubility of colloids in water which, therefore, made them immune to the laws governing the usual chemical solutions. He wrote, for instance, that

"Knowledge of the chemical and of the physiological role of colloids is linked to the study of the equilibria present in their solutions: the most important fact ruling these equilibria is that, some of the constituents of the colloid being totally absorbed by it, they are thus made incapable to affect further reactions in the mixture" (Duclaux, 1904b).

This was a convenient, lazy viewpoint: instead of detailed physico-chemical calculations, the colloidal scientist could content himself with descriptive statements in order to interpret the measurements he had made of various properties such as viscosity, the aptitude to undergo coagulation upon addition of salts, etc. With respect to proteins,

"there was immense resistance to the idea that a 'solution' of a protein or any other colloid, in water or any other solvent, could be regarded as a single phase. To the adherents of this school the attempts to treat protein solutions as if they consisted of definite molecules was a waste of time or worse" (Edsall, 1962).

Indeed, Wolfgang Ostwald was quick to seize upon the field of colloidal science as a discipline to create and to build up: in 1907 he became editor of *Zeitschrift für Chemie und Industrie der Kolloide* and he also edited *Kolloidchemische Beihefte* from 1909 onwards. He turned on powerful rhetoric to justify his endeavor, his leadership of a new school within chemistry. He was convinced that the colloidal state was a new state of matter, and that most of biology would come to be explicable by colloidal theory:

" . . . no one has yet tried to do justice to the importance of modern colloid chemistry *as a new science* . . . " (Ostwald, 1912).

he wrote in 1912. Ostwald also believed that the colloidal state served to make the transition between homogeneous solutions and heterogeneous, polyphasic systems.

The colloidal school he started can be viewed, in a way, as the continuation of one of the two strands inherent in the physical chemistry of his father, Wilhelm Ostwald, the 'chemical' and the 'physicalist' tradition. I borrow here from Servos his apt characterization:

"... the germs of both traditions may be found in the research school Ostwald founded at Leipzig in the 1880s (. . .) a tension existed within Ostwald's own thinking about the specialty which was reflected in his use of the terms *allgemeine Chemie* [General Chemistry], as in the title of his major textbook, and *physikalische Chemie*, as in the title of his journal" (Servos, 1982).

Wolfgang Ostwald thus, in setting up a school of colloidal science, was fulfilling his father's ambition of developing 'an interdisciplinary entity situated between the traditional fields of chemistry and physics' (Servos, 1982). Wolfgang Ostwald was helped, in the starting of this new school, in no inconsiderable way, by the appearance of Wilder D. Bancroft's *Journal of Physical Chemistry*. This was a new journal, started in the summer of 1896 at Cornell, the first English-language periodical in the field, meant in part to stem the hegemony of German physical chemistry. The *Journal of Physical Chemistry* had aims not unlike those of Wolfgang Ostwald. To use again a judicious remark of Servos, it appealed not only to the pride but also to the prejudices of its audience, 'their prejudice against modern physical theory' to put it concisely (Servos, 1982).

While other physical chemists, such as Harry Clary Jones (Johns Hopkins) and Arthur Amos Noyes (MIT) worked at exploring the limits of the Van 't Hoff–Arrhenius theory of solutions, Bancroft (1905), who derided their endeavor ('Physical chemistry', he wrote scathingly, 'should not become the chemistry "of slightly polluted water" '), charged that these chemists were constructing a narrow specialty confined largely to the study of dilute aqueous solutions instead of following Ostwald's bold injunction to develop a new *allgemeine Chemie*. (. . .) Bancroft championed the study of heterogeneous equilibria, catalysis, photochemistry, and colloid chemistry (Servos, 1982).

Thus was launched the influential and powerful colloidal school:

"For a time the prevailing view in many circles was that all colloids were more or less random aggregates not true chemical compounds, and that the attempt to investigate their properties as if they were genuine chemical substances was really a waste of time" (Edsall, 1962).

This is not the place to address the interesting question of the psychological relationships of Wolfgang Ostwald to his father Wilhelm and to his 'surrogate scientific father' Jacques Loeb: was he a 'naughty boy', who knew by instinct the ropes of scientific eminence, and who yearned for the prestige but shied away from the dull, patient, hard work of inductive scientific research? Did he unconsciously aim at making himself his father's equal by launching a new discipline equally powerful, as encompassing as physical chemistry? Did he react to Loeb's sometimes naive assertion of simplicity?

These seem legitimate questions to ask, and they are pertinent to the rest of the story: Jacques Loeb was an innovator, but one who followed in the footsteps of Arrhenius; whereas Wolfgang Ostwald, who reacted to disassociate himself from too close an identification with Wilhelm Ostwald and with Jacques Loeb, from hindsight seems to have been mostly an operator, self-aggrandizer, and propagandist.

4. Physicochemical studies on proteins

Jacques Loeb devoted the last six years of his life (1918–1924), at the Rockefeller Institute, to physicochemical studies on proteins: 'death came when he was actively engaged in what he regarded as *the most fundamental investigation of his life*' (Osterhout, 1928; emphasis added).

Loeb entered the protein field by chance, by way of studying interactions between ions and proteins, a topic on which he started work just before the turn of the century, and this was the topic with which he continued to be involved mostly after World War I. The accidental discovery occurred in the following way:

"One day, washing some eggs of *Fundulus* on a filter to free them from adhering salt solution, the idea occurred to him that he might treat proteins in the same way to get rid of the excess of substances which did not combine with them" (Osterhout, 1928j).

When much later Jacques Loeb returned to this idea, he found that insufficient attention had been given to the pH and to controlling the pH. Thus, he was able to show that many properties (such as the osmotic pressure, the viscosity, the 'power of swelling') went through a minimum at the isoelectric point. He could explain such a behavior by the Donnan equilibrium, assuming that the protein ionizes in aqueous solution. Jacques Loeb went on to predict and to find a membrane potential, whenever a protein is maintained in a compartment enclosed by a membrane, permeable to the aqueous solution of salts on the outside but impermeable to the protein. Starting in 1918, Jacques Loeb published his protein work in numerous publications and in an influential book, *Proteins and the Theory of Colloidal Behavior,* which came out the year before his death (Loeb, 1923).

The point has been made that in his work on proteins, as well as in some of his other interests, the aims of Jacques Loeb are better understood by playing down his proclaimed mechanistic, reductionist philosophy, and by bringing to the fore aims of a more 'biotechnological' slant (Pauly, 1981). I tend to agree in part with this reassessment, which helps to explain why Loeb's work, despite its immense influence at the time, did actually so little to clarify the nature of proteins as well-defined molecular objects. As Edsall assesses justly:

"Loeb's work made no contribution to determination of the molecular size of proteins. Indeed most of his work was carried out with gelatin, one of the most heterogeneous of protein preparations; but gelatin was quite sufficient for him to clarify many hitherto confusing characteristics of protein solutions in terms of the Donnan equilibrium and the net charge on the protein molecules" (Edsall, 1962).

Loeb, because of his forceful personality, gave a powerful push to the *chemical* conception of proteins. But most of the physicochemical studies on proteins, during the first quarter of the century, came from other isolated investigators.

First, the work of Thomas B. Osborne should be mentioned as that of a pioneer with a long-range influence mostly through his former student Edwin J. Cohn. Osborne worked on very many vegetable proteins which he crystallized, "preparations of crystalline globulins from Brazil nut, oat, hempseed, castor bean, squash seed, and flax seed. Osborne's photomicrographs leave no doubt of the high quality of the crystals he obtained, with well-defined faces and regular, often octahedral, forms" (Edsall, 1962).

Another, perhaps the most important contribution from hindsight was that of Sørensen (1917), a 'major landmark' according to Edsall (1962).

"Indispensable to his achievement was the definition of the pH scale, and the development of means to measure it, which Sørensen himself had worked out some years before. Solubility studies on proteins could not be reproducible until the necessity of specifying pH, as well as salt concentration and temperature, was fully appreciated. With his rigorously careful technique in purifying and handling the protein, Sørensen could now show, to a close approximation, that the solubility of ovalbumin in ammonium sulfate solutions could be defined by the phase rule of Gibbs, and that the protein in aqueous salt should thus be thought of as a true solution" (Edsall, 1962).

"(Sørensen) showed that the solubility was constant, and independent of the amount of crystalline phase present, if temperature, pressure, pH, and salt concentration were all fixed. Thus the protein behaved as a true pure compound, according to the phase rule of Gibbs" (Edsall, 1981).

In this manner, Sørensen (1917) could determine the molecular weight of egg albumin as 34 000, a value which, if somewhat low, is nevertheless close to the modern value.

Edsall offers this balanced evaluation of the relative impact of the contributions by Jacques Loeb and by Sørensen:

"When Jacques Loeb took up the study of proteins, he was not only a scientific investigator but (in this as in other matters) a crusader and a propagandist. He believed that the colloid chemists were causing confusion and retarding progress, and he wanted to set matters straight, as he saw them. This is made clear in his book *Proteins and the Theory of Colloid Behavior*, McGraw Hill, 1923 (. . .) based on a considerable series of papers in the *Journal of General Physiology*. The book had a great influence, it was quite exciting to me when I read it at the age of 20. I realize now that Loeb oversimplified, here as elsewhere; for instance his belief that the Hofmeister series was an artifact that would disappear, if only one takes account of pH, was quite wrong. But he had a great influence, and on the whole it was good, and helped to clear away much of the fuzzy thinking in which some of the colloid chemists indulged. But really the work of Sørensen, especially his *Studies on Proteins* (Sørensen, 1917) was in the long run more influential than Loeb's work, in helping to establish proteins as definite chemical compounds that obeyed the phase rule" (Edsall, 1983).

5. The controversy: proteins as molecules or as colloids?

I have chosen to refer briefly in this section to a few examples of the dispute between the colloidalists, who gave lip-service to physicochemical theory, and the adherents of strict physicochemical principles. The first instance picked goes back to landmark work by W.B. Hardy (1905) and by John Mellanby (1905). Both had studied 'proteids' in aqueous solutions, to which they added acids, bases, and salts; and both had used 'globulin' (bovine serum albumin) for this purpose. Hardy set about to 'deal with the phenomena of colloidal solutions (. . .) from a frankly chemical standpoint' (Hardy, 1905). He concluded that globulins have both acidic and basic functions, that they consist of large ionized molecules, and that their interactions with electrolytes are best described as chemical. John Mellanby posed quite similar questions and reported in his paper, published back-to-back with Hardy's, that the 'amount of globulin dissolved by a given percentage of neutral salt is directly proportional to the strength of the original globulin suspension', i.e. that the 'amount of protein in solution (serum globulins) increased with the amount of undissolved protein in the system' (Edsall, 1983). Hardy had also made similar observations.

A dozen years later, Wolfgang Ostwald drew upon these 'well-known and very important papers of W.B. Hardy and J. Mellanby' (Edsall, 1983) for his critique of Sørensen's solubility studies (Sørensen, 1917): he pointed out that the Hardy–Mellanby results were inconsistent with the presence of a pure compound in the protein solutions since, on the addition of more solid material saturation to a fixed value of the concentration would be reached (Ostwald, 1927). This was a good point. We now know that, of course, the behavior which Hardy and Mellanby observed in their simultaneous pioneering papers, was due to chemical heterogeneity: 'the globulin system they studied was no chemical individual but a highly complex mixture of more or less related proteins' (Edsall, 1962). But Ostwald himself felt justified in building upon these observations: he thought such behavior to be characteristic of colloids in general.

Ostwald asserted that

"Sørensen was wrong in treating solubility of proteins as if they were like simple small molecules that formed true crystals, and he criticized some weak points in the data of Sørensen.

Sørensen was then led to do further work, and the result was a very long paper in three parts (Sørensen, 1930) on proteins as reversibly dissociable component systems" (Edsall, 1983).

This argument between Ostwald and Sørensen was not the only one, and Sørensen had already earlier encountered the hostility of the colloidalists toward his ideas. As writes Edsall (1962):

"The conclusions Sørensen drew were not accepted easily by many of the colloid chemists. I well remember a story told me by Edwin J. Cohn, who had received his training in protein chemistry by Osborne and Sørensen, shortly after I had come as a student to work in Cohn's laboratory. At a scientific meeting shortly after 1920, Cohn had an exchange of remarks with Wilder D. Bancroft, the well-known colloid chemist, which went somewhat as follows:
Cohn: Sørensen has measured the osmotic pressure of egg albumin, and finds a molecular weight of 34 000.
Bancroft: Yes, yes, I understand. He is measuring a system of molecular aggregates. That is the molecular weight of the aggregate.
Cohn: But the tryptophan content and sulfur content of ovalbumin have also been determined, and they give a *minimum* molecular weight of 34 000.
Bancroft: Ah! Then in that case I would say that the aggregation factor is unity!

Although this little story may be partly legendary, I believe I have told it essentially as Cohn told it to me and *I think it does illustrate correctly the spirit of that period*" (emphasis added).

A third example of dissent flaring up between the colloidalists and the physical chemists on the factors determining the swelling of gelatin is instructive. It opposed Martin H. Fischer and his co-workers (Fischer et al., 1918) to Lawrence J. Henderson and Edwin J. Cohn.

Fischer and Hooker had studied the swelling of gelatin, that of fibrin too, in polybasic acids and their salts. They gave their results the general important interpretation that 'edema (was) a state of excessive hydration of the body colloids' (Fischer and Hooker, 1918). Martin Fischer's prime assumption was that a colloidal substance such as gelatin was "capable of existing in different degrees of association or polymerization depending upon the temperature and upon other changes in its environment, like the presence of acids and of alkalies. The degree of association and hence the size of the particles of which the gelatin is composed may be greatly varied" (Fischer and Coffman, 1918).

Fischer further introduced what he saw as a fundamental distinction for proteins (the emphasis is that of the original):

"the phenomena of hydration (swelling) and of 'solution' while frequently associated are essentially different. Hydration is to be regarded as a change through which the protein enters into physicochemical combination with its solvent (water); 'solution', as one which can be most easily understood at the present time as the expression of an increase in the degree of dispersion of the colloid".

And Martin Fischer concluded with a jab at the biologists, in the form of an epistemological remark which, were it not for the muddleheadedness of his whole approach, would have been well taken:

"These experiments emphasize again the necessity of interpreting in the simpler language of colloid-chemistry the mass of experimental material now collected by the biologists under the heading of 'permeability' studies. (. . .) it would look like a better conservation of mental energy to concentrate further study upon the colloid-chemical principles here to the front than to continue indefinitely the mere restatement of biological problems in the terms of biology itself" (Fischer and Coffman, 1918).

(Replace in the above quotation 'colloid-chemistry' by 'physical chemistry', and the Martin Fischer statements become the rejoinder by Henderson and Cohn, to which we now turn.)

In their attack on Martin Fischer, Henderson and Cohn scolded him for not taking into account pH changes, and for ignoring important recent work:

"It is simply not permissible to disregard the accurate quantitative measurements which have been accumulated in many laboratories of biological chemistry, of physiology, and of clinical medicine during the last decade. On the basis of Fischer's speculations the theory of the regulation of breathing, one of the most splendid achievements of modern physiology, which has stood the test of aviation, of gas warfare, and of military medicine, is almost meaningless. Yet it is a fact that the composition of the blood determines the activity of the respiratory centre so as to adjust the hydrogen-ion concentration of the blood at a constant point (Haldane, 1917; Barcroft, 1914)" (Henderson and Cohn, 1918).

Henderson and Cohn stressed that 'the acid or alkali content of the protein colloids is a function of the hydrogen ion concentration of the solution, *a theory which has perhaps never been doubted by any one but Professor Fischer'* (Henderson and Cohn, 1918; emphasis added), a point which they emphasized graphically by plotting the Fischer and Hooker data (1918) as a function of pH: in this manner the results of Fischer and Hooker were shown to conform with earlier results of Henderson and his co-workers (Henderson et al., 1914). And Henderson and Cohn could rail at Martin Fischer for neglecting to submit his results to the 'necessary quantitative mathematical analysis' (Henderson and Cohn, 1918).

Martin Fischer was not amused. He whipped out a reply, in which he tried to ridicule a number of the statements of Henderson and Cohn. Clearly on the defensive, he ended with this appeal to authority:

"This is Henderson's opinion. Among those of another persuasion in this matter of therapy are Arthur D. Dunn (1912), Albert J. Bell (1912), Paul G. Wooley (1914), Edgar G. Ballenger and Omar F. Elder (1914), Rufus Southworth (1914), Gordon F. McKim (1914), H. Lowenburg (1914), Geo T. Grinnan (1916), H.B. Weiss (1917a,b), J. Mitchell Clarke (1917), Herbert Brown (1917), W. de B. MacNider (1916; 1917a,b), A.W. Sellards (1917), V. Pleth (1917) and many more" (Fischer, 1918).

It is ironical that Martin Fischer, who had earlier derided the biologists, should now rush to enlist the help of physicians to his viewpoint! Lawrence Henderson fired the final shot in this volley (Henderson, 1918):

"Professor Fischer must demonstrate very much wider variations of hydrogen-ion concentration than are now known to occur within the organism or very much greater changes in swelling *within physiological ranges* than are now known, before there can be any ground for accepting his theories".

Proper documentation of the various flare-ups between colloid and physical chemists during the heyday of the former should also include the controversy between Wolfgang Ostwald and Joseph Barcroft on binding of dioxygen to hemoglobin, which occurred in 1907–1909: Ostwald, supported by the chemist Manchot, claimed non-specific absorption of oxygen to be concurrent with heme iron binding. This was refuted by Barcroft and Rudolph Peters, who carefully re-examined the Hb–O_2 binding curve (Edsall, 1972). Likewise, Edwin J. Cohn had to put to rest (Cohn, 1925) the view asserted by R.A. Gortner that, superimposed upon chemical bindings at well-defined groups (His, Arg, Lys, etc.), proteins would bind protons also non-specifically.

I have already referred in this book to another eminent colloid chemist, The Svedberg: his experiments with the ultracentrifuge, however, gave the kiss of death to colloid theory!

We shall come now to another litigant, Jacques Loeb. He entered the battle late at the end of his career. Colloid chemistry had taken over the whole ground for about a generation. Reclaiming it was an uphill fight. Jacques Loeb was psychologically cut to enter, enjoy and suffer scientific controversies.

"Often dogmatic in expressing his views" (Osterhout, 1928h), "he always remained the tense and excitable figure of his youth, short, lean, and spikey in appearance, with pointed nose and chin and dark glinting eyes that could stab an opponent but also break up in laughter. He was (. . .) hungry for recognition and never satisfied with what he got, assailed by a perpetual sense of isolation from his peers, biting and dogmatic in his pronouncements but prepared to seek an opponent out with an ungrudging, 'You are right, I was wrong!' " (Fleming, 1964c).

By the time of the near-hegemony of colloid chemistry, and before he entered into confrontation with colloid chemists, Jacques Loeb had already joined the fray in many a controversy; and often had won these contests.

In 1891, Hans Driesch (1867–1941) had obtained *complete* embryos from just *one* of the first two dividing cells from a frog's egg. Driesch could not interpret this experiment in mechanistic terms: are they machines able to reproduce themselves and to split into two identical machines? he asked. This had led Driesch, from 1895 on, to become a staunch vitalist. Jacques Loeb had lucidly analyzed Driesch's result, had seen that only apparently did it threaten biological mechanism: it sufficed that the cell separated by Driesch contained 'a representative selection of the physicochemical components of the fertilized egg' (Fleming, 1964f) to explain Driesch's observations.

Loeb had entered another battle, with his admired Wilhelm Ostwald, the physical chemist, who in the early years of the 20th century, went so far as to repudiate atomic theory and mechanism.

"Ostwald's alternative to a mechanistic conception of the universe was a science of
pure energetics, concerning itself solely with transfers of energy and rigorously avoid-
ing the invocation of any mechanistic or material basis for these transactions" (Fleming, 1964g).

Fortunately for Loeb, Jean Perrin had convincingly demonstrated
the existence of atoms and molecules by his work on Brownian
motion, and Wilhelm Ostwald gave up the fight, and accepted in the
fourth edition of his physical chemistry textbook (1909) the atomis-
tic-mechanistic viewpoint.

Having thus come out on top in each of these two major challenges
to his reductionist philosophy, Jacques Loeb felt justified to volun-
teer as the mouthpiece, the prophet as it were, for the mechanistic
viewpoint. In Hamburg, at the First International Congress of Mo-
nists in September 1911, 'in a state of fierce intoxication with his
vision, he poured out "The Mechanistic Conception of Life" to an
audience of several thousand' (Fleming, 1964h). Thus, Jacques Loeb
was in every way well prepared to fight the colloidalists. It is even a
little surprising that he bided his time so long before doing so, which
happened in the early 1920s.

As the reader will recall, for colloid chemists proteins consisted of
aggregates whose degree of association (or polymerization) de-
pended upon the circumstances of pH and ionic strength. In solution, proteins existed as biphasic suspensions. Thus, they were
unable, for instance, to form salts upon titration. Wilder D. Bancroft
made himself one of the chief propagators of this viewpoint, especially with a book he published on the topic (Bancroft, 1921).

Jacques Loeb countered this viewpoint with a molecular description of proteins; but this was a drawn-out fight, and even after a
decade of controversy full victory was yet to be achieved. I shall
resort here to Servos' excellent account of this polemic (Servos,
1982):

"Loeb maintained that proteins did in fact form salts and enter the state of solution.
Some proteins, he observed, could be crystallized like salts. More significantly, his

careful study of the osmotic pressure of gelatin in solution convinced him that protein solutions obeyed van 't Hoff's law. This, according to Loeb, constituted proof that proteins were substances of definite composition - similar to the familiar compounds of inorganic and organic chemistry except for their great size (Loeb, 1924).

Bancroft rejected Loeb's conclusions as not necessarily following from his evidence. For every protein Loeb could point to as being capable of crystallization, Bancroft could point to two that resisted crystal formation; whereas Loeb believed he was measuring the osmotic pressure of gelatin salts, Bancroft maintained that he was encountering a swelling effect, unrelated to osmotic pressure (Bancroft, 1921). Loeb had already decided to ignore Bancroft's position by 1923. Writing to Arrhenius of the 'great confusion which existed under the name colloid chemistry', Loeb linked Bancroft's name with medical quackery:

'In the protein industries in the United States the new work on proteins [Loeb's own research] has been accepted. Of course, the medical quacks probably will continue to stick to the old fashioned colloid chemistry which they find remunerative. One occasionally finds also a physical chemist like Wilder Bancroft sticking to that confusion, but I do not think Bancroft has ever done an experiment in his whole life, and since he is neither a mathematician nor capable of rationalistic [sic] thinking, I think he can be ignored' (Loeb, 1923).

Despite Loeb's dismissal of Bancroft, the controversy over proteins was not universally perceived to be as settled as Loeb suggested in his letter to Arrhenius. Loeb's position gradually won adherents during the 1920s, but Bancroft's view held the support of a significant number of reputable American and European investigators, and the colloid movement itself thrived as evidence mounted on both sides of the protein issue. Only after 1930, as evidence drawn from X-ray studies of protein crystals and ultracentrifuge studies of protein solutions became available, did Bancroft's view come to seem untenable to most informed scientists (Munro, 1936)".

Certain expressions are so well chosen! *drags on*; a house *divided against itself*; not to be *on the same wavelength* or, as the French say, *un dialogue de sourds*, i.e. a dialogue between deaf persons.

All these are applicable to the seemingly endless discussion about the molecular or colloidal status of proteins, during the first three decades of this century. Science advances by even aggressive discussions between the participants. It can be trying, but the tension between differing and sometimes opposed interpretations can be quite fruitful. In general, no one quarrels with such trite statements.

In the present case, however, the scientific historian is hard put to find merit, to ascribe any positive value to this fight between the colloid and the physical chemists. If I may be forgiven this metaphor, it is as if, in exploring an artifact sunk to the sea bottom the former, by their wild gestures and posturing, had stirred a cloud of mud which quickly totally masked the issue. This serves to explain my title for this chapter. I hasten to say though that mine is quite an unoriginal stand. Many historians of biochemistry have likewise blamed the colloidal school of chemistry for the delay biochemistry suffered till it became the molecular science now so familiar.

A balanced appraisal is that of Robert E. Kohler:

"The 'colloid' concept emerged as a general explanatory theory almost simultaneously with the enzyme theory in the early 1900's and soon surpassed it in popularity. By 1914 enzymes were widely regarded as simply a special kind of biologically active colloid, and in the 1920's, the heyday of the colloidal school, many biochemists and biologists appealed to colloid chemistry for answers to the problems of life. This remarkable episode has generally been dismissed by biochemists as a period of confusion, an aberration, a temporary throw back to vitalism – 'the dark age of biocolloidology' (Florkin, 1972). But at least two generations of biochemists were profoundly influenced by the colloidal concept: one inspired by its promise, and the next chastened by its failure. (. . .)

Like enzymes and protoplasm, the colloid concept had a double aspect. On one side, colloid theory appeared to be rigorously physical-chemical, drawing heavily on molecular physics. Its 'scientific' rigor was a principal reason why those with a reductionist bent preferred it to the more vaguely biological notion of an enzyme. On the other side, some colloid chemists pointed out that the properties of colloidal surfaces seemed quite different from the properties of ordinary matter. Those of a biologistic bent saw in these 'new laws of nature' a general explanation of the special properties of 'living' matter. The important point here is the family resemblance between the concepts of protoplasm, enzyme, and colloid: all were attempts to explain life in terms of entities that were more than 'molecular' and less than morphological; all appealed to both reductionist and vitalist sentiment. This was the slippery middle ground between chemistry and biology to which biochemistry laid claim" (Kohler, 1975).

To return to this last point made by Kohler (1975), there is a periodically recurring tendency to privilege supramolecular over molecular viewpoints; see for instance the recent article by Cohen (1984). However, in the long-term view of the historian it is obvious that, during the first half of this century, molecular concepts had to overrule supramolecular (colloidal) concepts, if progress was to be made.

To close this section with an illustrative example of the kind of high-faluting gibberish which the colloid chemists were wont to utter, the Frenchman Jacques Duclaux could still write in 1934:

"It is altogether doubtful that there exists in the living cell anything that can be represented with a constitutional formula; it is doubtful that anything occurs in it that can be represented with a reaction equation. Furthermore, it is a logical inference that there exists in the cell neither a molecule, nor an atom, nor a valence apart from a few exceptions. One can even go a step further and say that, as soon as a substance becomes amenable to our symbols of representation, living matter considers it as a foe and eliminates or neutralizes it" (Duclaux, 1934).

Accordingly, Duclaux in his book (1934) reverts constantly to the vitalistic metaphor of life processes as a whirlpool, one of the two traditional approaches to the physical basis of life according to the suggestive essay by Scott F. Gilbert (1982).

A comparison is apt. During the first three decades of the century, colloid theory enjoyed the same sort of hegemony, intellectual respectability, seemingly universal applicability, as psychoanalytic theory had till the 1970s or the 1980s at the very least. The reason for the simile is simple: neither theory is testable, neither theory is falsifiable, as was pointed out by a number of philosophers of science, such as Popper (1969) or Medawar (1967). Clearly, even in the 1920s, propositions such as the following (see Erwin, 1984) would have drawn close to unanimous support: (A) the experimental evidence supports certain parts of colloidal theory; (B) the present state of scientific research suggests that colloidal theory is of great promise; (C) colloidal theory is, and will continue to be, of great heuristic value to biochemists.

To end this chapter, I wish now to set in perspective the materialistic creed and the mechanistic bias of a Jacques Loeb.

6. Philosophical coda: mechanism versus vitalism

The militant version of reductionist materialism espoused by Jacques Loeb, which he disseminated in his influential pamphlet of 1912, *The Mechanistic Conception of Life,* is extremely well expressed in this statement made by another Frenchman - more lucid and more readable than Duclaux, though – also during the 1930s:

"Mechanism enjoys presently an extremely firm position, and one does not grasp readily what should be answered to it when, on the strength of its daily successes, it simply requests more time to complete its endeavor, namely *to explain life completely without recourse to life*" (Rostand, 1939).

This triumphant viewpoint, the 'hard-core' materialism of a Jacques Loeb, remained somewhat simplistic. To borrow from Allen's useful definitions, if *philosophical materialism* has these four aspects:
(a) existence of material reality outside human perception;
(b) ideas about the world are derived from our interaction with material reality;
(c) all change in the universe is a result of matter in motion;
(d) all non-physical forces or mystical causes are inadmissible as explanation of any phenomena,
then *mechanistic materialism* can be summarized in these five propositions:
(e) the parts of a complex whole are distinct and separate from one another;
(f) the proper method of study (analysis) is to break the whole into its component parts;
(g) the whole amounts to nothing but the sum total of its component parts;
(h) systems evolve due largely to the imposition from the outside of constant forces;
(i) 'the mechanistic world view is basically *atomistic*. Mechanists tend to see all phenomena in terms of a mosaic of separate, interacting, but ultimately independent parts' (Allen, 1983).
These statements (e–i) formed the crux of the controversy opposing physical chemists, such as Jacques Loeb and Lawrence Henderson, to the colloid chemists, such as Wolfgang Ostwald and Wilder Bancroft. Continually, the latter would summon to their help biological phenomena. They would allude to organization, differentiation, development as biological characteristics which escaped the grasp of the naive form of mechanism, of what I have just labeled as 'hardcore' mechanistic materialism.

This problem of accounting materialistically and mechanistically for biological development goes back a long way, and I have referred elsewhere in this book to the partial solution offered for it by Kant, which was later eagerly seized upon by Goethe, Blumenbach, and by a group of predominantly German biologists and physiologists in the first part of the 19th century. The key concepts for this *holistic materialism* - to continue using here Allen's terminology (1983) - are: (*i*) interconnection between the parts of a complex whole; (*ii*) necessity to devise new techniques to study the interaction of the parts; (*iii*) 'the whole is more than the *mere sum* of its parts'; (*iv*) 'processes in the world are dynamic and developmental' (Allen, 1983).

Perusing the few scattered notes which Marcel Florkin made in preparation of this volume, it is clear to me that he himself was such a holistic materialist. One of his notes reads: 'At the antipodes of Molecular Biology. (The organism results from the whole and not from a molecular order.) Write *The Molecular Order*' [this was one of the titles he toyed with for the present volume]. Let me reproduce here, and despite their sketchy nature, another two of Florkin's preparatory notes; the first appears to have been meant for the Introduction:

"The Sixth and last Part of this History is devoted to modern molecular correlates of biology. The author's thesis belongs to recurrent history and can be epitomized as demonstrating that the features of contemporary biochemistry, in the field considered, derive of a chemical progress, the introduction in the 'epistemological laboratory' of biochemistry of a new chemical chapter, concerned with ordered polymers. Before this knowledge was acquired, the concept of macromolecule was introduced, replacing the current views on aggregates with which we have been dealing in Chapters [30-14 inter alia]. The thesis proposed by the author, that of the seminal effect of an acquisition of organic chemistry, the concept of ordered polymers, on scientific explanation in the field of biochemistry, reaching a number of explanatory devices including the correlates of heredity through the double helix of DNA".

Elsewhere, Florkin writes:

"It was one of the teachings of modern biochemical developments that it does not pay to go to the primitive methods of experimenting on whole organisms, on isolated organs, on tissue slices, etc. further than to orient towards the important substrates and enzymes involved and to approach the problem immediately at their level by the detour of synthetic compounds. This relative sterility of the traditional approaches is illustrated by the teachings of horizontal history. The spectacular advances realized in biochemistry since [there is a blank here in the manuscript] is (sic) consequently based on advances in organic chemistry".

Nowadays, after the triumph of molecular biology, vitalism in biology is, at least temporarily, on the wane. Anti-mechanistic, anti-reductionist schools of thought, such as that of the colloid chemists, have acquired a bad reputation. Colloid chemistry is being blamed, quite generally and with good justification, for the protracted emergence of molecular concepts in biochemistry. Arguments for the irreducibility of biology to physics and chemistry periodically spring up, to be put down rapidly on various grounds, most often that of circularity (Schaffner, 1967).

The case *for* irreducibility has retreated to either a statement of equivalence between the teleological and the non-teleological viewpoints (Nagel, 1979), or to pragmatic use in certain sectors of biology for methodological reasons only (Schaffner, 1967).

I wish to give here the last word to Whitehead, in order to elaborate further upon the (key)point (b) above (p.154):

"In considering knowledge we should wipe out all these spatial metaphors, such as 'within the mind' and 'without the mind' (. . .) Natural philosophy should never ask, what is in the mind and what is in nature. To do so is a confession that it has failed to express relations between things perceptively known, namely to express those natural relations whose expression is natural philosphy. It may be that the task is too hard for us, that the relations are too complex and too various for our apprehension, or are too trivial to be worth the trouble of exposition. It is indeed true that we have gone but a very small way in the adequate formulation of such relations. But at least do not let us endeavour to conceal failure under a theory of the byplay of the perceiving mind. (. . .) The nature which is the fact apprehended in awareness holds within it the greenness of the trees, the song of the birds, the warmth of the sun, the hardness of the chairs, and the feel of the velvet. The nature which is the cause of awareness is the conjectured system of molecules and electrons which so affects the mind as to produce the awareness of apparent nature (. . .)

The recourse to metaphysics is like throwing a match into the powder magazine. It blows up the whole arena. This is exactly what scientific philosophers do when they are driven into a corner and convicted of incoherence. They at once drag in the mind and talk of entities in the mind or out of the mind as the case may be. *For natural philosophy everything perceived is in nature. We may not pick and choose" (Whitehead, 1920).*

REFERENCES

Allen, G.E. (1983) Am. Zool., 23, 829–843.
Arrhenius, S. (1884) Recherches sur la Conductibilité Galvanique des Electrolytes, Stockholm.
Ballenger, E.G. and Elder, O.F. (1914) J. Am. Med. Assoc., 62, 197.
Bancroft, W.D. (1921) Applied Colloid Chemistry : General Theory, McGraw-Hill, New York.
Bell, A.J. (1912) Am. J. Med. Sci., 144, 669.
Brown, H. (1917) personal communication from Flanders to Fischer, M.H. (1918) received Sept. 1, 1917.
Clarke, J.M. (1917) Br. Med. J., 2, 239.
Cohen, E.B. (1984) J. Theor. Biol., 108, 369.
Cohn, E.J. (1925) Physiol. Rev., 5, 349.
D'Arcy Wentworth Thompson (1942) On Growth and Form, Cambridge University Press, Cambridge, 2nd ed. (1st ed. 1916), pp. 347–349.
Duclaux, J. (1904) Recherches sur les Substances Colloïdales, Laval, (a) p. 2; (b) p. 101 (my translation).
Duclaux, J. (1934) L'Analyse Physicochimique des Fonctions Vitales, Paris, p. xi (my translation).
Dunn, A.D. (1912) Lancet-Clinic, 108, 8.
Edsall, J.T. (1962) Arch. Biochem. Biophys., suppl. 1, 12.
Edsall, J.T. (1972) J. Hist. Biol., 5, 205.
Edsall, J.T. (1981) TIBS, 6, 336.
Edsall, J.T. (1983) letter to Pierre Laszlo of June 13.
Erwin, E. (1984) Br. J. Phil. Sci., 35, 115.
Fischer, M.H. (1918) J. Am. Chem. Soc., 40, 862.
Fischer, M.H. and Benzinger, M. (1918) J. Am. Chem. Soc., 40, 292.
Fischer, M.H. and Coffman, W.D. (1918) J. Am. Chem. Soc., 40, 303.
Fischer, M.H. and Hooker, M.O. (1918) J. Am. Chem. Soc., 40, 292.
Fleming, D. (1964) Introduction to Jacques Loeb's The Mechanistic Conception of Life, Harvard University Press, Cambridge, MA, (a) p. xii; (b) p. xv; (c) p. xxi; (d) p. xiii; (e) p. xiv; (f) p. xxv; (g) p. xxvi; (h) p. xxviii.
Fleming, D. (1970–1980) 'Jacques Loeb', in Dictionary of Scientific Biography (Gillispie, C.C., Ed.), 14 vols., Scribner, New York, pp. 445–447.
Florkin, M. (1972) A History of Biochemistry (Comprehensive Biochemistry), (Florkin, M. and Stotz, E.H., Eds.), Vol. 30, ch. 14, Elsevier, Amsterdam.
Fruton, J.S. (1976) Science, 192, 327.
Gilbert, S.F. (1982) Persp. Biol. Med., 26, 151.
Graham, T. (1861) Phil. Trans. Roy. Soc. London, 151, 183 (emphasis added).
Grinnam, G.T. (1916) Virg. Med. Semi-Month., 20, 523.
Hardy, W.B. (1905) J. Physiol., 33, 251.
Henderson, L.J. (1918) J. Am. Chem. Soc., 40, 867 (emphasis added).

158   REFERENCES

Henderson, L.J. and Cohn, E.J. (1918) J. Am. Chem. Soc., 40, 857.
Henderson, L.J., Palmer, and Newburgh, (1914) J. Pharmacol., 5, 449.
Kohler, R.E. (1975) J. Hist. Biol., 8, 275.
Loeb, J. (1912) The Mechanistic Conception of Life, University of Chicago Press,
    Chicago (repr. 1964; Fleming, D., Ed., Harvard University Press, Cambridge,
    MA).
Loeb, J. (1916) The Organism as a Whole, Putnam, New York, pp. 299-302.
Loeb, J. (1923) Proteins and the Theory of Colloidal Behavior, McGraw-Hill, New
    York.
Loeb, J. (1923) letter of December 14 to Svante Arrhenius. Box 1, Jacques Loeb
    Papers, Library of Congress, Washington. Quoted by Servos (1982).
Loeb, J. (1924) Proteins and the Theory of Colloidal Behavior, McGraw-Hill, New
    York, 2nd ed., esp. pp. 10-18, 163-170, 367-371.
Lowenburg, H. (1914) J. Am. Med. Assoc., 63, 1906.
McKim, G.F. (1914) personal communication to Fischer, M.H. (1918).
McNider, W. de B. (1916) J. Exp. Med., 23, 171.
McNider, W. de B. (1917a) J. Exp. Med., 26, 19.
McNider, W. de B. (1917b) Proc. Soc. Exp. Biol. Med., 14, 140.
Medawar, P.B. (1967) The Art of the Soluble, Methuen, London.
Mellanby, J. (1905) J. Physiol., 33, 338.
Munro, L.A. (1936) J. Chem. Ed., 13, 462.
Nagel, T. (1979) The Structure of Science, Mechanistic Explanation and Organismic
    Biology, Harcourt, Brace and World, New York, p. 398 et seq.
Osterhout, W.J.V. (1928) J. Gen. Physiol., 8, ix–xcii. (a) ix; (b) xiii; (c) xv; (d) xxi
    (emphasis added); (e) lv; (f) liii; (g) xxvii; (h) lvi–lvii; (i) li; (j) xlv.
Ostwald, W. (1912) The New Developments of Colloid Chemistry; quoted by Hauser,
    E.A. (1955) J. Chem. Ed., 32, 2 (emphasis added).
Ostwald, W. (1927a) Kolloid Z., 41, 163.
Ostwald, W. (1927b) Kolloid Z., 43, 249.
Pauly, P. (1981) Jacques Loeb and the Control of Life : An Experimental Biologist in
    Germany and America, 1859-1924. Dissertation submitted to Johns Hopkins Uni-
    versity, Baltimore.
Pleth, V. (1917) personal communication to Fischer, M.H. (1918).
Popper, K.R. (1969) Conjectures and Refutations, Routledge and Kegan Paul, Lon-
    don, 3rd ed., p. 34 et seq.
Rostand, J. (1939) La Vie et ses Problèmes, Flammarion, Paris, p. 155 (my transla-
    tion and emphasis; also quoted by Canguilhem, J. (1965) La Connaissance de la
    Vie, Vrin, Paris, p. 94).
Rubin, L.P. (1980) J. Hist. Med., xxxv, 397.
Schaffner, K.F. (1967) Science, 157, 644.
Sellards, A.W. (1917) Acidosis and Clinical Methods, Cambridge.
Servos, John W. (1982) ISIS, 73, 207.
Sørensen, S.P.L. (1917) C.R. Trav. Lab. Carlsberg, 12.
Sørensen, S.P.L. (1930) Kolloid Z., 53, 102, 170, 306.

Southworth, R. (1914) Lancet-Clinic, Sept. 5.
Weiss, H.B. (1917a) J. Am. Med. Assoc., 68, 1618.
Weiss, H.B. (1917b) Ohio State Med. J., 13, 595.
Whitehead, A.N. (1920) The Concept of Nature (Tarner Lectures, Trinity College, 1919) Cambridge University Press, Cambridge, pp. 28–32 (emphasis added).
Wooley, P.J. (1914) J. Am. Med. Assoc., 63, 596.

PART II

Proteins as Molecules

(Chapters 9–17)

Chapter 9

The Name 'Protein'

In 1838, the Dutch scientist G.J. Mulder (1802–1880) coined a new term for a single substance which he had isolated from a variety of organic sources. To quote from the original article in French (Mulder, 1838):

"La matière organique, étant un principe général de toutes les parties constituantes du corps animal (...) pourrait se nommer *Protéine* de πρωτειος, primarius". (The organic matter, being a general principle of all the constituent parts of the animal body (...) could be termed *Protein* from πρωτειος, primarius.)

Thus, far from being trivial, this denomination raises a number of far-reaching questions. Why was Mulder interested in reducing the diversity of albuminoids to a *single* entity? What kind of arguments and experimental methods did he resort to? How was the new concept welcomed by the contemporary chemists? Does the present language of biochemistry owe to Mulder's definition its emphasis upon the primacy of proteins?

Mulder had trained himself to become a physician-apothecary. As such, he

"over-exerted himself during the cholera epidemic of 1832–33, and finally had to give up the medical profession. He became a lecturer at the Bataafsch Genootschap voor Proefondervindelijke Wijsbegeerte (Batavian Society for Experimental Philosophy) and the Clinische School (Clinical School), both at Rotterdam. A grave insomnia gave him opportunity to work during twenty hours a day, which afforded him ample time for chemical experimentation. Untouched by any fear of contagion, he began analyzing diarrhoea, vomit, bile, blood and gastric juice from cholera patients, and found that the symptoms of the disease were brought about by an expulsion of serum protein to the digestive tract. During his lecturership he analyzed plant pigments, gossamer, silk and much else" (Glas, 1983).

[163]

Two features in his character, probably related to one another, are thus proeminent: his single-minded application to science, and his personal courage.

Mulder was working within the experimental program, which dominated organic chemistry during the whole first half of the 19th century, of determining compositions of purified substances. This was a systematic survey of organic compounds according to their formulae. The great compiler Berzelius, for several decades, gathered them in the successive editions of his *Treatise of Chemistry* – which can thus be seen as the germ for, or likened to our present-day *Chemical Abstracts*. Indeed, Mulder was active in translating Berzelius' textbook; and conversely Berzelius helped Mulder to obtain his appointment as professor at Utrecht (Glas, 1983).

Let us here recall a few discoveries prior to those of Mulder. The German chemist Meinecke had found in 1817 that three different substances, starch, gum arabic, and sugar, shared an identical $C_2H_2O_2$ formula. His solution (Meinecke, 1817) was to distinguish for each of these three cases differing sub-systems, which he wrote $2 C + 2 HO$ for starch, $CH + CO + HO$ for gum, and $2 C + 2 H + 2 O$ for sugar, thus accounting for the differences in properties and chemical reactivity of these three compounds. This is the idea that two substances can be identical in constitution, i.e. have the same elemental composition, yet, differ in the arrangement of the parts; as Humphry Davy had written in his *Elements of Chemical Philosophy* (1840):

"There is, however, no impossibility in the supposition that the same ponderable matter in different electrical states, or in different arrangements, may constitute substances chemically different".

This is the concept of *isomerism*, which Berzelius would formally define and incorporate in his textbook in the early 1830s.

Interestingly, for what is to follow, another German chemist, Liebig shared this idea, and wrote in 1825 that:

"there is no doubt that the same elements, precisely in the same amounts, can often produce substances with very different properties, and that the chemical nature of the result depends on the way these elements are combined with one another" (Liebig, 1825).

Among the above-mentioned chemists, at that time (ca. 1825), Humphry Davy, Meinecke, and Liebig were all sharing a joint world-view, that of *Naturphilosophie*.

Another strand in the weaving of the protein concept, as it came to crystallize in the elucidation by Mulder, was, following the proposals by Kant in his *Critique of Judgment* (1790), the work of vital materialists such as Johann Friedrich Blumenbach (1824), Johann Christian Reil, and especially Karl Ernst von Baer. This embryologist in his lectures on anthropology (1824)

"adopted the view that the basic material constituents of the body: albumen, fibrin, gluten, etc. were all slightly altered chemical combinations of carbon, hydrogen, nitrogen and oxygen, the order and arrangement of these constituents being the principal determinant of the properties and ultimately the function of these substances" (Lenoir, 1981).

These various strands formed the basis for Mulder's experimental work. In order to perform elementary analysis of organic substances, he was extremely careful in their purification so that he would be dealing with the real chemical as it occurs in vivo, rather than an experimental artifact from the isolation technique. He was also careful in his choice of minerals for the oxygen source, heat-resisting lead salts, so that contamination of the organic compounds with volatile sulfur and phosphorus oxides be avoided (Glas, 1983). Mulder found that sulfur was the element occurring in the smallest proportion (0.38%) in albumin. Assuming that this quantity corresponded to one atom per molecule, Mulder derived from his analytical determinations the number of atoms of the other elements (C, H, O, N).

"Mulder also prepared albuminate of lead and silver, as well as albumin sulphate, taurate and chlorite. In these compounds, the 'saturation capacities' (proportions between the inorganic and albuminous moieties) turned out to be fixed and multiple. So it seemed possible to 'translate back' the macroscopic multiplicities to proportionalities at the molecular level (. . .) Mulder also made the crucial discovery that the saturation capacity of protein was ten times greater than that of albumin or fibrin. From this result it followed that protein was indeed a subunit of albuminoid, and not just an isomeric transformation product. Neither was it a purer form of albuminoid. The chemical relations between the albuminoids could now be expressed quantitatively. Putting 'Pr' for 'protein', serum albumin became $Pr_{10}S_2P$, globulin $Pr_{15}S$, keratin PrS_2, etc." (Glas, 1983).

Hence, Mulder viewed albuminoids as composed of identical subunits, made of protein, and linked by sulphide bridges.

"Physiological transformations, as for instance the conversion of vegetables into flesh and blood, had to be comprehended as rearrangements of intact protein blocks through reversible dissolution of the sulphide linkages" (Glas, 1979).

Since Mulder's discovery of the protein substance gave rise, almost immediately, to a bitter controversy in which his opponents interpreted his find in various ways, let us emphasize Mulder's definition of the word 'protein' by borrowing from the Oxford English Dictionary:

"Name given by Mulder to a complex residual nitrogenous substance, of tolerably constant composition, obtained from casein, fibrin, and egg albumin, to which he assigned the formula $C_{40}H_{62}N_{10}O_{12}$, and which he regarded as *the essential constituent* of organized bodies, animal or vegetable" (emphasis added).

Florkin (1972) has already indicated Mulder's indebtedness to Berzelius for the word 'protein'; and he has stressed that Mulder's theory of the composition of albuminoid substances was first spelled out in 1838 in a letter to Liebig, who published it in his house journal, which he had founded in 1831, *Annalen der Chemie und Pharmacie.*

    The initial reception of the Mulder find by Liebig was a mix of agreement with the facts, and of mild disagreement with the theory.

"In the period 1838–42, extensive analyses of albuminous substances were undertaken in the Giessen laboratory [of J. Liebig]. The results were in excellent agreement with Mulder's data (...) Liebig's eventual formulae resembled Mulder's like two peas in a pod" (Glas, 1983).

In 1842, following publication the preceding year of his book on *Plant Chemistry*, J. von Liebig came out with his book *Animal Chemistry*. In this book, Liebig based himself on Mulder's discovery

(Glas, 1983). But for him, the central feature of albuminoid substances was their interconversion. And Liebig found the Mulder
protein concept to be a convenient operational definition for these
mutual transformations: perhaps the protein molecule was a common intermediate in these interconversions. This was a cautious
position.

This is how Liebig presented the protein (1842):

"This body (composed of nitrogen, carbon, hydrogen and oxygen) can be considered
as the starting point for all the other products of animal economy, since all these
products are derived from blood. Basing himself on this point of view, Mr. Mulder has
given to the body described here the name of *protein*, from the Greek πρωτευω, I stand
in the first place. Blood, or rather the principles of blood would therefore be combinations between this protein and variable amounts of other inorganic substances
(. . .).
    We shall assume as a consequence that the nitrogen-containing principles in the
animal economy, however different their compositions, all derive from *the* protein, by
the effect of either fixation or elimination of oxygen or of the elements of water, or
also as the effect of a molecular resolution".

However, Liebig who had learned chemistry from analysts of the
French School such as Gay-Lussac and Thénard, was fond of specu-
lating from a database of quantitative elementary analysis. In *Animal Chemistry*, thus, he started from the postulate that only plants
can synthesize organic substances from the elements; animals, who
receive these pre-formed organic substances as food, 'metamor-
phose' them. Liebig then proceeded to compare, by careful book-
keeping, the elemental analyses of those substances which he
believed to be thus interrelated by way of animal physiology, natural
products from plants as the initial state, and their 'metamorphosed'
animal counterparts as the final state (Florkin, 1972). Florkin
(1972) gives a number of examples of 'the fantastic metabolic calculations piled up by Liebig's extreme "analysm" in the second section
of his *Animal Chemistry*. Accordingly, Berzelius in his Yearbook
for 1843 was quick to condemn Liebig for his speculative theories of
animal physiology.

Stung by this criticism, Liebig turned his wrath upon Mulder.

The following year, in 1844, he denounced him as a iatrochemist, in
a violent diatribe (Liebig, 1844):

"The iatrochemist knows a protein trioxide and bioxide, he determines the molecular
weights of fibrin, albumin and casein from their combinations with hydrochloric acid
and lead oxide, it is he who seeks to establish the absolute numbers of the atoms of the
elements and enters on debates about the formulae. This is the present-day iatro-
chemistry".

Also because of the Berzelius criticism, in 1846 Liebig back-tracked
and withdrew from the second edition of *Animal Chemistry* the
whole section on animal 'metamorphoses'.

The stage was now set for a public controversy between Liebig and
Mulder – rivals who incidentally shared a very similar background in
elementary analysis, and whose institutes in Giessen and in Utrecht
were organized in like manner, strongly hierarchized and with the
teaching of chemistry firmly rooted in practical laboratory work!

Mulder had made his discovery of the protein, as a single substance, within the framework of the basic *unity of organic and inorganic chemistry*, an idea from *Naturphilosophie* which Berzelius
started espousing in the 1830s. In 1845, Liebig was won over (Liebig,
1845) to the Dumas view of the *specificity of organic substances*:
contrary to inorganic compounds, organic substances owe their
properties to their architecture or 'type' rather than to their elemental composition. 'The very idea of protein subunits with a chemical
character of their own, in Liebig's view violated the new ideas in
organic chemistry' (Glas, 1983). Hence, Liebig set about to refute
Mulder's discovery of the protein. Starting in 1846, he got one of his
students (Laskowski) to go through the experiments of Mulder, with
a view to find them lacking in some respect: indeed Laskowski failed
to isolate a sulfur-free protein. Instead of blaming Laskowski for his
poor experimental procedure (he had used boiling instead of lukewarm alkali), Liebig challenged Mulder to produce a detailed procedure for the isolation of the 'protein' (Glas, 1983). The following
year, 1847, saw the success of another of Liebig's co-workers, Fleitmann, who demonstrated that proteins contained both tightly
bound 'thio'-sulfur (which could only blacken lead salts after treatment with concentrated mineral acids) and weakly bound 'sulfide'

sulfur (present after treatment with lukewarm alkali; Fleitmann, 1847). Ironically, and as is often the case in the history of science, both Liebig and Mulder found in this experiment full justification of their opposing views! Glas (1983) makes an interesting analysis of the Liebig-Mulder debate, in the context of Liebig's prominent position in the German scientific-industrial establishment which he was so instrumental in setting up, in the 1830s and 1840s.

The Liebig-Mulder controversy had enduring value. In its aftermath, caution prevailed in the definition of protein concepts during the period 1850-1870. In 1851, we find W.B. Carpenter, in his *Manual of Physiology* writing thus (Carpenter, 1851):

"Proteine and Gelatine are remarkable, not only for containing four elements, but for the very large number of atoms of these components which enter into the single compound atom of each".

Thomas Huxley expresses himself in 1869 in guarded terms (Huxley, 1869):

"All forms of protoplasm (. . .) yet examined, contain the four elements carbon, hydrogen, oxygen, and nitrogen, in very complex union . . . To this complex combination, the nature of which has never been determined with exactness, the name of *Protein* has been applied".

A modern biochemist could subscribe almost entirely to this descriptive statement apart from the omission of sulfur.

This brief summary of the origins of the word 'protein', and of some of the attendant disputes, shows the key role of Berzelius in both framing the grand design of a unitary chemistry, and in proposing this name of 'protein' to Mulder. It shows also the rival viewpoints, occupying the scene in the 1830s and which would only be unravelled in our century: are chemical properties determined predominantly by the nature of the component atoms, or by molecular architecture? This latter view, that of Dumas which came to be espoused by Liebig, would lead much later, in the late 1930s and in the 1940s, to the Pauling theories of protein structure (see Chapter 13). In a more philosophical vein, the search by Mulder for a single vital substance or principle is to be set in parallel with other attempts of the same ilk, with respect to protoplasm for instance (notice in passing the resemblance between the terms '*prot*ein' and '*prot*oplasm'). Finally, there is an evident convergence in aims between the discovery of amino acids as building blocks and Mulder's discovery of 'protein' as the basic unit in albuminoid substances.

REFERENCES

Berzelius, J.J. (1843) Jahresber. XXII, 535 (quoted by Glas, 1983).
Carpenter, W.B. (1851) Manual of Physiology, 2nd ed., p. 9.
Compact Oxford English Dictionary, Oxford University Press, Oxford.
Davy, H. (1840) Elements of Chemical Philosophy, in The Collected Works of Sir Humphry Davy, Bart, LL.D., F.R.S. (Davy, J., Ed.), Smith, Elder, Cornhill, London, 9 vols., vol. 4, p. 364.
Fleitmann, Th. (1847) Ann. Chem. Pharm., LXI, 121-126.
Florkin, M. (1972) A History of Biochemistry, Part II, Elsevier, Amsterdam, (a) pp. 123-125; (b) pp. 149-158.
Glas, E. (1979) Chemistry and Physiology in their Historical and Philosophical Relations, Delft, pp. 31-40.
Glas, E. (1983) Stud. Hist. Phil. Sci., 14, 39-57.
Huxley, T.H. (February 1869) Physiological Basis of Life, Fortn. Rev., 135.
Lenoir, T. (1981) Stud. His. Phil. Sci., 12, 293-354.
Liebig, J. von (1825) Kastner's Arch. ges. Naturlehre, 6, 95.
Liebig, J. von (1842) Chimie Organique Appliquée à la Physiologie Animale et à la Pathologie, Fortin, Masson, Paris, pp. 111-113 (my translation).
Liebig, J. von (1844) Ann. Chem. Pharm., L, 334.
Liebig, J. von (1845) Ann. Chem. Pharm., LIII, 1.
Meinecke, J.L.G. (1817) Erläuterungen zur Chemischen Messkunst, Halle; Leipzig.
Mulder, G.J. (1838) Bull. Sci. Phys. Néerl., III.

Chapter 10

The Conceptualization of Albumins as a Class

One might distinguish three stages in the appearance of a new concept: its emergence, when a pioneer deems it necessary to recognize or devise a new intellectual entity; its crystallization, which leaves its trace in the vocabulary simply by the inclusion of a new term (or of a novel meaning for an already existing word); and its flourishing, when the germ thus seeded has generated an epidemic and the concept finally meets with its generalized acceptance.

Of course, intellectual history shares with history itself similar sudden accelerations of change after long periods of apparent immobility. Hence, the three stages just defined are not always clear-cut and easily picked out. Often, two of these (or even all of them) may be fused into a single well-characterized intellectual event.

The concept of proteins is better behaved! It offers almost a test case for the intellectual historian, since the three stages we have mentioned are well-separated in time. Indeed, long before Mulder crystallized the concept in the late 1830s by giving these substances the name 'protein', these species had been recognized as sharing a set of joint characteristics: one may already discern in the 17th and the 18th century evidence for the emergence of such a concept.

We shall quote from the *Encyclopédie* of Diderot et d'Alembert in order to document this statement in unambiguous manner. In the entry 'Albumineux', we find the following assertions (Diderot et al., 1751–1780):

"Egg white has almost the same properties as lymph; it is because of this resemblance that Mr. Quesnay has resorted to the word *albuminous* to refer to lymph and to other humors of the same kind. Lymph is intermediate between blood and aqueous humors (. . .) it resembles blood by the ease with which it coagulates with heat, and even more from the mix with acidic and vinous spirits. Heat alone, raised to 150°F (. . .) thickens lymph, and makes it into a jelly; the above-mentioned spirits do likewise".

This somewhat lengthy quotation of a text, from the early 1750s since it appears in the second volume of the *Encylopédie*, published from 1751 till 1780, is nevertheless useful: it shows that conceptualization of what we now name 'proteins' and was known then as 'albumins' or 'albuminous substances' was soundly based on a property recognizable from experiment, that of coagulation, clotting or curdling induced either by heat or by acid or by alcohol. Because the 18th-century scientists already knew these three types of what we would term nowadays agents for the denaturation of proteins, we are justified in assuming that, in the second half of the 18th century already, proteins (known then as 'albumins') were already an *operational concept*, i.e. an object firmly within the fold of Baconian experimental science.

François Quesnay (1694–1774), to whom paternity of the word 'albumineux' is attributed in this entry from the *Encyclopédie*, was a physician and surgeon practising in the hospital (Hôtel-Dieu) at Mantes, a small town about 30 miles west of Paris. He is best known for his epochal work in economics. As the founding father of the physiocrats, he gave a *Tableau Economique* of France in 1758 which is widely recognized as an important foundation of modern economics.

Quesnay was struck by the common features of egg white and lymph to such an extent that he termed as '*albuminous* lymph and other humors of the same kind': he conceived of these substances as sharing the same intrinsic *whiteness*. Color, or the absence thereof, was important to chemists of the time who were distinguishing, in their investigation of living tissues, between colored and colorless substances.

After it had been evolved by Quesnay, this concept of albumins as organic substances which heat, acid or alcohol could coagulate was fully crystallized. Thus a name had already been given and it gained general acceptance, firstly on the strength of the *Encyclopédie*, and also because Macquer included it (1777) in his influential writings, of which the *Dictionnaire de Chimie* is best known.

REFERENCE

Diderot, D. et al. (1751-1780) Encyclopédie ou Dictionnaire Raisonné des Sciences, des Arts et des Métiers (35 vols.) entry ALBUMINEUX, vol. 2, p. 29; signed 'H.D.G.', Neuchâtel.

Chapter 11

Crystallization of Proteins and the Molecular Weight of Hemoglobin

The importance of achieving crystallization, for the conceptualization of proteins as molecules, cannot be underestimated. To obtain
crystals, even more to see them under the microscope, is a decisive
gesture, an all-important step when dealing with the molecules of
life. Very often indeed, as a testimony to this crucial operation, one
finds microphotographs of crystals (e.g., those of pepsin) adorning
the proud reports of their first crystallization.

Crystallization provides a fairly homogeneous proteic material,
usually of high purity. Let us consider the attendant thermodynamics, when a protein crystal is at equilibrium with its aqueous saline
solution. Usually, there is acid or base present, unless the protein is
strictly isoelectric. Hence, in the sense of Willard Gibbs's phase rule
$v = c + 2 - \varphi$, there are c = four components (water, protein, salt, acid
or base), and φ = two phases, crystalline solid and solution. The
system therefore has a variance v of four: 'if salt concentration, pH,
pressure and temperature are all fixed, the chemical potential of the
protein should also be determined and it should give a definite solubility, independent of the amount of protein in the crystalline solid
phase' (Edsall, 1979). ' . . . If more of the same solid is added no
change in concentration will occur. If any other substance is added,
however, a change will occur. The test is exceedingly specific, even
more so than the serological tests as Landsteiner and Heidelberger
pointed out. (. . .) From the experimental point of view it has the
great advantage that the test may be carried out in concentrated salt
solution, in which the proteins are most stable' (Northrop, 1946).

Protein crystals are also needed for x-ray work: in 1912, the x-ray
diffraction patterns of the very first crystal (copper sulfate) were

[177]

obtained by Friedrich, Knipping and von Laue (Friedrich et al., 1912); the following year already, Japanese pioneers were concluding from x-ray diffraction patterns obtained from fibers such as natural silk, that there is molecular order in these materials (Nishikawa and Ono, 1913).

But perhaps the most important aspect in the crystallization of a protein is a more philosophical (or perhaps ideological) point, since many outstanding scientists were convinced for decades that such a feat was impossible in principle. Graham had postulated in 1861 a mutual exclusion between crystalloids and colloids (see Chapter 4), a distinction according to which proteins were noncrystallizable substances. The widespread conviction that proteins could not be crystallized is indicated by Pasteur's statement in 1883 (Pasteur, 1886): "You know that the most complex molecules of plant chemistry are the albumins. You know besides that these immediate principles have never been obtained in the crystalline state. May one not add that they probably cannot crystallize" ' (Fruton, 1976). It is ironical that, during the same period (1870–1900 approximately) where there was general agreement about the impossibility of crystallizing proteins, crystallized proteins were already known or being made available! Within 15 years of the Pasteur statement, as indicated by Fruton (1976), 'albumins had been crystallized from several sources, including egg white and blood plasma'.

We shall refer here to the example of hemoglobin, for its historical importance: the molecule was one of the first, if not the first, protein to be crystallized (1830 and 1840); to have its molecular weight determined (1871 and 1886) and to be shown as *isodisperse* in the first ultracentrifuge experiments of Svedberg (1926).

Perhaps the first recorded occurrence of hemoglobin crystals is in the report by Baumgaertner in 1830. This was confirmed 10 years later, when Hünefeld (1840) saw the growth of hemoglobin crystals from blood held between glass plates on his microscope. More than a century later, his experiment was repeated successfully in Oxford by Barbara Low and Dorothy Crowfoot (Hodgkin, 1979). Apparently, while Hünefeld had observed under a light microscope crystalline tabular (*tafelförmig*) crystals separating from samples of either human or pig blood when placed under glass plates and partly dessi-

cated (Edsall, 1979, quoting Horace Freeland Judson), the experiment Dorothy Crowfoot-Hodgkin refers to was performed on blood
from the earthworm (Hodgkin, 1979).

The next step was the empirical discovery, shortly after 1850, of
the salting-out technique by Panum (1852), Virchow (1854) and
Claude Bernard (see Robin and Verdeil, 1853):

"A protein is precipitated by adding a salt at high concentration – commonly sodium
chloride or sodium sulfate; later ammonium sulfate became a favorite salting-out
agent because of its high solubility in water. (. . .) Qualitatively salting-out represents an attraction of the highly polar water molecules for the ions of the salt; the
water molecules crowd around the ions, and 'squeeze out' less polar molecules like
acetone W or proteins X" (Edsall, 1979).

Plasma proteins were thus salted out (Denis, 1859). Another technique, inducing crystallization by addition of alcohol in the cold,
allowed Preyer (1871) to crystallize hemoglobin from some 40-odd
animal species, including mammals, birds, fish, reptiles and amphi-
bians.

Preyer (1871) determined from the iron analysis the molecular
weight of hemoglobin as 13 000. This estimate was later refined by
Zinoffsky (1885), who was working in the laboratory of Gustav
Bunge, to 16 700 based on an iron content of 0.336%. Zinoffsky also
found a 2:1 sulfur-to-iron ratio 'a figure fully in accord (. . .) with
the most recent measurements on this protein' (Edsall, 1979).

An important and clean determination, also indirect, of the molecular weight of hemoglobin was that by Barcroft and Hill (1910). It
was based upon the distinction between calorimetric and van 't Hoff
enthalpies: the calorimetric experiments of Barcroft and Hill gave a
heat of oxygenation of 1.85 cal per g of hemoglobin at oxygen saturation, and their equilibrium measurements at constant p_{O2}, in the
temperature range 16–49°C, gave a van 't Hoff enthalpy change of 28
kcal.mol^{-1}. 'Since the ratio, 15 100, agreed within experimental
uncertainty with the molecular weight of 16 700 deduced from iron
determinations they concluded that in the equation:

$$(Hb)_n + nO_2 \rightleftarrows (HbO_2)_n$$

$n = 1$. They used hemoglobin extensively dialysed against distilled
water, which was presumably monomeric' (Sturtevant, 1980).

In 1925, Adair, from osmotic pressure measurements, reported a molecular weight of 67 000, four times that deduced from the iron analysis (Adair, 1925). His report met with some disbelief, especially from the side of the older chemists who believed strongly in the dictum of Emil Fischer that organic compounds with molecular weights above ca. 5000 could not exist (see Chapter 4), 'but all doubts were soon swept away when Svedberg and Fåhraeus (1926) reported the same value quite independently, from measurements of sedimentation equilibrium in the ultracentrifuge. Moreover, the concentration gradient at equilibrium indicated that the molecular weight was the same at all points in the cell, and therefore that the molecules were all of the same size. This was a point that could not be established from osmotic pressure alone, for this yielded only what we would now call a number average molecular weight. A year later, Svedberg and Nichols (1927) studied hemoglobin by the sedimentation velocity method, which provided an even more rigorous test of homogeneity by the observation of the form of the single sedimenting boundary' (Edsall, 1979).

We have already drawn attention to the epochal significance of the 1926 paper by Svedberg and Fåhraeus (1926), because of their demonstration of the *isodisperse* nature of the hemoglobin protein. Their paper not only helped in removing the widespread belief in the colloidal nature of proteins, it also 'left no doubt that the hemoglobin molecule consisted of four subunits' (Haurowitz, 1979). The earlier osmometric determinations had diverged about the molecular weight of hemoglobin: Adair (1925) reported a value of 67 000, while Hüfner and Gansser (1907) had much earlier found a value of $M_r \cong$ 15 000 for horse hemoglobin. As Dr. Felix Haurowitz (1979) recalls, 'this was quite a shock for Professor von Zeynek in whose department I worked. Von Zeynek had been a student of Hüfner and knew that Hüfner was a very careful and precise worker; he asked me to prepare crystalline horse oxyhemoglobin and to redetermine its molecular weight by osmometry. I found the same value as Adair. At about the same time Svedberg (1926) had developed his ultracentrifuge and confirmed the value obtained by Adair (1925)'.

The following table summarizes the history of the determination of the molecular weight of hemoglobin:

TABLE V

Determinations of the molecular weight of hemoglobin

Investigator(s)	Date	M_r	Method
Preyer	1871	13 000	iron content
Zinoffsky	1885	16 700	iron content
Hüfner and Gansser	1907	15 000	osmometry
Barcroft and Hill	1910	15 000	heat of oxygenation
Adair	1925	67 000	osmometry
Svedberg and Fåhraeus	1926	67 870	sedimentation equilibrium
Svedberg and Nichols	1927	68 350	sedimentation speed

Clearly, from examination of the above table, during more than 50 years, there were investigators who knew or behaved as if hemoglobin was a bona fide molecule, with a well-defined molecular weight. This, 'at a time when protein chemistry was laboring under the view that proteins are colloids in the older sense – that is that they are dispersed as lumps of a wide range of sizes, lumps of glue in the case of proteins, probably of indefinite and variable composition' (Sturtevant, 1980). Contemporary with the first determination of the molecular weight of hemoglobin by Preyer, we find chemists excluding proteins from the realm of organic chemistry, as in this gem culled by Fruton (1976): 'albuminoid substances . . . properly speaking, do not constitute a chemical species; they are organs, or the remains of organs, whose history belongs to biology rather than to chemistry' (Naquet, 1868). One should be fair, however: such examples of short-sightedness should be complemented with the converse: examples of far-reaching long-sightedness; these also existed: for instance, Barcroft and Hill could state in the introduction to their 1910 paper, with daring and utmost clarity:

"In the present papier, we propose (1) to set forth some additional evidence for supposing the union of oxygen with hemoglobin is a chemical one, (2) to press this sup-

position to its logical conclusion and on thermodynamical principles deduce the heat
generated when one molecule of hemoglobin unites with oxygen, (3) by actual determinations of the heat produced by the union of one gram of hemoglobin, to calculate
the molecular weight of hemoglobin".

Other workers, such as T.B. Osborne (1924) in New Haven who
'prepared dozens of crystalline proteins and carefully studied their
chemical composition and their physical properties, (also) continued quietly to proceed on the assumption that proteins were definite
large molecules' (Edsall, 1979).

Let us now return to the main theme of the crystallization and
purification of proteins, in general. Wilhelm Kühne (1884) used
salting-out procedures [Na_2SO_4, $(NH_4)_2SO_4$] to purify proteins.
Also in the 1880s, Hofmeister (1887–1890) crystallized ovalbumin
from egg white by salting out, and 'listed the effectiveness of certain
salts in salting out, in the following order: Na_2SO_4 > K_2HPO_4 >
Na_3 citrate > $(NH_4)_2SO_4$ > $MgSO_4$ > NaCl > KCl' (Edsall, 1979),
i.e. that of the so-called Hofmeister series. 'Hopkins, in 1898, had
greatly improved the method, by careful acidification of the solution
(. . .) at high salt concentration. (. . .) In 1901, F.N. Schulz gave a
valuable survey of the many protein crystals that had been obtained
up to that time' (Edsall, 1979). As a result, at the turn of the 20th
century, very many protein crystals had been obtained. For hemoglobin alone, 'by 1909, a great volume of 600 microphotographs of
hemoglobin crystals from 106 species had been collected by E.T.
Reichert and A.P. Brown' (Crowfoot-Hodgkin, 1979).

We return now, in closing this chapter, to the question of the
homogeneity of crystallized proteins. Work by Sørensen and Høyrup (1917) had demonstrated from solubility studies on egg albumin
that this was a pure chemical component. Later, in 1930, Sørensen
re-examined the question and 'demonstrated that the solubility of
(. . .) apparently pure proteins depends (in fact) on their concentration. He concluded correctly that the dissolved protein consists of a
mixture of reversibly dissociating particles. We would call them
today polymers and monomers, or molecules and subunits' (Haurowitz, 1979).

This brief account of the history of the crystallization of proteins,

during the period of the hegemony of colloidal chemistry and biochemistry, is highly interesting for epistemological reasons. Instead
of the Romantic view of the upheaval by a pioneer (in this case
Hermann Staudinger or The Svedberg) of the previous paradigm in
the Kuhnian sense (here, the tenets of colloid chemistry), we find
rather surprisingly a relatively easy coexistence, during very many
years, between the colloidal concept and the molecular concept of
proteins.

REFERENCES

Adair, G.S. (1925) Proc. Roy. Soc. London (A) 108, 627–637.

Barcroft, J. and Hill, A.V. (1910) J. Physiol., 39, 411.

Baumgaertner, K.H. (1830) Beobachtungen über die Nerven und das Blut, Freiburg, p. 25 (Table V).

Denis, P.S. (1859) Mémoire sur le Sang, Paris.

Edsall, J.T. (1979) Ann. N.Y. Acad. Sci., 325, 53–73 (footnote, on p. 55).

Friedrich, W., Knipping, P. and von Laue, M. (1912) Sitz. Ber. Math. Phys. Klasse. Bayer. Akad. Wiss. München, 303.

Fruton, J.S. (1976) Science, 192, 327–334.

Haurowitz, F. (1979) Ann. N.Y. Acad. Sci., 325, 37–47.

Hodgkin, D.C. (1979) Ann. N.Y. Acad. Sci., 325, 121–145.

Hofmeister, F. (1887–1888) Arch. Exp. Pathol. Pharmakol., 24, 247.

Hofmeister, F. (1887–1888) Arch. Exp. Pathol. Pharmakol., 25, 1.

Hofmeister, F. (1887–1888) Arch. Exp. Pathol. Pharmakol., 27, 395.

Hüfner, G. and Gansser, E. (1907) Arch. Anat. Physiol., 31, 209.

Hünefeld, F.L. (1840) Die Chemismus der Thierischen Organisation, Brockhaus, Leipzig, pp. 160–161.

Kühne, W. and Chittenden, R.H. (1884) Z. Biol., 20, 11.

Naquet, A. (1868) Principles of Chemistry Founded on Modern Theories (Stevenson, T., Ed.), Renshaw, London, p. 721 (transl. M. Cortil).

Nishikawa, S. and Ono, S. (1913) Proc. Math. Phys. Soc. Tokyo, 7, 113.

Northrop, J.H. (1942–1962) Nobel Lecture, December 12, 1946; in Nobel Lectures. Chemistry, Elsevier, Amsterdam, p. 124.

Osborne, T.B. (1924) The Vegetable Proteins, Longmans Green, London.

Panum, P.L. (1852) Virchow's Arch. Pathol. Anat., 4, 419–467.

Pasteur, L. (1886) Conférences Faites à la Société Chimique de Paris en 1883–1886, Bureau des Deux Revues, Paris, p. 36.

Preyer, W. (1871) Die Blutkrystalle, Mauker's Verlag, Jena.

Reichert, E.T. and Brown, A.P. (1909) The Differentiation and Specificity of Corresponding Proteins and other Vital Substances in Relation to Biological Classification and Organic Evolution, Carnegie Institute, Washington, DC.

Robin, C. and Verdeil F. (1853) Traité Chim. Anat., 3, 299.

Schulz, F.B. (1901) Die Krystallisation von Eiweissstoffen und ihre Bedeutung für die Eiweisschemie, Fischer, Jena.

Sørensen, S.P.L. (1930) Kolloid Zeitschr., 53, 102.

Sørensen, S.P.L. and Høyrup (1917) C. R. Trav. Lab. Carlsberg, 12, 213–261.

Sturtevant, J.M. (1980) in Bioenergetics and Thermodynamics: Model Systems (Braibanti, A., Ed.), Reidel, Dordrecht, pp. 17–22.

Svedberg T. and Fåhraeus (1926) J. Am. Chem. Soc., 48, 430–438.

Svedberg, T. and Nichols, J.B. (1927) J. Am. Chem. Soc., 49, 2920–2934.

Virchow, R. (1854) Virchow's Arch. Pathol. Anat., 6, 572–579.

Zinoffsky, O. (1885) Z. Physiol. Chem., 10, 16–34.

Chapter 12

Polypeptidic Nature of Proteins

1. A history of slow progress: isolation of amino acids from hydrolysates of proteins

The 20 common amino acids were isolated in the span of a little over a century! This seemingly mundane statement deserves closer scrutiny. The extreme slowness of the identification of individual amino acids isolated from proteins can be considered as an empirical fact. What can be read from it? As we shall find out, the answer is multiple. It has to do with the dominance of *Naturphilosophie* on the concepts of physiological chemistry, when this science was still in its infancy; with the later difficulties of physiological chemistry to establish itself as a respected academic discipline in German universities; and, last but not least, with analytical problems in the separation of amino acids from one another.

I shall take as my starting point the year 1813, and a fine book of science popularization written by a woman which, in those times, she could not think of publishing under her name.

One finds in the anonymous *Conversations on Chemistry* (1813) the statement that animal bodies contain carbon, hydrogen, oxygen and nitrogen which 'form the immediate materials of animals, which are *gelatine, albumen* and *fibrine*'. Shortly thereafter the first two amino acids, leucine and glycine, were isolated from proteins by the Frenchmen Proust (1819) and Braconnot (1820). The latter tried to find out whether the hydrolysis of gelatin, like that of cellulose, would yield a sugar. Since glycine had indeed a sweet taste, he named it accordingly (Gr. *glykys*, sweet). I point out this etymology, not for its anecdotal flavor, but because it underlines Braconnot's misconception of gelatin as a sugar-containing compound.

TABLE VI

Isolation of amino acids from protein hydrolysates

Amino acid	Author (date) of discovery
Leucine	Proust (1819)
	Braconnot (1820)
Glycine	Braconnot (1820)
Tyrosine	Liebig (1846)
	Bopp (1849)
Serine	Cramer (1865)
Glutamic acid	Ritthausen (1866)
Aspartic acid	Ritthausen (1868)
Alanine	Schützenberger and Bourgeois (1875)
	Weyl (1888)
Phenylalanine	Schulze and Barbieri (1881)
Lysine	Drechsel (1889)
Arginine	Hedin (1895)
Histidine	Kossel (1896)
	Hedin (1896)
Cystine	Mörner (1899)
Valine	Fischer (1901)
Proline	Fischer (1901)
Tryptophan	Hopkins and Cole (1901)
Hydroxyproline	Fischer (1902a)
Isoleucine	Ehrlich (1903)
Methionine	Mueller (1929)
Threonine	Meyer and Rose (1936)

The other common amino acids came upon the stage grudgingly, in slow succession (Table VI). Florkin (1972) has reproduced in the first volume of this *History of Biochemistry* the plot made by Cohn (1925) of the number of known proteic amino acids N as a function of time t between 1820 and 1918. Interestingly, this plot closely approximates an exponential growth curve $N = \exp(a.t)$, with a $\simeq$ 0.028 if t is measured in years (Table VII). Assuming that this empirical finding goes beyond an accidental coincidence, its significance is obviously the proportional rate of discovery of new amino

TABLE VII

Number of known proteic amino acids as a function of time, calculated as an
exponential growth (see text)

Year	t	calc.	exp.
1819	0	1	1
1864	45	3.5	3
1900	81	9.5	12
1906	87	11.2	15
1929	100	16.1	18

acids dN/dt to the size of the existing pool of amino acids already
known at time t: $(dN/dt) = A.t$.

In other words, were one to rely upon this single fact, one would
gain the impression of a 'band-wagon effect' spanning four or five
generations of scientists all sharing a joint paradigm.

Before doing so, however, let us recall (see Chapter 9) that both
Mulder and Liebig had, in the late 1830s and early 1840s, quite
another notion, that of the existence in the animal economy of a
single parent albuminoid substance, for which Mulder coined
accordingly the term 'protein' in 1838. From this standpoint, the
Proust–Braconnot discovery of 1819–1820 could be safely ignored as
some sort of artefact. I submit that the paradigm of *the protein* as
the sole basic constituent of animal bodies, which became embodied
in the name 'protein', was effective during the whole period
1820–1840 to discourage scientists from testing the compound
nature of what we now know to be proteins. Indeed, one could argue
that discovery of individual amino acids from proteins could be
resumed only after 1844, when the prestigious and influential Justus
von Liebig turned his back on this paradigmatic notion of a single
protein. Indeed Liebig himself discovered in 1846 the next amino
acid, tyrosine (Table VII), thus starting the hunt for other amino
acids in proteins, a research program which could then proceed un-
abated during the next three quarters of a century. More detailed

accounts of the identification of individual amino acids from proteins can be found in the *Biochemisches Handlexikon* (Abderhalden, 1911), in the publications by Plimmer (1908), Vickery and Schmidt (1931), Schmidt (1938) and Vickery (1941).

This last proposition now comes fairly close to the actual course of history: the Proust and Braconnot double find occurred in the *pre*history of isolation of amino acids from proteins. That their number went up exponentially after Liebig had isolated tyrosine in 1846, in consonance with the harmonious development within an homogeneous Kuhnian paradigm, can now be stated with greater assurance. What is striking, if one examines Table VI more closely, is that indeed most of the scientists who discovered the proteic amino acids not only were Germans, but also published their finds in a small number of periodicals, in journals such as *J. Prakt. Chem., Z. Physiol. Chem.*, and *Ber. Chem. Ges.*

I have thus provided a first, tentative answer to the question of why isolation of amino acids from proteins was such a protracted, drawn-out activity: it was delayed by some 20 years due to the influence of *Naturphilosophie*.

One word now about the experimental procedure by which most of the amino acids listed in Table VI were isolated. The protein chosen for study is dissolved in an aqueous solution of a strong mineral acid, typically hydrochloric (20%) or sulfuric (35%) acid. This mixture is then brought to boiling point and refluxed at that temperature. After cooling down, the aqueous solution consists of a protein hydrolysate. Needless to say, such a drastic procedure can lead to artefacts: ornithine, for instance, gave rise to discussions of this type, regarding its actual presence in a native protein.

This procedure of acid hydrolysis, which by and large all the physiological chemists responsible for isolation of amino acids from proteins resorted to well into the 20th century, provides us with another, complementary (although totally positivistic) answer to the question of why it took so long to find out the 20 natural amino acids (Fieser, 1956): starting with Proust and Braconnot, in the early part of the nineteenth century, 'investigation of protein hydrolysates was largely a matter of chance observation'. 'The isolation of leucine', writes Fieser (1956) 'from muscle fiber and that of tyrosine from casein was accomplished at an early period (1820-49) since both substances are sparingly soluble in aqueous solutions of low acidity'. And Fieser implies that all the other proteic amino acids were discovered subsequently more or less by accident, during the period 1820-1922, 'before introduction of efficient modern methods for separation of protein hydrolysates'. This notion of the accidental discovery of the amino acids finds added support in the indifference of the contemporary scientists, even when these were competent. From hindsight, the isolation of the various amino acids from proteins had enormous importance. But the contemporaries were, to a large extent, blind to the significance of these results. As Fruton indicates (1979), 'one of the striking features of the 19th century speculations about protein structure is their lack of emphasis on amino acids as structural units of proteins'.

Fruton (1979) gives the example of a textbook (the academic swan song as it were) by Neumeister. Richard Neumeister published his *Lehrbuch der Physiologischen Chemie* in 1897. This was a critical year for him: he had just taken a difficult decision, that to abandon chemistry. Neumeister held the chair of chemistry in Jena since the suicide in 1889 of the first incumbent, Friedrich Krukenberg, who had become desperate because he had not been given proper support. While Richard Neumeister did not take his life, similar material difficulties led him to resign his academic position in 1897 in favor of medical practice and amateur philosophy (Kohler, 1982). This by way of background for the discrepancy between the statement made in Neumeister's book, when 11 amino acids had at that time been actually identified in proteins (Table VI), that only three (tyrosine, leucine, and aspartic acid) were regular products of the hydrolysis of proteins (Neumeister, 1897). Did he reflect, as implied by Fruton (1979), a generally held prejudice? Or was Neumeister unconsciously downgrading achievements in a discipline he himself had been unable to contribute to, because of the harsh material circumstances? Or was he simply ignorant, because of his own scientific isolation in Jena, of the breakthroughs made by physiological chemists elsewhere? I believe that the first two factors best explain his attitude.

Neumeister's unsuccessful career can indeed be contrasted with

those of some of his contemporaries who had performed the isolation of proteic amino acids which Richard Neumeister chose to
ignore in his textbook. For instance Edmund Drechsel, who
reported the isolation of lysine in 1889, was a militant physiological
chemist who was responsible for the introduction of this discipline
in Leipzig where he started in 1878 (he became afterward Nencki's
successor in Bern, in 1891, where his chair combined physiological
chemistry and pharmacology). Albrecht Kossel, to whom we owe the
first isolation of histidine, was likewise a chemical physiologist who
held the chair of physiology in Marburg (Kohler, 1982).

Hence, all three men (Neumeister, Drechsel, Kossel) worked during the difficult, resisted, and not fully recognized rise of the new
discipline of physiological chemistry. Neumeister did not succeed
personally, but he contributed to the public support of the field with
the publication of his book (Neumeister, 1897). Kossel worked from
an established position in physiology which he slanted toward physiological chemistry. Drechsel, I have already indicated, pushed
strongly for recognition of physiological chemistry as a respectable
academic discipline. This was at the time when Hoppe-Seyler was
launching from his institute in Strasbourg his journal, the *Zeitschrift für Physiologische Chemie*. This periodical, started in 1877,
was for 30 years the only one of its kind in Germany (Kohler, 1982).
It should be no surprise then that no less than five of the amino acids
isolated from protein hydrolysates were first reported in the
Zeitschrift für Physiologische Chemie (Table VI).

An example will serve to convey the pattern of the studies of amino acids by German physiological chemists, during the last quarter
of the 19th century: histidine was isolated simultaneously in 1896 by
Kossel from sturine (a sturgeon-sperm protein) and by Hedin from
several protein hydrolysates, among which were those of casein and
egg-white albumin. Kossel and Hedin both reported their findings
in the *Zeitschrift für Physiologische Chemie*. The structure determination, which was immediately started, was reported also in the
Zeitschrift für Physiologische Chemie (Pauly, 1904). This structure
was confirmed in the same year (Knoop and Windaus, 1904). Windaus went on to synthesize histamine (Windaus and Vogt, 1907)
which Best et al. (1927) much later characterized as by far the most
important product of histidine.

2. A discourse on the hybrid status of physiological chemistry in Germany during the second half of the nineteenth century

In order to set the stage for Emil Fischer's seminal contribution, I have to amplify what I have just obliquely referred to, the uneasy status of physiological chemistry in Germany during the second half of the 19th century. This question has been addressed in a recent useful book (Kohler, 1982) from which I shall summarize the most salient points: as part of chemistry, physiological chemistry played second fiddle to organic chemistry. As part of physiology, physiological chemistry remained subservient to the medical physiologists. The very slow identification of amino acids as proteic constituents (Table VI) follows with almost embarrassing clarity the very slow emergence of physiological chemistry as an academic discipline, the forerunner of biochemistry.

Justus von Liebig had deliberately set as the aim for his well-known school in Giessen the foundation of such a new discipline of physiological chemistry. His studies of animal and of agricultural chemistry were astutely framed within the traditional alliance between organic chemistry and pharmacy. Liebig's school of experimental chemistry indeed started in the Giessen school of pharmacy. But Liebig was much more ambitious. He wanted to further increase the academic respectability of his new discipline, and brought it into the philosophic faculty. In order to make this graft succeed, Liebig conceived of a program for chemistry as a combination of an applied and a pure science (a program whose inspirer, I believe, is none other than Humphry Davy). Also, as Kohler (1982) writes, 'it is no accident that Liebig took up physiological and agricultural chemistry after losing the lead in structure theory to the French school of Dumas'.

While Liebig made a huge impact on the popularization of chemistry, as lecturer and author of best-selling books, academically he failed to secure a stronghold for physiological chemistry as the discipline he had attempted to conceptualize anew and to remould according to his ideas. As nicely put by Kohler (1982), 'the failure of physiological chemistry to become a discipline in the period from

1840 to 1860 is the reverse side of the success of organic chemistry, seen from the loser's point of view'. At the time when Liebig moved from Giessen to Munich (1850), he had been trounced in his effort to establish an autonomous discipline of physiological chemistry. On the contrary, most chairs of chemistry in Germany became the province of general or organic chemists.

Pushed out by the success of organic chemistry, the infant discipline of physiological chemistry sought shelter in the medical schools, under the mantle of physiology. By doing so, it renounced its autonomy as a new discipline and remained throughout the second half of the 19th century secondary to physiology and to physiologists, whom it helped to set their own discipline apart from anatomy. Or it daringly proclaimed its aim at a takeover of physiology within medical faculties, in which case expulsion quickly followed. Physiological chemists were ridiculed: clearly it was impossible to master such a wide field extending from medicine to chemistry. The demise of Babo, that of his rival Latschenberger in the Freiburg Medical Faculty at Leipzig, that of Johann von Scherer's chair of medical chemistry in Würzburg, at the death of their incumbents, are all illustrative of this resistance of the medical faculties to the new interdisciplinary program of physiological chemistry (Kohler, 1982). The one great exception is that of Felix Hoppe-Seyler, in Tübingen first, where he had sympathetic chemist colleagues, and later in Strasbourg, for historical reasons: after the victory of Prussia against France in 1870, Germany had meant the University of Strasbourg to be a showcase for the efficiency of Prussian methods of education and the excellence of German science. Nevertheless, Felix Hoppe-Seyler's famous Strasbourg Institute, whose creation was decided on in 1872 but which was not founded for 11 years, did not survive him when he died in 1895. The university administration 'decided to split the chair, giving the bulk of its responsibilities and perquisites to a new chair of hygiene' (Kohler, 1982).

All these institutional reasons explain why, at the turn of the 20th century, Emil Fischer could revolutionize in a few years this research field of the chemical composition and structure of proteins: because as an organic chemist, he was an outsider, but one with an extremely powerful base within his own well-established discipline;

and because the question he took up, the role of amino acids in proteins, had built up over so many years that the analytical information was almost fully complete (Table VI) in order for Emil Fischer to start his synthetic work: perfect timing!

A few words now about Emil Fischer and his achievements prior to his studies of the structure of proteins. Emil Fischer, deemed by his father too stupid for business, started studying chemistry at the University of Bonn in 1871. The following year, he accompanied his cousin Otto Fischer to the newly established University of Strasbourg where he came under the spell of Adolf von Baeyer, with whom he prepared his Ph.D., which he obtained in 1874, at the age of 22. The same year, still in Strasbourg as an assistant instructor, he made the chance discovery of phenylhydrazine. Emil Fischer followed von Baeyer to Munich where Fischer became a Privatdozent in 1878, and an Associate Professor in Analytical Chemistry in 1879. Fischer's career then moved to Erlangen whose call Fischer accepted in 1881; to Würzburg (1888–1892); and finally to the University of Berlin, till his death in 1919.

In Erlangen Fischer studied the active principles of coffee, cocoa, and tea; he identified and synthesized caffeine and theobromine. This led him to find a whole family of natural compounds, the purine bases as we now term them. He named the parent compound, then still hypothetical, purine in 1884, and he synthesized it in 1898. His monumental work on sugars was performed during the period 1884–1894. Fischer established the aldehyde formula of glucose and, by forming a common osazone, he showed the relationship of glucose, fructose, and mannose (1888). He then went on, based upon the Le Bel and van 't Hoff theory of the asymmetric carbon (1874), to establish the stereochemical relationships of all the known sugars. He interrelated chemically the hexoses and, by degradation or synthesis to the hexoses, the heptoses and pentoses. This sugar work was capped in 1890, shortly before Emil Fischer moved to the prestigious chair at the University of Berlin, by the synthesis of glucose, fructose and mannose from glycerol as their joint precursor. This outstanding work on the sugars rapidly became exemplary of what biochemists should try to emulate (Kohler, 1982), and Emil Fischer's institute of organic chemistry in Berlin attracted large numbers of physiological chemists from Germany and from abroad. In 1902, Emil Fischer was awarded the Nobel Prize in Chemistry for his work on the sugar and purine syntheses.

His Nobel Lecture makes interesting reading (Fischer, 1902b). Fischer starts it with a retrospective look on the history of chemistry. He notes that, at the start of the 19th century, when organic chemistry became established as a discipline separate from inorganic chemistry, it quickly escaped from its original frame of reference (Fischer, 1902b):

"Strange to relate, organic chemistry, as the new discipline was termed, did not remain for long within its original terms of reference. It found the exploration of new avenues more worthwhile. It replaced the animal and vegetable substance by many artificial products such as the hydrocarbons and cyano compounds, wood tar and coal tar, wood alcohol, etc. and by pressing into its service the synthetic methods of inorganic chemistry it appropriated the fundamental problems of our science at the same time".

Fischer then remarks that organic chemistry, during the same period and following Friedrich Wöhler's urea synthesis of 1828, became the testing ground for chemical theories. Fortunately, after having thus strengthened itself, chemistry returned to

"the paths which it was following at the beginning of the 19th century under the guidance of Berzelius, Gay-Lussac and Davy. *A necessary consequence of this reorientation must be the reversion of organic chemistry to the great problems of biology*" (my emphasis; Fischer, 1902b).

Fischer then explains in this remarkable text his work on the purines, and he offers this vivid (or rather morbid) analogy (Fischer, 1902b):

"To determine the structure of the molecule the chemist proceeds in a similar way to the anatomist. By chemical actions he breaks the system down into its components and continues with this division until familiar substances emerge".

One comment is in order: Fischer is describing the familiar Cartesian methodological rule of breaking down complex reality into smaller and simpler entities; yet at the same time his reference to anatomy must be seen in the context, referred to above, of the break-away of physiology as a field, and with the active support of physiological chemists, from the rule of anatomists. Emil Fischer gives this rhetorical tribute to anatomy when, at the same time, he is influential in detaching physiological chemistry from anatomy to lure it back into organic chemistry!

After summarizing his work on sugars, Emil Fischer states near the end of his Nobel Lecture (Fischer, 1902b) his next goal: despite his success with carbohydrates,

"the chemical enigma of Life will not be solved until organic chemistry has mastered another, even more difficult subject, the proteins, *in the same way as it has mastered the carbohydrates*" (emphasis added).

Which way? Does Emil Fischer have a magical formula for mastering Nature? The modern reader turns back to this text in the hope that perhaps Fischer, who is so nicely articulate, has stated his general strategy in it. Indeed, this is what Emil Fischer says of his way of tackling the sugars, which in my opinion is a cogent summary of how he had already attacked the protein problem. He writes of the sugars that:

"more than a century elapsed from the elucidation of their elementary composition by Lavoisier before science prepared them by artificial means. The reason for this slow progress lies firstly in the peculiar difficulties which those substances pose for experimental treatment, and secondly in their great profusion of forms which also necessitates a rather complex systematology".

When Fischer uttered these words in 1902, he had already started working in the protein field since 1899, and he must have had the proteins also in mind when he made this dual comment about both the experimental and the conceptual difficulties of the proliferation of like structures.

3. The polypeptide syntheses and theory of Emil Fischer

Among the experimental difficulties which Fischer had to vanquish, the purification of the amino acids obtained from hydrolysis of pro-

teins indeed was a prime problem. The melting points of amino acids
are very high (e.g. glycine, m.p. 289–292° C), in the neighborhood of
300° C as befits these dipolar substances. Hence, they could not be
purified by distillation without incurring some decomposition.
Fischer's solution was to turn to derivatives and to prepare methyl
or ethyl esters which could be distilled, at much lower temperatures
and without decomposition (Cohn, 1943). Fischer introduced this
technique of esterification of a protein hydrolysate and fractional
distillation of esters of the type $RCH(NH_2)CO_2CH_3$ or
$RCH(NH_2)CO_2C_2H_5$ in 1901. 'Since the component acids are of
rather low molecular weight, the relatively large difference in molecular weight between homologs results in a fair spread of boiling
point adequate for separation by fractional distillation' (Fieser and
Fieser, 1956). This single technical improvement, almost routine for
an organic chemist in Fischer's time, led him immediately to the
discovery in protein hydrolysates of valine, proline, and hydroxyproline (Table VI); single-handedly, Emil Fischer added to the store
of amino acids from proteins a full one fourth of those already
known.

Of course, he also evolved syntheses of the various amino acids.
Historically, the first amino acids to have been synthesized were
glycine and alanine. Perkin and Duppa made glycine in 1858 by the
α-halogen acid method, displacing the halogen-leaving group by
ammonia (Fieser and Fieser, 1956). Alanine was first synthesized by
Strecker in 1850 (25 years before it was identified in protein hydrolysates): addition of hydrogen cyanide, in the presence of ammonia,
to the carbonyl group of an aldehyde RCHO gives first the cyanohy-
drin RCH(OH)CN which is then transformed by ammonia to the
corresponding aminonitrile $RCH(NH_2)CN$ and which, on hydrolysis, gives the amino acid $RCH(NH_2)COOH$. Fischer, in 1902, prepared serine $HOCH_2CH(NH_2)COOH$ in this manner from glycolal-
dehyde (Fieser and Fieser, 1956). He then used serine as the starting
material for his syntheses in 1908 of the sulfur-containing amino
acid, cystine, in approx. 25% yield (Fieser and Fieser, 1956).

More useful yet, Emil Fischer devised a general synthesis of amino acids by the malonic ester method. For instance, he obtained
leucine in 37% yield in 1906 by condensation of isobutylbromide

with the sodium salt of malonic diethylester, followed by hydrolysis to the diacid, bromination (Br_2) in the remaining activated position, decarboxylation through heating to α-bromoisocaproic acid, an α-halogen acid which was then converted to leucine by the Perkin method.

This scheme:

$$RBr + {}^-CH\begin{smallmatrix}COOC_2H_5\\COOC_2H_5\end{smallmatrix} \longrightarrow RCH\begin{smallmatrix}COOC_2H_5\\COOC_2H_5\end{smallmatrix} \xrightarrow{H_3O^+} RCH\begin{smallmatrix}COOH\\COOH\end{smallmatrix}$$

$$\xrightarrow{Br_2} RCBr\begin{smallmatrix}COOH\\COOH\end{smallmatrix} \xrightarrow{\Delta} RCBr\begin{smallmatrix}COOH\\H\end{smallmatrix} \xrightarrow{NH_3} RCNH_2\begin{smallmatrix}COOH\\H\end{smallmatrix}$$

$$R = CH(CH_3)_2$$

is quite nice because, depending on the nature of the starting alkyl halide, a variety of amino acids can indeed be generated.

Having now in his possession extensive knowledge about the amino acid composition of a number of proteins (see Table VI) and having mastered the syntheses of the individual amino acids, Emil Fischer could now turn to combining amino acids into oligo- and polypeptides, to see if he could obtain compounds with properties similar to those of proteins. This is the program on which he started to work at the turn of the 20th century.

The field was his for the taking, this land had been untilled (with the insignificant exception of the accidental formation of benzoyldiglycine when Curtius in his Munich laboratory condensed hippuric acid with glycine (Curtius, 1881)).

The reader will remember that Emil Fischer had turned to methyl esters for the fractional distillation and separation of amino acids from protein hydrolysates. Fischer then observed, early on (1901) with Fourneau, that these methyl esters of amino acids would form linear condensation products upon gentle heating. For instance, at 100°C, methyl α,β-diaminopropionate undergoes dimerization:

$$2 \; H_2NCH_2CH(NH_2)COOCH_3 \xrightarrow{\Delta} H_2NCH_2CH(NH_2)CO\text{-}HNCH_2CH(NH_2)COOCH_3$$

Fischer took advantage of this observation for making higher oligomers and synthesized pentaglycylglycine by this method, with 76% yield, from diglycylglycine methyl ester. The yields, however, dropped when he attempted further lengthening of the linear chain (Fieser and Fieser, 1956). Fischer was thus led to look for another condensation method.

In 1903, he reported such a method, in which the peptide bond is formed by condensing the acid chloride of an amino acid with the amino group of another. The problem of self-condensation was circumvented, again by recourse to the same old Perkin–Duppa strategy: Fischer used the acid chloride of an α-halo acid, replacing the α-halo (leaving) group with an amino group (by addition of ammonia) *after* the peptide bond had already been formed. The schemes below illustrate how Emil Fischer obtained diglycylglycine and leucylglycylglycine (Fieser and Fieser, 1956):

(1) $ClCH_2COCl+H_2NCH_2CO\text{-}NHCH_2COOC_2H_5 \xrightarrow{27\%}$

$\qquad ClCH_2CO\text{-}HNCH_2CO\text{-}NNCH_2COOC_2H_5 \xrightarrow[100\%]{NaOH}$

$\qquad ClCH_2CO\text{-}HNCH_2CO\text{-}HNCH_2COOH \xrightarrow[25\%]{NH_4OH}$

$\qquad H_2NCH_2CO\text{-}HNCH_2CO\text{-}HNCH_2COOH$

(2) $(H_3C)_2CHCH_2CHBrCOCl+H_2NCH_2CO\text{-}HNCH_2COOC_2H_5 \xrightarrow{76\%}$

$\qquad (H_3C)_2CHCH_2CHBrCO\text{-}HNCH_2CO\text{-}HNCH_2COOC_2H_5 \xrightarrow[100\%]{NaOH}$

$\qquad (H_3C)_2CHCH_2CH(NH_2)CO\text{-}HNCH_2CO\text{-}HNCH_2COOH \xrightarrow[63\%]{NH_3}$

$\qquad (H_3C)_2CHCH_2CH(NH_2)CO\text{-}HNCH_2CO\text{-}HNCH_2COOH$

During this extremely productive period, 1899–1908, Emil Fischer synthesized many oligopeptides. It is hard for us to believe, in the age of Merrifield's automated resin-supported peptide synthesis, that a full half-a-century earlier, Emil Fischer had achieved in 1907 the almost impossible feat of synthesizing, with the normal synthetic organic methods he had himself devised, an octadecapeptide:

$$\text{Leu-(Gly)}_3\text{-Leu-(Gly)}_3\text{-Leu-(Gly)}_9$$

Emil Fischer was very much aware of the epistemological significance of this work. In a lecture he gave in 1906, he states, after describing his new separation method for the monoamino acids through the use of esters (Fischer, 1906a):

"More significant seem to me to be the similarly found methods for conversion of the amino acids to their amide-like anhydrides, for which I have chosen the collective name 'polypeptides'. The higher members of this group of synthetic materials are so similar to the natural peptones in their external properties, in some color reactions, and in their behavior toward acids, alkalis and ferments that one must consider them to be very close relatives, and that I may describe their preparation as the beginning of the synthesis of the natural peptones and albumoses".

The next year, 1907, the target seemed to Emil Fischer within sight. He had succeeded in making a peptide composed of 14 glycine and leucine units, and this polypeptide gave a strong biuret reaction and was precipitated by protein reagents. Emil Fischer became convinced that, if he could synthesize a duodecapeptide, he would have made an artificial protein (Fruton, 1979). However, the same year 1907, when he had prepared the octadecapeptide of the constitution previously indicated, with a molecular weight of 1213, Emil Fischer decided not to proceed with the synthesis of yet higher homologs. He rationalized his decision as follows (Fischer, 1907, quoted by Fruton, 1979):

"If one imagines the replacement of the many glycine residues by other amino acids, such as phenylalanine, tyrosine, cystine, glutamic acid, etc. one would soon attain two to three times the molecular weight, hence values assumed for some natural proteins. For other natural proteins the estimates are much higher, up to 12 000–15 000. But in my opinion these numbers are based on very insecure assumptions, since we do not have the slightest guarantee that the natural proteins are homogeneous substances. From the experience thus far, I do not doubt that the synthesis can be continued by means of the same methods beyond the octadecapeptide. I must however provisionally waive such experiments, which are not only very laborious but also very expensive".

Let me note also in this context, of taking one's last bow before leaving the scene of a great scientific success, that Emil Fischer had published in 1906 a book on his work with amino acids, polypeptides and proteins (Fischer, 1906b).

Emil Fischer thus capitalized, in this incredibly short time for such a productive period (1899-1908), on the inferior status of physiological chemistry as a discipline which he had intellectually conquered with his earlier work on the carbohydrates; on the slow accumulation of knowledge about the amino acids present in proteins (Table VI) which just awaited someone like him to grasp it as a whole and use it as a firm foundation; on his design of a simple albeit powerful analytical technique; and likewise on his design of both simple and powerful synthetic methods.

Yet, and even though Emil Fischer had a clear notion of the peptide bond, it was too early for him in the early 1900s to accept the idea of proteins as polypeptides of enormous (macromolecular) size (see Chapter 1, section 4). One should note in this respect that Emil Fischer was also responsible for the conservative notion of the impossibility for organic molecules to have molecular weight beyond approx. 5000 - a dogma which Staudinger had to put aside in 1920 (and which was not collectively challenged till the epochal meeting of the Society of German Naturalists and Physicians in Düsseldorf, in 1926). Another mental reservation of Emil Fischer is worth mentioning. Fischer questioned peptide bonds as being 'the only mode of linkage in the protein molecule. On the contrary,' he writes in the same lecture already referred to (Fischer, 1906a) 'I consider it to be quite probable that on the one hand it contains piperazine rings, whose facile cleavage by alkali and reformation from the dipeptides or their esters I have observed so frequently with the artificial products, and on the other hand the numerous hydroxyls of the oxyamino acids are by no means inert groups in the protein molecule. The latter could be transformed by anhydride formation to esters or ether groups, and the variety would increase further if poly-oxy-amino acids are assumed to be probable protein constituents'.

4. Vindication of the polypeptide theory

History of science is made of seemingly endless standstills punctuated with sudden, dizzying accelerations. We have witnessed the slow progress throughout the whole of the 19th century in the characterization of the amino acids present in proteins. Approximately 100 years were spent identifying the 20 or so amino acids found in proteins. After this lengthy analytical period, and as a counterpoint to it, we have just admired the breath-taking speed of Emil Fischer who, in less than 7 years, managed not only to discover three new natural amino acids but also to provide efficient syntheses of oligopeptides of quite impressive size. The long, drawn-out period of analysis had been followed by this burst of synthetic activity. Then what? You have guessed right: the field returned to its dormancy! Nearly half a century of relatively little activity followed. Till 1943, when, in a seemingly nice Hegelian dialectical scheme, overturning again the ruling paradigm, Frederick Sanger started analytical work on the determination of the sequence of amino acids in a protein, insulin.

One may ask: why such a delay, between the tantalizing oligopeptide syntheses of Emil Fischer in the early 1900s and the first establishment of a sequence, completed by Sanger in the mid-1950s? Before I venture to answer this new question, let me try to capture the atmosphere in the early 1950s. It was easily summarized: the peptide theory of protein structure was at that time an attractive working hypothesis. Nothing more, and nothing less. *A tout seigneur, tout honneur*: let us turn to Sanger himself who declared in 1952 (Phillips, 1978):

"As an initial working hypothesis it will be assumed that the peptide theory is valid, in other words, that a protein molecule is built up only of chains of α-amino (and α-imino) acids bound together by peptide bonds between their α-amino and α-carboxyl groups. While this peptide theory is almost certainly valid, it should be remembered that it is still a hypothesis and has not been definitely proved. Probably the best evidence in support of it is that since its enunciation in 1902 no facts have been found to contradict it".

When E.T. Edsall summed up the Royal Society Discussion Meeting convened on May 1, 1952, about the structure of proteins, he asked this question (Phillips, 1978):

"Is it reasonable to hope that the endless variety of proteins found in nature, and their extraordinarily diverse and specific interactions with one another and with other substances, can be explained on the basis of a few relatively rigid general patterns, simply by varying the nature and sequence of the side-chains attached to the fundamental repeating pattern?'.

This question partly answers the question of the half-a-century delay between the Fischer breakthrough and the Sanger milestone achievement: the deceptive simplicity of the polypeptide theory may have appeared to many as incongruent with the profusion and diversity of proteins. Another reason, besides the almost mandatory pause and regrouping after Emil Fischer had set a new paradigm with his peptide-bond hypothesis, was very simply the enormous difficulty of the work ahead which we should not, from our vantage position of hindsight, underestimate. Even today there remains within organic chemistry a sort of a 'no man's land': molecules with molecular weights in between 500 and 5000 are unfamiliar and unwieldy, many techniques applicable either to small or to large molecular objects do not extend to this intermediate range – which had to be spanned in order to establish continuity between the molecular concepts of organic chemistry and the molecular concepts of biochemistry. That Frederick Sanger took approximately 12 years to unravel the sequence of the amino acids in insulin gives an idea of the daunting task he set himself when he started work on this problem in 1943.

Why did he choose insulin as his target? Because its amino acid composition was known accurately, was rather simple (trytophan and methionine were entirely missing), and because of a high content of free α-amino group (Chibnall, 1942). Insulin was even known (Jensen and Evans, 1935) to possess a terminal phenylalanine: 'at that time this was the only case where the position of an amino acid in a protein was known' (Sanger, 1958).

Perhaps this singularity also led Sanger to conceive the general technique he used to decipher the amino acid sequence in insulin. Consider a peptide fragment, and label it with a dye marker affixed exclusively on the N-terminus. Then subject the peptide to acid hydrolysis, or to cleavage by proteolytic enzymes, and repeat the labeling procedure: there should be enough redundancy in a set of such experiments to reconstruct from their results the desired sequence.

Sanger chose a yellow-colored marker, 1,2,4-fluorodinitroben-zene, whose reaction with free amino groups of a peptide or protein gives dinitrophenyl (DNP) derivatives. These can be separated from the unreacted amino acids by ether extraction, fractionated with the powerful method of partition chromatography on paper which had just been introduced (Gordon et al., 1943), identified by comparison with synthetic DNP derivatives, and quantitated by colorimetry.

In this manner, Sanger could identify first the end groups of insulin (N-terminal) as phenylalanine and glycine. Cleavage of the disulfide bridges with performic acid gave him two chains, the A chain with 21 residues and a Gly terminus, the B chain with 30 residues and a Phe terminus. The A chain had four Cys and the B chain had two Cys residues.

Then it became a matter of patient, accurate experimental analysis, in order to produce peptide fragments from which to recompose the actual sequences of the A and of the B chains, as illustrated in Table VIII for a partial acid hydrolysate of fraction B (Sanger, 1958). In this manner, Sanger could identify (to continue with the example of the B fraction) the following five sub-structures:

(1) Phe.Val.Asp.Glu.His.Leu.CySO$_3$H.Gly

(2) Gly.Glu.Arg.Gly

(3) Thr.Pro.Lys.Ala

(4) Tyr.Leu.Val.CySO$_3$H.Gly

(5) Ser.His.Leu.Val.Glu.Ala

(partial sequence 1 is the N-terminal). Taken together, these account for all but four of the amino acid residues of fraction B. This information from the acid hydrolysates was then supplemented with information from the enzymatic hydrolysates (using proteolytic

TABLE VIII

Cysteic acid peptides identified in a partial acid hydrolysate of fraction B (Sanger, 1958)[a]

	CySO$_3$H.Gly	CySO$_3$H.Gly
	Val.CySO$_3$H	
		Leu.CySO$_3$H
	Val.(CySO$_3$H, Gly)	
		Leu.(CySO$_3$H,Gly)
	Leu.(Val,CySO$_3$H)	
	Leu.(Val,CySO$_3$H,Gly)	
Sequences deduced:	Leu.Val.CySO$_3$H.Gly	Leu.CySO$_3$H.Gly

[a]The inclusion of residues in brackets indicates that their relative order is not known.

enzymes such as pepsin, trypsin or chymotrypsin). In this manner Sanger and his group could establish the sequence of chain B as: Phe- Val- Asp- Glu- His- Leu- CySO$_3$H- Gly- Ser- His- Leu- Val- Glu- Ala- Leu- Tyr- Leu- Val- CySO$_3$H- Gly- Glu- Arg- Gly- Phe- Phe- Tyr- Thr- Pro- Lys- Ala. Similar methods led to identification of the sequence for the A chain as: Gly- Ile- Val- Glu- Glu- CySO$_3$H- CySO$_3$H- Ala- Ser- Val- CySO$_3$H- Ser- Leu- Tyr- Glu- Leu- Glu- Asp- Tyr- CySO$_3$H- Asp.

Sanger and his collaborators then established the position of the three disulfide bridges, which gave them the complete structure of insulin. The work was completed in 1955 (Sanger et al., 1955; Ryle and Sanger, 1955; Ryle et al., 1955).

Besides the awesome experimental work of peptide fragmentation, labeling, separation, identification, and the patient build-up, piece by piece as in a jigsaw puzzle, of the insulin sequence, Sanger also had to circumvent a conceptual difficulty which could have crippled the work. When he started, insulin was (erroneously as it turned out) believed to have a molecular weight of approx. 12 000. And this was the accepted value during most of his work. Only when he was looking for the position of the disulfide bridges, the correct molecular weight of ca. 6000 was established (Harfenist and Craig, 1952).

Let me emphasize what I wrote above, prior to this extremely concise and simplified summary of the insulin sequencing by Sanger: in the early 1950s, the polypeptide theory of protein structure, as it had been formulated by Emil Fischer in 1902, was by no means fully accepted. Rival theories flourished. When Sanger started his work, 'the most widely discussed theory was that of Bergmann and Niemann who suggested that the amino acids were arranged in a periodic fashion, the residues of one type of amino acid occurring at regular intervals along the chain. On the other extreme there were those who suggested that a pure protein was not a chemical individual in the classical sense but consisted of a random mixture of similar individuals' (Sanger, 1958). Hence, Sanger's contribution was crucial, because it clinched the case conclusively and demonstrated cleanly the polypeptide structure of proteins.

From an operational viewpoint, obviously Sanger's work was dependent upon the new technique of paper chromatography. The coincidence is far from being accidental: as soon as Gordon, Martin and Synge reported their invention in 1943, Sanger elected it for separation of the amino acids (and their DNP derivatives) from insulin hydrolysates.

The point has been made that, in the development of paper chromatography, 'necessity, in the guise of zero budget, (was) the mother of invention. (. . .) When during World War II, the budgets for biochemical research in Great Britain diminished markedly, the scientists developed and exploited paper chromatography, which proved to be a very powerful but remarkably inexpensive research technique' (Stetten, 1980). Of course, such a statement cannot be taken at face value (and I am not being defensive here merely as an active scientist arguing for the largest possible budgets from granting agencies!). It has to be complemented with these other logical propositions: (i) not all research run on a shoestring gives rise to cute little (or large) inventions; far from it; (ii) some research projects with very large money allocations lead to important inventions; (iii) not all research projects with considerable funding strike pay-dirt!

All these truisms – common experience is the justification for theorem (i); the Manhattan project is an example of (ii); and the failure of the large task force, assembled during World War II in American university and industrial laboratories for the synthesis of penicillin, is an example for (*iii*) – are reduced to a single rule for research funding: 'there is no rule for achieving the most efficient research funding'. Enough for this ironical aside.

To conclude this section on the polypeptidic nature of proteins, the analyses performed by Emil Fischer and by Frederick Sanger shared a joint methodology: these two scientists – whose accomplishments complement so nicely – evolved techniques for the characterization of each individual amino acid through derivatization, formation of alkyl esters on the carboxyl end, formation of 2,4-dinitrophenylamines on the amino end, respectively.

REFERENCES

Abderhalden, E. (1911) Biochemisches Handlexikon, Berlin, p. iv.
Best, C.H., Dole, H.H., Dudley, H.W. and Thorpe, W.W. (1927) J. Physiol., 62, 397.
Bopp, F. (1849) Annals, 69, 16.
Braconnot, H. (1820) Ann. Chim. Phys., 13, 113.
Chibnall, A.C. (1942) Proc. Roy. Soc. London, B131, 136.
Cohn, E.J. (1943) Proteins, amino acids and peptides, in Proteins, Amino Acids and
    Peptides as Ions and Dipolar Ions (Cohn, E.J. and Edsall, J.T., Eds.), ACS Mono-
    graph Series, Reinhold, New York, Ch. 15, pp. 338–369.
Conversations on Chemistry in which the Elements of that Science are Familiarly
    Explained and Illustrated by Experiments (1813) Longman, Hurst, Rees, Orme
    and Brown, London, 4th ed. (2 Vols.), Vol. 2, p. 271.
Cramer, E. (1865) J. Prakt. Chem., 96, 76.
Curtius, T. (1882) J. Prakt. Chem., 24, 239–240.
Drechsel, E. (1889) J. Prakt. Chem.N.F., 39, 425.
Ehrlich, F. (1903) Z. Ber. Deut. Zucker Ind., 53, 809.
Fieser, L.F. and Fieser, M. (1956) Organic Chemistry, Reinhold, New York, 3rd ed.,
    pp. 418–420.
Fischer, E. (1901) Z. Physiol. Chem., 33, 151.
Fischer, E. (1902a) Ber. Chem. Ges., 35, 2660.
Fischer, E. (1902b) Syntheses in the purine and sugar group. Nobel Lecture, Decem-
    ber 12, 1902, in Nobel Lectures. Chemistry. 1901–1921, Elsevier, Amsterdam, pp.
    15–39.
Fischer, E. (1906a) Ber. Chem. Ges., 39, 530–610.
Fischer, E. (1906b) Untersuchungen über Aminosaüren, Polypeptide und Proteine,
    Springer, Berlin.
Fischer, E. (1907) Ber. Chem. Ges., 40, 1754–1767.
Florkin, M. (1972) A History of Biochemistry. Part I. Proto-Biochemistry, Part II.
    From Proto-Biochemistry to Biochemistry, in Comprehensive Biochemistry
    (Florkin, M. and Stotz, E.H., Eds.), Elsevier, Amsterdam, vol. 30, p. 285.
Fruton, J.S. (1979) Ann. N.Y. Acad. Sci., 325, 1–18.
Gordon, A.H., Martin, A.J.P. and Synge, R.L.M. (1943) Biochem. J., 37, 79.
Harfenist, E.J. and Craig, L.C. (1952) J. Am. Chem. Soc., 74, 3087.
Hedin, S.G. (1895) Z. Physiol. Chem., 20, 186.
Hedin, S.G. (1896) Z. Physiol. Chem., 22, 191.
Hopkins, F.G. and Cole, S.W. (1901) J. Physiol., 27, 418.
Jensen, H. and Evans, E.A. (1935) J. Biol. Chem., 108, 1.
Knoop, F. and Windaus, A. (1904) Beitr. Chem. Physiol. Pathol., 7, 144.
Kohler, R.E. (1982) From Medical Chemistry to Biochemistry. The Making of a
    Biomedical Discipline, Cambridge University Press, Cambridge.
Kossel, A. (1896) Z. Physiol. Chem., 22, 176.

Liebig, J. von (1846) Annals, 57, 127.

Meyer, C.E. and Rose, W.C. (1936) J. Biol. Chem., 115, 721.

Mörner, K.A.H. (1899) Z. Physiol. Chem., 28, 595.

Mueller, J.H. (1929) Proc. Soc. Exp. Biol. Med., 19, 161.

Neumeister, R.E. (1982) Lehrbuch der Physiologischen Chemie, Fischer, Jena, p. 334.

Pauly, H. (1904) Z. Physiol. Chem., 42, 508.

Phillips, D.C. (1978) Molecular Biology 1952-1977, in Highlights of British Science, Royal Society, London, pp. 193-202.

Plimmer, R.H.A. (1908) The Chemical Constituents of the Proteins, London.

Proust, M. (1819) Ann. Chim. Phys., 10, 29.

Ritthausen, H. (1866) J. Prakt. Chem., 99, 4, 454.

Ritthausen, H. (1868) J. Prakt. Chem., 103, 233.

Ryle, A.P. and Sanger, F. (1955) Biochem. J., 60, 535.

Ryle, A.P., Sanger, F., Smith, L.F. and Kitai, R. (1955) Biochem. J., 60, 541.

Sanger, F. (1958) The Chemistry of Insulin. Nobel Lecture, December 11, 1958, in Nobel Lectures. Chemistry. 1942-1962, Elsevier, Amsterdam, pp. 539-556.

Sanger, F., Thompson, E.O.P. and Kitai, R. (1955) Biochem. J., 59, 509.

Schmidt, C.L.A. (1938) The Chemistry of the Proteins, Thomas, Springfield, IL.

Schulze, E. and Barbieri, J. (1881) Ber. Chem. Ges., 14, 1785.

Schützenberger, P. and Bourgeois, A. (1875) C. R. Acad. Sci., Paris 81, 1191.

Stetten Jr., D.W. (1980) Perspectives in Biology and Medicine, 357.

Vickery, H.B. (1941) Ann. N.Y. Acad. Sci., 41, 87.

Vickery, H.B. and Schmidt, C.L.A. (1931) Chem. Rev., 9, 169.

Weyl, T. (1888) Ber. Chem. Ges., 21, 1407.

Windaus, A. and Vogt, W. (1907) Ber. Dtsch. Chem. Ges., 40, 3691.

Chapter 13

Dorothy Wrinch: the Mystique of Cyclol Theory or the Story of a Mistaken Scientific Theory

We turn now into a blind alley, Dorothy Wrinch's cyclol theory of protein structure. One might discard it – and more often than not, it has been discarded in historical accounts – with the implicit dictum: 'It is a capital mistake to theorize before one has data' (Conan Doyle, 1891).

However, there is a didactic value of error – if not for individuals, certainly when studying the history of science. Immersion in the actual ebb and flow of science in the making, watching half-baked theories rise buoyed only by incomplete or inaccurate data, is a healthy safeguard against the heresy of Whig historiography. This (it will be recalled) is the writing of the history of any science as that of its blossoming; as a fairy-tale story jumping from success to success, all these landmarks on the road to full understanding as finally achieved, at long last, in our privileged time.

Moreover, before being recognized as mistaken from hindsight and eliminated from consideration, any scientific interpretation drew a measure of support during some period. In Dorothy Wrinch's case, a substantial segment of the scientific community came, however briefly, under its spell. That such a distinguished scientist as Irving Langmuir lent his prestige to her theory of protein structure is an eloquent testimony to the attention it received for a while.

The study of scientific errors from the past is indeed worthwhile. It can be an incisive tool for resurrecting distant conceptual strata, for reconstructing the complex sets of attitudes, of beliefs, what might be termed as the dominant paradigms active at various times in the past.

There is yet a third potential benefit from focussing the historian's attention, not only on the royal road of success as we can chart it from our privileged position, but also on the other avenues which Father Time, by touching them with his sickle, made into by-roads: they tell us how to choose the proper vehicle for traveling along the main road. In the case at hand, by studying first the admittedly erroneous theory of protein structure imagined by Dorothy Wrinch, we shall become better equipped to report the essentially exact theory of protein structure devised by Linus Pauling.

Accordingly, this chapter and Chapter 15 should be read as companion chapters (somewhat in the manner of *Parallel Lives* as written by an author of Antiquity).

1. To be young, gifted and a woman

I will not sketch here a biography of Dorothy Wrinch, but only give a few indications on her remarkable personality to help us understand her attack on the structure of proteins during the second half of the 30s. Dorothy Wrinch was born (in 1895) and educated in Britain. She was a clever young woman of her times, and she became a mathematician. From her early twenties, she started contributing to mathematical journals. A second strong interest was philosophy, she wrote articles in such renowned periodicals as *Mind*. Thirdly (and not least), she was a militant emancipated young woman and feminist. Dorothy Wrinch even published, under a pseudonym (Jean Ayling), a little pamphlet entitled *The retreat from parenthood* (Ayling, 1940) in which she advocates the collective upbringing of children (her ideas are somewhat similar to those made familiar to us by the kibbutz in Israel). These three elements are, of course, extremely close to the trail brilliantly blazed by Bertrand Russell; and there is little doubt that Dorothy Wrinch was inspired by his example.

Dorothy Wrinch was a student of Russell's and she was a friend of his second wife, Dora Black (they were both Fellows of Girton College): it was Dorothy Wrinch who introduced them to one another in 1919. Dorothy Wrinch was intellectually extremely close to Russell. For instance, when the genius philosopher Ludwig Wittgenstein sent Bertrand Russell the manuscript of what was to become the *Tractatus Logico-Philosophicus*, he discussed it both with the young French philosopher Jean Nicod and with Dorothy Wrinch (Russell, 1968). Not only did the young Dorothy Wrinch lead the life of an academic scholar, she also escaped repeatedly to the big city, to London. There is a letter to Bertrand Russell from his brother Frank exclaiming, in early 1921: 'I have not seen the elusive little Wrinch again although she seems to spend as much time in London as at Girton. I did not know a don had so much freedom of movement in term time' (Russell, 1968).

Apart from the influence of Bertrand Russell, it should be pointed out that Dorothy Wrinch belonged to the same generation as other young women intellectuals who, in the 1920s and 1930s, with considerable stamina and a measure of courage went into careers of their own. Dorothy Wrinch was a contemporary of Rebecca West, she was younger than Virginia Woolf (by 13 years) and older than Simone Weil (by 14 years).

What is striking about Dorothy Wrinch is, not only that she had the mathematician's clarity of thought, with well-marked distinctions and fully logical argumentation; she added to it the breadth of her interests, running from mathematics to biology, and also including philosophy. Furthermore, she was extremely well-read. For instance, in her elegant paper of 1937 in the *Proceedings of the Royal Society* (Wrinch, 1937b) she has this footnote: 'I have found this deduction from Euler's theorem in the edition of Legendre's *Eléments de Géométrie* published in 1809, and it may well have been made long before that date'.

Dorothy Wrinch came to be interested in the major problems of biology as a consequence of a tragedy in her personal life (Crowfoot-Hodgkin, 1984):

"(Dorothy Wrinch) had come to Oxford on her marriage to John Nicholson, a very brilliant physicist, Fellow of Balliol. As students we greatly admired Dorothy Wrinch Nicholson – she was the first woman to take a D.Sc. at Oxford and she had talked to Einstein. (. . .) When I went to Cambridge (1932) I began to know her as a friend both of J.D. Bernal and of Margery Fry, ex-principal of Somerville, then in London. It was through Margery I began to know Dorothy Wrinch as a troubled person; Nicholson had become mentally unbalanced and removed to the Warneford. Dorothy felt she wanted to change her life and what she was working on and particularly to understand biology. She courageously set out to take elementary biology courses at Oxford and elsewhere - also chemistry. At the same time she was drawn into speculative biology discussions - the [J.H.] Woodger group - Bernal, the Needhams, Piries, etc. also took part. Speculation about linear sequences of amino acids or nucleic acids being involved in the genetic process were common to them - some are preserved in Bernal's file at Cambridge".

Thus, Wrinch did not become interested in the structure of proteins all of a sudden, but by way of, rather remarkably, the problem of the structure of nucleic acids. She addressed it in a letter to *Nature* (Wrinch, 1934). Her starting point was the belief, commonly held at the time, that 'chromosomes consist substantially of aggregates of protein molecules in association with nucleic acid'. Hence, she thought that the 'specificity of a given chromosome may be regarded as an expression of its particular protein pattern'. With the polypeptide chain written as:

...-NH-CO-CP-NH-CO-CQ-NH-CO-CR-NH-...

Wrinch postulated that 'the specification of a chromosome will be in terms of the side chains ... *P Q R* belonging to molecules consisting of certain numbers of amino acid residues' (Wrinch, 1934).

This note about *Chromosome Behaviour in Terms of Protein Pattern* already shows some arresting features which would remain characteristic of Dorothy Wrinch's scientific style: she saw that the 'fundamental approach to the problem is clearly the study of the molecular structure of chromosomes by X-rays'. And in the same breath, she wrote of her own irresistible temptation to guess at the answer: 'Pending the necessary technical developments we must pursue our inquiries inductively'. Her whole work on proteins, from 1936 on, would be conducted under the same constraints and, ultimately, limitations.

We have an idea of the reception of her 1934 paper on the structure of chromosomes from a letter of Michael Polanyi to her (Polanyi, 1935), in which he wrote:

"Dear Dorothy,
May I try to sum up the position as I think people see it here now, with regard to your work.
1. That the chromosome must have a fibre structure since the organism developed from it is fibrous is certainly a fundamental and true idea.
2. To connect this fibre structure with the genetic functions is an equally good idea.
3. To attribute to the side chains [of the aggregated proteins] the genetic identity of the chromosomes is also a brilliant idea, but one cannot feel just as sure that it is true as with respect to 1. and 2.
4. To take some sort of longitudinal structure linked up by some sort of annular structure is obviously a trustworthy leading idea; though tentative to some extent.
(...)
6. Swelling as the basis of the splitting of chromosomes is certainly something to hold on to; but remember the various models of amoeboid motion (Rumbler) based on surface tension which never became useful.
(...)
Speaking as a distant relative of the gypsies, I might add, as a piece of fortune-telling, that you will live to great recognition of your vision".

I have quoted extensively from this letter because it gives an impression of the reception of Wrinch's ideas: her fellow-scientists were impressed with the cleverness and insight but were not totally convinced. Even when they shared the speculations, 'the scientists tended to treat them as more speculative and less publishable than Dorothy did' (Crowfoot-Hodgkin, 1984). Let us now move on to Wrinch's work on protein structure, which started being published in early 1936. It was for her the high point of her scientific career.

2. The protein research problem in 1936

We come now to a description of the field Dorothy Wrinch entered brilliantly in the second half of the 1930s. What was known about proteins then?

The polypeptide theory of Emil Fischer was held to be probable; but, as detailed in the previous section, it was yet unproven. The notion of proteins as macromolecules, the concept for which Staudinger was responsible, was also accepted; there were a number of molecular weight determinations, the most reliable of which came from the work of The Svedberg with the ultracentrifuge (see Chapter 2), which made this clear. Another contribution by The Svedberg was the demonstration of proteins as isodisperse biopolymers.

Then there was a tantalizing bit of evidence. The first x-ray pictures on protein crystals had been taken by Dorothy Crowfoot and J.D. Bernal on pepsin in 1934 (Bernal and Crowfoot, 1934). These showed that proteins had highly ordered structures. Furthermore, while the pepsin crystals showed hexagonal symmetry, the insulin crystals which Dorothy Crowfoot and J.D. Bernal started examining in 1935 had trigonal symmetry. Dorothy Wrinch was well aware of this feature from her informal contacts with both Bernal and Crowfoot (Hodgkin and Riley, 1968).

To complete the description of the protein structure problem as it stood in 1936 when Dorothy Wrinch proposed her cyclol theory, another three questions which were very important at the time must be mentioned.

(*i*) The folding of globular proteins: because of the overall shape of these molecular objects, as indicated by hydrodynamic measurements, folding of the polypeptide chain was necessary. Which were the forces responsible for folding? How did the chain arrange itself? In the words of Dorothy Crowfoot-Hodgkin (1984), 'we all speculated on protein structures - particularly that to go to three-dimensional structures from two-dimensional chains required a further link'.

(*ii*) pH-dependent changes: the molecular weight of proteins had been observed to depend upon pH and ionic strength. There was a possibility that cluster formation - protein oligomerization - accounted for the observations. Hence, what was the nature of the groups on the protein surface capable of entering into intermolecular interactions?

(*iii*) The Svedberg had obtained a not inconsiderable body of data on the molecular weights of various proteins. The molecular weight, far from being random, seemed in 1935-1936 to run in multiples of a basic ca. 35 000 unit. (To be fair, this last result was transformed into a 17 500 basic unit in 1938-1939, after Dorothy Wrinch had published her theory of protein structure.) What was the explanation for these observations?

Hence, we find that the letter to *Nature* which Dorothy Wrinch sent from the Mathematical Institute in Oxford, where she was still working in early 1936, starts with this admirable summary of the extant knowledge about the structure of proteins (Wrinch, 1936a):

"(1) The molecules are largely, if not entirely, made up of amino acid residues. They contain -NH-CO linkages, but in general few -NH$_2$ groups not belonging to side chains, and in some cases possibly none.

(2) There is a general uniformity among proteins of widely different chemical constitution which suggests a simple general plan in the arrangement of the amino acid residues, characteristic of proteins in general. Protein crystals possess high, general trigonal, symmetry.

(3) Many native proteins are 'globular' in form.

(4) A number of proteins (Svedberg, 1934) of widely different chemical constitution, though isodisperse in solution for a certain range of values of pH, split up into molecules of submultiple molecular weights in a sufficiently alkaline medium".

3. Nature of the secondary bonds

After she had thus reviewed the properties of proteins, Dorothy Wrinch went on to state the structural characteristics which therefore could be presumed for proteins (Wrinch, 1936a):

"The facts cited suggest that native protein may contain closed, as opposed to open, polypeptides, that the polypeptides, open or closed, are in a folded state, and that the type of folding must be such as to imply the possibility of regular and orderly arrangements of hundreds of residues".

I find this a neat, handsome, and percipient statement of the sort of structure which, at the time, several scientists were postulating for proteins.

The next logical step – the one which turned out to be flawed – consisted in the identification of the secondary forces or bonds responsible for the folding of the polypeptide chain. This missing link was supplied to Dorothy Wrinch by none other than J.D. Bernal (Wrinch, 1936a). He recalled that, during the discussion at a lecture given by W.T. Astbury to the Oxford Junior Scientific Society in 1933, Dr. F.C. Frank had suggested condensation of amino acid residues according to:

$$\text{\large$\diagdown$C} = \text{O} + \text{H} - \text{N}\diagdown \quad \rightleftharpoons \quad \text{\large$\diagdown$C}\diagup^{\text{OH}}_{\diagdown\text{N}\diagdown}$$

This is a superficially plausible tautomeric equilibrium, arising from nucleophilic addition of the nitrogen lone pair to the carbonyl. This addition is analogous to formation of the cyclic forms of sugars as hemiacetals by addition of the hydroxy group from an alcohol function to the carbonyl group of an aldehyde. Dorothy Wrinch, as soon as she heard this possibility mentioned by J.D. Bernal, started drawing possible protein structures embodying these secondary bonds between amino acid residues.

At this stage, in 1936, when she embarked upon her speculative search for protein structure, Wrinch considered also another possibility for the secondary interactions between amino acids responsible for folding of the polypeptide chain: hydrogen bonds. She was seemingly led to doing so by the appearance of the Mirsky and Pauling paper (Mirsky and Pauling, 1936). She immediately sent off a letter to *Nature* (Wrinch and Lloyd, 1936) whose gist is that: (1) the suggestion by Mirsky and Pauling had little originality and had already been made by many others, especially by Jordan Lloyd (Lloyd, 1932; Lloyd and Marriott, 1933) – a jibe which surely did not endear her to Linus Pauling! – and (2) that hydrogen bonds are very closely related to the lactam-lactim tautomerism she had earlier proposed:

$$\text{\large$\diagdown$C} = \text{O} \ldots \text{H} - \text{N}\diagdown \quad \rightleftharpoons \quad \text{\large$\diagdown$C(OH)} - \text{N}\diagdown$$

In her words (Wrinch and Lloyd, 1936): 'In putting forward the two postulates [the lactam-lactim tautomerism, and hydrogen bonding[we would stress the fact that there is no conflict between them. Indeed the formation of a hydrogen bond may be an intermediate step in the lactam-lactim transformation'.

4. Model building

As soon as she addressed a biological problem having to do (she believed) with protein structure, Wrinch opted for model building. This is what she wrote in her paper on chromosome structure (Wrinch, 1934):

" . . . a molecular model – however inadequate – must be constructed".

This was what she set out to do as soon as it dawned upon her that the lactam-lactim tautomerism, 'a simple *working hypothesis* for which no finality is claimed' (Wrinch, 1936a), could be responsible for the folding of the polypeptide chain of a protein. Dorothy Wrinch had learned from Dorothy Crowfoot how to build models (Crowfoot-Hodgkin, 1984):

"I returned to Oxford to teach chemistry myself in 1934 – and fit in research as best as I could. Dorothy Wrinch welcomed me with red roses in my room on arrival and an invitation to come round any evening I liked to discuss biochemical problems. I used to drop in quite often – to talk over different things that I was working on and advise her how to begin making models. I made myself some at Cambridge in vacation to bring back to Oxford and Dorothy had them copied for her use. Actually model building became usual at this time as we began to know more exact interatomic distances. I sometimes think I must have made more wrong models myself than anyone else".

Being thus coached into model building, nevertheless Dorothy Wrinch went on, as the good mathematician she was, to make the distinction between a topology and a metric:

"If and when the actual values of these angles and distances are known, these structures will presumably lend themselves to modification by a uniform or systematic deformation which leaves the topology unchanged. As will be seen (. . .), a considerable part of the argument is concerned wholly with topological considerations and so is independent of any metrical data" (Wrinch, 1937b).

For this topology, Wrinch hit upon hexagonal arrangements – which she termed *cyclols* – for the polypeptide chain. A very simple arrangement is the 'cyclol 6' depicted below:

As she wrote:

"Closed polypeptide chains consisting of 2, 6, 18, 42, 66, 90, 114, 138, 162 . . . (18 +
24n) . . . residues form a series with three-fold central symmetry. (. . .) The hexagonal folding of polypeptide chains, open or closed, evidently allows the construction of
molecules containing even hundreds of amino acid residues in orderly array, and
provides a characteristic pattern, which in its simplicity and uniformity agrees with
many facts of protein chemistry" (Wrinch, 1936a).

In this same paper, Wrinch considers briefly, to dismiss almost
immediately, other families of cyclols with twofold, and with sixfold
central symmetry. And she goes on, describing tesselation (= full
coverage) of a plane by cyclols with threefold central symmetry,
recognizing that these form a sheet 'the thickness of which is one
amino acid residue. *Since all naturally occurring amino acids are of
laevo type* (Lloyd, 1932) this fabric is dorsiventral, having a front
surface from which the side chains emerge, and a back surface free
from side chains. Both front and back can carry trios of hydroxyls
normal to the surface in alternating hexagonal arrays' (Wrinch,
1936a). And Wrinch feels that this may well be the right structure
for a protein, on two counts, because of an analogy with inanimate
materials and because of a well-established feature:

"Linkage by means of hydroxyls recalls the structures of graphitic oxide and mont-
morillonite, etc. Such a structure suggests a considerable capacity for hydration, an
outstanding characteristic of many proteins" (Wrinch, 1936a).

The testimony of Dorothy Crowfoot is again extremely valuable to
the historian (Crowfoot-Hodgkin, 1984):

"Dorothy's first cyclol model looked very nice – hexagonal sheets – till I realised it was built of alternate d and l residues. We rebuilt it together: when she changed it to all l, I ceased to like it any more – the residues were too crowded on the surface, though the fact they were all on one side of it had certain advantages. At that time, I had just made the first measurements on insulin. It was natural Dorothy should try to tailor her molecule to fit my data. She suggested a three layer structure in essentially cylindrical close packing. The size relations were plausible but the very limited data I had, showed no evidence of a layer structure".

Another difficulty with the cyclol structure arose nearly immediately. The *Nature* issue of August 8, 1936 carried two back-to-back communications on the thermochemical problem with cyclols, the first by D.M. Wrinch (Wrinch, 1936b) and the second by her colleague from the Engineering Laboratory at Oxford, F.C. Frank (Frank, 1936). Dr. Wrinch used the following bond energies which she had taken essentially from Pauling (Pauling and Sherman, 1933):

$$(kcal.mol^{-1}) \quad \begin{array}{llll} C - O & 82.8 & C = O & 177.8 \\ C - N & 61.5 & C = N & 119.6 \\ O - H & 109.5 & N - H & 83.3 \end{array}$$

With these, she calculated the energy balance for the two equilibria:

$$NH \quad O = C \; \rightleftarrows \; N - (HO)C \quad (1)$$

and

$$N \quad HO - C \rightleftarrows N - (HO)C \quad (2)$$

and found that the 'formation of cyclol 6 from a closed polypeptide consisting of six residues requires 21.9 kcal if the polypeptides are in amide form throughout; if the three links concerned are already in the imide form, energy amounting to 10.3 kcal is emitted' (Wrinch, 1936b). Thus, this back-of-the envelope calculation made cyclol formation from a genuine polypeptidic chain *endothermic*. However, Dr. Wrinch was not too concerned: firstly, she had not taken into account resonance which, she notes, 'amounts to 20 kcal for acetamide' – a grievous oversight, as it turned out – and the chain might consist of amino acid residues in the imide rather than in the amide form. Another way out was pointed by Frank (1936); very lucidly, he wrote in his companion letter that 'the "cyclol" molecule in itself offers no chance of constructing a resonant system', whereas equilibria (1) and (2) above are endothermic by about 27 and 10 kcal.mol^{-1}, respectively due to resonance in the lactam form on the left of eqn. (1) and (2): 'though these are not free energies, and in any event not very accurate', F.C. Frank continued, 'they appear certainly too large to allow the equilibrium to rest close to the right hand of the equation, as the [Wrinch] theory requires, so that we must either abandon the theory or find some compensating source of energy' (Frank, 1936). This extra stabilization was conveniently provided by hydration, especially that of the hydroxyls sticking out from the cyclol surface into the water solution.

Having thus faced squarely the energetics of the situation (and swept them somewhat under the rug), Wrinch proceeded to state the three-dimensional problem (Wrinch, 1937a):

" . . . can a cyclol network be drawn, not on a plane but on the surface of some polyhedron such that the faces at any edge crossed by the network abut at the tetrahedral angle?".

In this manner, cyclols which had previously been considered only with a planar topology could become 'closed (that is, space-enclosing)' (Wrinch, 1937a). This would be obviously a structural feature congenial to the properties of soluble globular proteins. Indeed, when a three-layer cyclol structure failed to account for Crowfoot's data on insulin, Dorothy Crowfoot made this very suggestion to Dorothy Wrinch (Crowfoot-Hodgkin, 1984):

"I suggested she folded her framework on the surface of a solid. She said - impossible - Euler's theorem. I said: but what about the ultramarines? and showed her the picture in Bragg's book. She realised then that the clue was to have different sized holes in the framework and Euler's theorem only referred to networks with one kind of hole".

Let me interject at this point, before we examine Dorothy Wrinch's proposal for the structure of globular proteins, a modest conjecture. It should be prefaced by noting that, as soon as Wrinch had proposed the planar cyclols, she rushed into print with a molecular explanation of cancer (Wrinch, 1936c): she found that the structures of various carcinogens, constituted by polycyclic aromatics such as 1,2-benzpyrene, were nicely superimposable on the cyclol tesselation or mosaic. In the fall of 1936, she crossed the Atlantic and gave lectures on her cyclol theory at Harvard, Yale and George Washington University Medical School and at the Medical Center of Columbia University, in October and November 1936. Probably someone in the audience, during one of these seminars, also called her attention to the inability of her theory to provide for space-enclosing three-dimensional cyclol sheets with closed structures which could account for the structure of globular proteins.

In any case, this is a move for which we know from her testimony that Dorothy Crowfoot is personally responsible (she even feels a measure of guilt for contributing innocently to launch Dorothy Wrinch into her destiny!) (Hodgkin, 1979).

So Dorothy Wrinch the mathematician turned to solving this problem (Wrinch, 1937a):

"To do so, it is necessary to investigate the conditions under which a cyclol fabric can bend about a line. Evidently it is permissible for two abutting median hexagons [such as the central one in the 'cyclol 6' structure represented earlier] to lie on different planes, if the angle between the planes is the tetrahedral angle. (. . .) among the regular and semi-regular polyhedra, only four satisfy the conditions. These are the truncated tetrahedron, the octahedron, the truncated octahedron and the skew triangular prism".

Dorothy Wrinch was especially interested in the truncated tetrahedron. The closed cyclol networks she could fold unto it (Wrinch, 1937a) 'form a linear series C_1, C_2, . . . C_n . . . which comprise 72288, . . . 72 n^2 . . . amino acid residues'. And there came, seemingly, a handsome experimental confirmation for her theory:

'It is found,' she wrote 'that the molecular weights of proteins are not distributed at random, but fall into a sequence of widely separated classes'. Not only were they such discrete clusters of molecular weights, but the actual values seemed to agree with the predictions from her theory (Wrinch, 1937a): 'It is (. . .) suggested for consideration (. . .) that the group of proteins with molecular weights ranging from 33 600 to 40 500 (Svedberg and Eriksson-Quensel, 1935-1936) are closed cyclol molecules of type C_2. (. . .) Svedberg has suggested that very many, possibly most, other proteins have molecular weights which are multiples of (say) 36 000'. And then came at the very end of the paper, as a note added in proof on May 28, 1937 and as the clinching, triumphant piece of evidence which seemed to prove her whole cyclol theory right, this (Wrinch, 1937a):

"Bergmann and Niemann (1937a) deduce from the chemical analysis of egg albumin that this molecule (. . .) consists of exactly 288 residues as predicted by the cyclol hypothesis".

This apparent success must have been the zenith of the cyclol theory of protein structure. To give an idea of how important it seemed to its discoverer, I shall quote from a letter of Dorothy Wrinch to Niels Bohr (Wrinch, 1938):

"I really think that there may be something about the cage structure which does represent the first step from the inorganic to the organic. Evidently in the phenomenon of death there is no instantaneous change of chemical composition. Nevertheless, there is an over-whelming change which does in fact occur. This, not being a change in chemical composition can only be a change in structure and this is, I suppose where the intrinsic importance of cages comes in".

The year 1937 was also marked for Dorothy Wrinch by publication of the elegant and carefully worded presentation of her cyclol theory in relation to the structure of globular proteins which she wrote and which Sir William Bragg presented to the Royal Society on April 22nd, 1937 (Wrinch, 1939). As an illustration I have selected Fig. 2 from that paper (see page 223).

It also includes several photographs, from various angles, of the C_1 protein (72 residues) and of the C_2 protein (288 residues).

A small point is worthy of note, out of loyalty to my institution and to a former colleague, the x-ray crystallographer Henri Brasseur of the University of Liège: Dorothy Wrinch refers there to his work with Rassenfosse on hydration of strontium platinocyanurate (Brasseur and Rassenfosse, 1936) in support of her picture of protein hydration. This reference to an out-of-the way periodical confirms how well-read a scientist Dorothy Wrinch was!

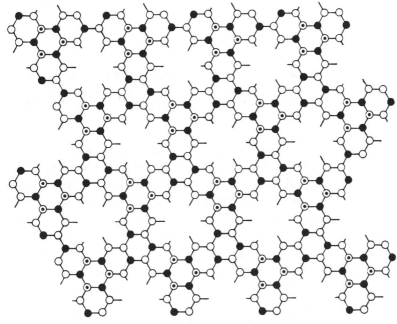

Fig. 6. The cyclol fabric. ●=N. ○=C(OH), peptide hydroxyl upwards. ◉=C(OH), peptide hydroxyl downwards. ○–=CH*R*, direction of side chain initially upwards. The median plane of the lamina is the plane of the paper. The lamina has its 'front' surface above and its back surface below the paper.

5. The collaboration with Irving Langmuir

During the same pivotal year 1937, the prestigious scientist Irving Langmuir started to work and publish together with Dorothy Wrinch (Langmuir et al., 1937). They and Schaefer jointly addressed the question of the structure of insulin (one might re-phrase this statement as: they jointly rushed to grab the structure of insulin). The insulin crystals, it will be recalled, had started being examined by x-ray in 1935 (Hodgkin and Riley, 1968). Dorothy Crowfoot reported a rhomboedral unit cell in 1935 (Crowfoot, 1935).

224   DOROTHY WRINCH - CYCLOL THEORY

In 1938, she was able to publish Patterson–Harker projections (Crowfoot, 1938), which Patterson (1935) and Harker (1936) had recently shown how to calculate:

"The Patterson–Harker maps represent not distributions of matter in space but of vectors erected at a common point or origin, each vector giving in magnitude and direction the distance between two atoms in the original structure" (Wrinch and Langmuir, 1938).

Before Dorothy Crowfoot had had the opportunity or wish to interpret these Patterson–Harker diagrams, Wrinch rushed to provide one: she pointed out that all the 18 peaks on Crowfoot's xy-plane projection were consistent with her C_2 cyclol structure, the one with 288 residues (Wrinch, 1938b, c). Wrinch and Langmuir went on to consider 80 well-defined points, nearly all of them with high intensities, from the Crowfoot diagrams (Crowfoot, 1938). They concluded that 'all the prominent features of the Crowfoot diagram are deducible from the C_2 structure whose octahedron [it was convenient to replace the C_2 truncated tetrahedron with an equivalent regular octahedron having almost exactly identical surface areas and volume] has a side 29.4 Å if a [the mean between the C–C and the C–N bond lengths] = 1.50 Å. During the development of this analysis,' continued Wrinch and Langmuir with dashing rhetoric, 'we repeatedly found that as we introduced one by one the more delicate features the more perfect became the concordance between the Patterson–Harker diagrams and the Crowfoot pictures. We feel that *these X-ray data, in giving so perfect a picture of the C_2 structure, provide the experimental basis for the cyclol theory*' (Wrinch and Langmuir, 1938, emphasis added).

It is no surprise if Irving Langmuir became an eloquent advocate for the cyclol theory (Langmuir, 1939):

"The original idea of native proteins as long chain polymers of amino acid residues, while consistent with the facts relating to the chemical composition of proteins in general, was not a necessary deduction from these facts. Moreover, it is incompatible with the facts of protein crystallography, both classical and modern, with Svedberg's results which show that native proteins have definite molecular weights, and with the high specificity of proteins discovered in immunochemistry and enzyme chemistry.

All these factors seem to demand a highly organized structure for the native proteins, and the assumption that the residues function as two-armed units leading to long-chain structures must be discarded. The cyclol hypothesis introduced the single assumption that the residues function as four-armed units, and its development during the last few years has shown that this single postulate leads by straight mathematical deductions to the idea of a characteristic protein fabric which in itself explains the striking uniformities of skeleton and configuration of all the amino acid molecules obtained by the degradation of proteins. The geometry of the cyclol fabric is such that it can fold round polyhedrally to form closed cage-like structures. These cage molecules explain in one single scheme the existence of megamolecules of definite molecular weights capable of highly specific reactions of crystallizing, and of forming monolayers of very great insolubility. The agreement between the properties of the globular proteins and cyclol structures proposed for them is indeed so striking that it gives an adequate justification for the cyclol theory . . . ".

(As Cardinal Newman wrote (Newman, 1848): '*We can believe what we choose. We are answerable for what we choose to believe*'.)

6. The downfall of the cyclol theory

Even though it has not glossed over them, my brief account has made clear some of the flaws of the cyclol theory of protein structure. To use modern language, the nitrogen lone pair, which is delocalized by resonance to the amide oxygen in a peptide bond is thus made unreactive (unavailable) for nucleophilic attack on the carbonyl group from another residue. Here, the *caveat* by Dr. F.C. Frank was quite justified a posteriori. Besides this thermochemical impossibility, since the cyclol theory demanded the energy to go uphill, there were other difficulties. The cyclol theory implied the presence in a protein of *two kinds* of peptide bonds, normal -NH-CO- groups on the periphery, and tautomerized -N-C(OH) groups on the inside of each cyclol. Hence, there could be a dynamic interchange, an overall reshuffling of the structure which would run counter to the observations by x-rays of highly ordered protein structures. Dorothy Wrinch also ignored some of the basics of structural chemistry which, of course, were quite familiar to Linus Pauling: the planarity of all amide groups due to resonance had been discussed by Linus Pauling as early as 1932 (Sturdivant, 1968). Furthermore, as more and more data from various techniques came to be collected the

Svedberg-initiated belief in proteins having molecular weights multiple of 17 500 became less and less tenable. The cyclol theory also failed to explain *why* proteins did arrange spontaneously into such highly symmetrical (or pseudo-symmetrical) conformations.

Here is what Professor Crowfoot-Hodgkin (1984) wrote to me during the preparation of this book:

"In retrospect, I think she was unlucky that three scientific observations made on proteins at that time seemed to fit so well with the cyclol hypothesis. The first – that protein molecules were highly symmetrical – was only favoured by the first few we picked up which were hexagonal, trigonal and cubic. By '36, '37 we were making measurements on others, lactoglobulin, haemoglobin and chymotrypsin, and we knew they did not have to be. The second was the Bergmann–Niemann numbers [see Section 9, below]. Again by 1937 we guessed that in so far as these were anything, they were a product of the subunit structures of some of the protein molecules – I say as much in my 1937 paper on insulin. The third was that the molecular weights estimated by combined chemical analysis and the centrifuge came to exactly the predicted number 288 required by the cyclol for egg albumin. All three were deductions from too little knowledge and this was realised by most protein scientists already by 1937–38. In Oxford there was a great deal of inter-departmental talk about protein structure. F.L. Simon, Professor of Thermodynamics and a particular friend of mine said 'The great beauty of the cyclol hypothesis is that it does give a reason for such large molecules having definite molecular weights'. That argument one could only answer by saying, there must be some other reason. And again it wasn't till later we realised how different the molecular weights could be of different proteins and that even individuals might lose a residue or two with little effect".

Despite all these apparent initial successes, the cyclol theory was quickly challenged, and by none other than Linus Pauling. For one thing, he had a rival theory based upon hydrogen-bonding (Mirsky and Pauling, 1936) which was almost the exact contemporary to Wrinch's theory. When I put to Linus Pauling (in a letter of May 19, 1983) this question: 'How early did you become convinced that the cyclol theory of protein structure, conceived by Dorothy M. Wrinch, was wrong, and what were the precise points where you judged (it must have been in the late 30s) that she had made decisive mistakes?', he answered me in these words (Pauling, 1983):

"In the middle of the 1930s I read the papers by Dorothy Wrinch. Her structures had some appeal to me, as a crystallographer, but not to me as a chemist. It is my memory that I never had a feeling that there was enough probability that her ideas were right to justify my paying much attention to them. However, in the fall of 1937 Dr. Warren Weaver of the Rockefeller Foundation asked me if I would ask Dr. Wrinch, who had been brought to the United States by the Rockefeller Foundation, to come to Ithaca, New York, to talk with me. I was for one semester the George Fisher Baker Lecturer in chemistry at Cornell University.

I invited Dr. Wrinch to come and to give a seminar about her ideas. I was impressed by the fact that some of the arguments that she presented were mutually inconsistent, and that when I pointed out that they were mutually inconsistent she seemed not able to understand what I was saying – at any rate she did not clarify the situation for me. I made a report to the Rockefeller Foundation on her presentation of her ideas and on my various points of criticism. Then she published a paper together with Irving Langmuir in which they attempted to obtain support for the cyclol theory by interpreting the x-ray diagrams published by Dorothy Hodgkin. The fact that she had got Langmuir to support her ideas caused me to decide that the various points of criticism of the cyclol theory should be published. This was done by my younger colleague Professor Carl Niemann in a paper in JACS in 1939. So far as I am aware, Dorothy Wrinch did not attempt to reply to our paper".

The paper Pauling refers to is one he published with Niemann in 1939 (Pauling and Niemann, 1939); it was 'a strong critique of the cyclol hypothesis, and its popularity waned rapidly thereafter' (Fruton, 1979).

The paper by Pauling and Niemann indeed uses strong language ('there exists no evidence whatever in support of this hypothesis (...) proteins do not have such cage structures') to blast at the Wrinch cyclol theory. The refutation is twofold. It addresses specifically the Wrinch and Langmuir work on the structure of insulin and demolishes this fit of the Crowfoot x-ray data to a C_2 cyclol structure. And it considers more generally the cyclol theory to rule it out, mostly on thermochemical and structural grounds.

The criticism of the Wrinch–Langmuir insulin structure is severe: (i) 'the evidence adduced by Wrinch and Langmuir has very little value, because their comparison of the x-ray data and the cyclol structure involves so many arbitrary assumptions as to remove all significance from the agreement obtained (...) the authors had at their disposal seven parameters, to which arbitrary values could be assigned in order to give agreement with the data' (Pauling and Niemann, 1939); (ii) 'it has been pointed out by Bernal (1939) that the authors did not make the comparison of their suggested structure and the experimental diagrams correctly. They compared only a fraction of the vectors defined by their regions with the Crowfoot diagrams, and neglected the rest of the vectors. Bernal (1939) reports that he has made the complete calculation on the basis of their structure, and has found that the resultant diagrams show no relation whatever to the experimental diagrams' (Pauling and Niemann, 1939).

Pauling and Niemann thus attacked and destroyed the Wrinch cyclol theory. They stressed its implausibility on a thermochemical basis: taking into account resonance stabilization of the peptide bond 'the cyclol structure is less stable than the polypeptide chain structure by 27.5 kcal.mol^{-1} per amino acid residue. (. . .) *the cyclol structure cannot be of primary importance for proteins; if it occurs at all (which is unlikely because of its great energetic disadvantage relative to polypeptide chains) not more than about three per cent of the amino acid residues could possess this configuration* (. . .) the cyclol structure is so unstable relative to the polypeptide structure that it cannot be of significance for proteins' (Pauling and Niemann, 1939). These authors then proceeded to hurl structural arguments at their target. They recalled an objection which had already been raised (Huggins, 1939): the cyclol theory entailed atomic contacts sometimes smaller than the sum of the van der Waals radii, even ridiculously low in certain cases (hydrogen atoms 0.67 Å instead of 2 Å apart). They referred also to Haurowitz's failure to find experimentally in proteins the large number of hydroxyl groups implied by the cyclol hypothesis (Haurowitz, 1938; Haurowitz and Astrup, 1939). And they pointed out that 'the area provided per side chain by the cyclol fabric, about 10 sq. Å, is far smaller than that required' (Pauling and Niemann, 1939).

The refutation by Pauling and Niemann (1939) was entitled *The Structure of Proteins*. Dr. Irving Langmuir at that time a staunch support of the Wrinch cyclol theory, counter-attacked by criticizing the Mirsky and Pauling rival model. This is what he wrote to Linus Pauling (Langmuir, 1939b):

"In your paper with Mirsky [Mirsky and Pauling, 1936], on p. 442, you say that the characteristic properties of the native proteins indicate that the polypeptide chain is folded into a uniquely defined configuration. This implies that every atom in every molecule of the protein is definitely located. Why may not the same be true of the side chains in the Wrinch polyhedra? In fact, in the Wrinch theory it is natural that the necessarily long range actions which make the symmetrical fabric structure should also cause a symmetrical arrangement of side chains. In the polypeptide chain theory, however, even if you do postulate out of a clear sky a uniquely defined configuration, in spite of the normally free rotation about bonds in a long chain, you still have no reason for giving to these unique configurations any unique characteristics. (. . .) The theory that you and Mirsky propose fails completely to account for the striking fact that the high symmetry of the molecules of the globular proteins, as indicated by their crystals, is definitely correlated with the high specificity, discrete molecular weights and other properties which you have so well summarized in your paper with Mirsky as being characteristic of native proteins".

Then came World War II. Dorothy Wrinch had moved to the U.S.A., where she settled ('Dorothy took (her daughter) Pamela for safety to America'; Crowfoot-Hodgkin, 1984); this, compounded with the war-time difficulties of communication, decreased markedly her contacts with her friend Dorothy Crowfoot (who had met, through Dorothy Wrinch's indirect agency (Crowfoot-Hodgkin, 1984), and married Thomas Hodgkin in 1937). Hence, whereas Dorothy Wrinch had been very close to the x-ray work on proteins when she lived in Britain, she was now at a distance from prime information. However, Dorothy Hodgkin continued to supply her with the data she needed (Crowfoot-Hodgkin, 1984):

"I took a full series of x-ray photographs of wet insulin crystals just before the war and measured intensities by eye over bedside lamps in the summer of 1940 in the country, at Crab Mill, my parents-in-law's house to which I had taken Luke [her son] for safety. I was a bit reluctant to send them over to Dorothy when she asked for them. I feared they were too inaccurate for serious use, and I had to work mainly on other things myself. But I did send them and Dorothy had calculated the 3 D Patterson diagrams on IBM machines – I only calculated the zero section myself".

When reading the correspondence between Dorothy Wrinch and Dorothy Crowfoot–Hodgkin during the 1940s, one gets the distinct impression that the requests by the former for fresh information are gradually being skirted by the latter. Dorothy Crowfoot–Hodgkin, after it appeared that the Wrinch cyclol theory was untenable, had good reason not to provide her friend with good sound data which

Plate 4. Dorothy Crowfoot as she was in the late thirties, early forties.
(Courtesy Prof. Crowfoot-Hodgkin.)

might get prematurely disclosed in support of structures she (D.C. Hodgkin) did not believe in. All this little drama is echoed in their correspondence, otherwise extremely friendly.

In it, there is increasing obduracy on Wrinch's part in defence of the cyclol theory (true, very few scientists in the history of science accept to abandon a theory they have authored). For instance, she writes in a letter to Dorothy Crowfoot-Hodgkin, which apparently was never sent (Wrinch, 1943):

"Well to sum - all this new excitement of the diffraction patterns of all sorts of structures means just this to me - the really important thing for protein X-ray data interpretation in the future is that one should make clear for future workers (if by any sad chance neither you nor I nor others in this generation actually do find it possible to establish a structure for insulin) what characteristics of the data are clinching evidence for or against this or that (including the cyclol cage and the cage in general) type of structure".

Dorothy Crowfoot-Hodgkin, in her part of the correspondence, appears intent on preserving her relationship with Dorothy Wrinch, on trying not to injure her friend's feelings. Therefore, in most of her letters, when she does not write about her children and family, she is very cautious when talking shop. And when she does so, usually she avoids adroitly the issue of cyclol cages.

This brings up the last letter from one Dorothy to the other Dorothy which I shall quote here, a truly remarkable letter for its firm gentleness, for the attempt not to bruise the feelings of one's friend, for the delicate wording; I find it one of the most touching letters exchanged between two scientists, perhaps only a woman could have written it! (Hodgkin, 1951):

"The thing that inhibited my talking to you much about insulin and which has probably led to this feeling of suspicion is quite different, and perhaps it would be best if I said what it was. I was only worried for your sake since it seemed to me that you cared a great deal that the cyclol idea should be right and that it would make you most unhappy if it should be proved wrong. When I saw you first with Otto [Otto Glaser, D.M.W.'s second husband] I had just been through a very gruelling time with [Sir Robert] Robinson minding far too much his penicillin structure being disproved. I found myself extremely sorry for him - it quite spoiled my own pleasure - and I feared similar trouble for you. I do not at all like the sensation of being an instrument for bringing unhappiness.

I may, of course, be quite wrong about cyclols and proteins but I just felt our attitudes to the problem were so different that we should probably hurt one another by discussing it - and that therefore, perhaps it was best not to".

Professor Hodgkin was well advised to deal carefully, even skillfully with Dorothy Wrinch, who during the 1940s and 1950s became increasingly paranoid about the cyclol theory. Professor Wrinch convinced herself that recognition was denied her because of a conspiracy led by none other than Pauling.

It is rather pathetic, as one reads through her papers in the archives at Smith College, to see her getting more into a corner, becoming more and more isolated (intellectually speaking), and being gradually cut-off from research funding. To close the Wrinch story and before I attempt reading the lesson from it, I shall quote from yet another letter of Linus Pauling (Pauling, 1956):

"I do not know about recent developments of the Wrinch protein theory. The papers that Mrs. Wrinch has published during the last few years and that I have read have not been convincing; I remember in particular a paper in which she presented an argument to show that the alpha helix could not be present in some globular proteins – hemoglobin, I think. I remember that there was a serious flaw in her argument.

I do not know about Miss Anslow's cage model for insulin. There is X-ray evidence, by Arndt and Riley, strongly supporting the assignment of the alpha helix as the principal way of folding the polypeptide chains in insulin. I think that the infrared spectra of insulin and other globular proteins show that these proteins contain polypeptide chains. You mention that Miss Anslow's cage model on denaturation gives exactly Sanger's sequences; if she has an argument that leads exactly to Sanger's sequences, without these sequences having been put in as a postulate, it would be very interesting. So far as I know such an argument has not been given; everyone who has thought about the problem has ended up with the conclusion that the sequences cannot be predicted.

A number of years ago I made a thorough study of Mrs. Wrinch's ideas - the Rockefeller Foundation had asked her to come to talk with me, and I made a report to the Rockefeller Foundation. The conclusion that I reached then was that the ideas that she had at that time were in considerable part self-contradictory. When I discussed her theories with her, especially the self-contradictions and the contradictions with experiment, she abandoned item after item, until at the end there was almost nothing left. The contrast between the extensive claims that she had made in her papers and the very small amount that she was willing to defend caused me to decide not to make an effort to find the reliable parts of her later publications. There is, so far as I know, nothing wrong with her mathematical work on Fourier transforms.

The X-ray studies that have been made on silk fibroin and a number of other proteins have now been carried out so thoroughly as to leave no doubt of the existence of polypeptide chains in these proteins".

7. Some reflections on the cyclol theory

A first point has to be raised, even if it can be dispelled easily. This is the matter of *pathological science*: Irving Langmuir, who collaborated with Dorothy Wrinch on the correspondence between the x-ray data and the cyclol (288) structure for insulin, also made a detailed inquiry into several episodes in the history of science (such as the N-rays of Blondlot) in which astounding phenomena had, it was claimed, been observed. The Langmuir study is a little gem in the history of science: not only does the General Electric leading scientist of the time know how to tell a story, but he shows analytic skill and a synthetic grasp of his chosen topic (Langmuir, without date). Putting together cyclols and 'pathological science', since Langmuir took responsibility for both, it is thus natural to ask if cyclols qualified as pathological science! Fortunately, Langmuir and Wrinch are immune from such criticism. The criteria suggested by Langmuir for pathological science are: (*i*) minute effects which are seemingly independent of the magnitude of the causative agent; (*ii*) effects at the limit of detectability; (*iii*) fantastic accuracy is claimed in the measurements or observations; (*iv*) a ludicrous theory, contrary to experience, is proposed; (*v*) criticisms are met with ad hoc excuses; (*vi*) the ratio of supporters to critics rises, epidemic-like, rapidly to ca. 50% and then gradually dwindles into oblivion.

To cut it short, *none* of these six criteria can be leveled at the Wrinch cyclol theory.

Another type of criticism could perhaps be made of the cyclol theory, on methodological grounds (and this general criticism is implicit in the attitude taken toward Wrinch's theory by most of her contemporaries, such as Pauling and Hodgkin): she did not rely exclusively, despite her token claim earlier quoted (' . . . we must pursue our inquiries inductively' (Wrinch, 1934)), on Baconian induction, which would have led her ultimately to the true structure of proteins.

I do not feel that it is totally fair to blame Dorothy Wrinch's failure on this ground. For one thing, some very interesting computer experiments conducted very recently by Herbert Simon and his group in Pittsburgh suggest that scientific discovery is a mix of slow induction and of sharp deductive insights. For instance, the following paragraph applies nicely to theories of protein structure, not only successful ones such as Pauling's but also unsuccessful ones such as Wrinch's:

" . . . the experiments also show that *introducing pretheoretical constructs, like symmetry* and conservation, may reduce significantly the amount of search required to detect empirical regularities in data. In fact, if sufficiently strong hypotheses are available, *the roles of theory and data may be reversed, so that laws are now deduced directly from theoretical assumptions and subsequently tested by data*" (Bradshaw et al., 1983; emphasis added).

Granted that the cyclol theory was decidedly not an example of pathological science in the Langmuir sense, nor in the Simon sense, are there other criteria by which to condemn it? A philosopher of science belonging to the Popperian school might discard the cyclol theory because it sought verifications - as for instance in the comparison performed by Wrinch and Langmuir (1938) between the Crowfoot x-ray diagrams for insulin and a cyclol structure for this protein - rather than falsifications. But we have alluded already to the psychological unlikeliness for a founder of theory to try deliberately to ruin it; as writes Crowfoot-Hodgkin (1984): 'We're only human beings and become devoted to our brain children even when wrong'; and other scientists - Bernal, Huggins, Haurowitz, and Pauling and Niemann especially, in this case - are only too eager to wield the hatchet!

Still one feature comes up repeatedly in the Wrinch writings and has to be noted: wishful thinking. Clearly, she got carried away by the sheer esthetic beauty of the honeycombed cyclol structures she had imagined. Behind her stubborn defence of the cyclol theory, there is undoubtedly the notion that these molecular models are too pretty not to be adopted by nature.

At this point, the weary historian must interject a note of caution: we are dealing here, post facto, with a lost scientific cause. Who would question it, had it been a winning one? If Dorothy Wrinch had had the luck to hit upon a theory much closer to reality, then the initial imperfections of the theory, its various discrepancies with experiment or with the extant knowledge about structure of smaller organic molecules, instead of being raised as major objections, could very well have been smoothed away rather quickly. In other terms, from (say) an abstract Popperian position, one could equally well challenge two emerging theories of protein structure, a losing one such as Wrinch's and a triumphant one such as Pauling's; in their rough nascent state when the necessary adjustments have not yet been performed, still imperfect.

But one need not accept the points made in the previous paragraph. Even from a narrow pragmatic standpoint, adjudging Wrinch 100% wrong and Pauling 100% right, still the Wrinch cyclol theory had considerable merit because of the seeds for the future which it contained. These are merits begging to be recognized by the historian. She had the daring to try to conceive a protein structure, even at a time of sparse experimental evidence, in terms of molecular architecture. For this purpose, she made first a planar outlay of the polypeptide chain – mistakenly, it turned out, transmuting it first into a cyclol network – and then wrapped the resulting 'cyclol fabric' into three dimensions. When, a few years later, Pauling and Corey (working from a much more sound knowledge of chemical structure, both experimental and theoretical), evolved their structures from the α-helix and the β-pleated sheet, they did not do otherwise. Even if Dorothy Wrinch's theory turned out to be totally wrong, her operational strategy, the procedures she pioneered for conceptualization of the structural solutions to a structural problem for biomolecules – a methodology which betrays the trained mathematician she was – became, I submit (even if of course such a proposition cannot be proved), part and parcel of the structural biophysics from which ironically she became quickly excluded!

Indeed a parallel begs for attention. When, in the early 50s, two young scientists who had had a look at some preliminary x-ray pictures of an important molecule (DNA rather than insulin) started building molecular models like mad, in the process ignoring certain basic structural features held to be true by the organic chemists (po-

sition of the keto-enolic tautomerism for the nucleic bases), were
they behaving in any different way from Wrinch's in the 30s? The
only difference – empirically an important one, but methodological-
ly or philosophically a minor one – was that they met with success!
Their good luck – and Wrinch's bad luck. To go beyond the impossibility of stating in general how we should proceed to make true statements about natural things and phenomena would be intellectually
irresponsible.

8. The seminal influence of Kepler

I have referred above to the 'too-good-not-to-be-true' flavor of cyclol
theory, shown repeatedly in the Wrinch papers. It brings to mind a
very important episode in the history of science, when Kepler tried
to make sense of the world in terms of its underlying Platonic and
Pythagorean harmony of form and number.

It will be recalled that Kepler discovered the laws governing the
motions of the planets, still named after him, when he set about
inscribing and circumscribing the spherical surfaces of their orbits
within the five Platonic solids (the tetrahedron, cube, octahedron,
dodecahedron, and icosahedron). Before him, quite a number of
Renaissance scientists had already attempted relating planetary
motions to geometrical forms. They were convinced of 'a music of
the spheres', which meant that 'the configurations and motions of
the heavens were governed numerically by specific key integer relationships, namely 2:1, 3:2, and 4:3, the perfect harmonic ratios of the
octave, the fifth, and the fourth' (Schneer, 1983). They displayed
these graphically by correspondences between various regular polygons and astrological positions and conjunctions. In Schneer's
words (Schneer, 1983), 'the great inspiration that spurred the
youthful Kepler, at 24, to publish the *Mysterium cosmographicum*
(1595) was his replacement of the two-dimensional polygons with
the sequential inscription of the five regular solids of Plato'.

And Kepler did not content himself with such a geometrical description of the harmony within the cosmos. If celestial bodies con-
formed to such a scheme, marvelous for its simplicity and beauty of

proportions, other natural artefacts should also display similar, divinely inspired, harmony. Kepler chose to exemplify this belief with the snowflake. He wrote an entire small treatise, *Strena seu de nive sexangula* (1611) on this topic, in which he related the six-fold symmetry of the snowflake with the packing of the water spheres entering into its composition. This 'seminal suggestion of a correlation between the external solid forms of crystals and the Pythagorean concept of a physical form for numbers' (Schneer, 1983) was an insight of genius. It gave rise, as documented by Schneer in his well-argued article entitled *The Renaissance Background to Crystallography* (Schneer, 1983), to modern crystallography, with the contributions of the French school of mineralogy in the 1770s (Romé de Lisle) and the 1800s (Haüy).

There is more than an echo of Kepler in Wrinch's cyclol theory: the cyclol structures, such as the 'cyclol 6', the 'cyclol 42', etc., resemble snowflakes in their hexagonal framework; both snowflakes and the Wrinch cyclols are *fractals*, these mathematical objects recently recognized for their universal applicability and their importance through the work of Benoît Mandelbrot. Another analogy is in the use of regular polyhedra as supports for the cyclol fabric: yet another similarity is in the importance given to numerological considerations (see below, section 9).

It is even possible to point at a very plausible 'missing link' between Kepler and Dorothy Wrinch. This is the classic of biological science, the book *On Growth and Form* which D'Arcy Wentworth Thompson first published in 1917. This book was highly influential on the whole generation of biochemists to which both Dorothy M. Wrinch and Linus Pauling belonged. Linus Pauling read it in the 1920s (Pauling, 1983). As for Mrs. Wrinch, she was so much impressed and enthralled by this book, by its subject matter as well as by its approach, that, when a new edition was issued in 1942 she was asked by the editor of *Isis* (the leading periodical in the history of science) to review it (Wrinch, 1943). In this second edition of *On Growth and Form*, Kepler's name is cited nine times, admiringly: for his work on the snow crystals; for his deduction - alluded to in Wrinch's *Isis* review - that the honeycomb angles are those of the rhombic dodecahedron; for his fascination with the five regular Platonic polyhedra, and with the sequence of numbers associated with the Golden Section and known as the Fibonacci series. All these are questions close to Wrinch's own more or less obsessive concerns.

Indeed, Dr. Wrinch's review of Sir D'Arcy's book (Wrinch, 1943) is worth careful reading and re-reading for the insight it provides, somewhat obliquely, into her own philosophy of science. She praises the 'constructive iconoclasm' of D'Arcy Wentworth Thompson, which I find in line with her own intellectual courage. She writes also lucidly about the 'inevitable transition from the visible morphology of today to the underlying molecular morphology of tomorrow'. She mentions explicitly Kepler's name, in direct connection with the topics of such overwhelming importance to her, cyclols:

"Much little-known history of this fundamental problem of the angular quantities associated with space filling structures is contained in the passages on the bee's cell and the work and thoughts thereon of KEPLER, of MARALDI the astronomer, nephew of CASSINI, of REAUMER [sic], and of KOENIG. *Such matters lead easily to the general problems of polyhedral geometry and their relation to biological structures*" (Wrinch, 1943; emphasis added).

She even goes one step further and makes an explicit allusion to the protein structural problem:

"Nothing can illustrate the importance of a correct 'feel' for this (intricate and beautiful material that is an organism) better than *the present deplorable state of protein chemistry. Here the worker untouched by biological knowledge* [read Pauling, Wrinch's foe] *hammers away trying to force living material into the cast-iron moulds of classical chemistry*" (Wrinch, 1943; emphasis added).

As if this polemic were not enough self-revealing, Wrinch goes one step further (and there are a couple of spots in this review which are a little disturbing). Her self-defensive posture is such that she could not resist this other comment (Wrinch, 1943; emphasis added):

"Fundamental advances in our understanding of biological structures, whatever their order of magnitude, will require support from physics and mathematics - the two sciences in which structure problems *free from needless adventitia have made the greatest progress*".

Clearly, when writing of 'needless adventitia', Dorothy Wrinch must have had in mind the thermochemical and structural objections raised by Pauling and by others to her cyclol theory of protein structure! The affinity with Kepler helps us to understand the almost religious belief of Dorothy Wrinch in polyhedra-based cyclols as the very fabric of life.

Coming to the end of this chapter about Dorothy M. Wrinch's tragic failure to erect a durable theory of protein structure, I feel compelled to state the qualities which I admire in her work even though these very qualities led her badly astray. This was chiefly the quest for beauty (see Chandrasekhar, 1979). In this fine article, the Nobel Laureate for 1983 addresses the question raised by the literary critic Roger Fry, 'what one should make of a theory which is aesthetically satisfying but which one believes is not true' (Chandrasekhar, 1979). (This is precisely how one may feel about the cyclol theory of protein structure.) And Chandrasekhar proceeds to answer this question, quoting Weyl to this effect: 'My work always tried to unite the true with the beautiful; but when I had to choose one or the other, I usually chose the beautiful'. To which the Chicago cosmologist adds this other quote, from Heisenberg, which I find quite relevant to the very real achievements of Dorothy Wrinch to which I hope this chapter will help to pay justice:

"If nature leads us to mathematical forms of great simplicity and beauty – by forms, I am referring to coherent systems of hypotheses, axioms, etc. – to forms that no one has previously encountered, we cannot help thinking that they are 'true', that they reveal a genuine feature of nature . . . You must have felt this too: the almost frightening simplicity and wholeness of the relationships which nature suddenly spreads out before us and for which none of us was in the least prepared" (Chandrasekhar, 1979).

The original title for this chapter read 'Dorothy Wrinch, or a Daring Renaissance Scholar in an Age of Patient Data Collection'. It conflated unwillingly two quite distinct periods: in the 1930s and 1940s, before the advent of high-speed computers, there were quite a number of 'Renaissance scholars' among the scientists intent upon unraveling the structure of proteins; they numbered Desmond Bernal, of course, but also W.T. Astbury, I. Fankuchen, D. Crowfoot to consider only the group in Cambridge; the age of 'patient data collection' came after the war, when digital computers became generally available. Fortunately, Professor Dorothy Crowfoot-Hodgkin, to whom I had submitted this chapter in draft form, reacted strongly. She had these, among other comments (Crowfoot–Hodgkin, 1984):

"I thought to myself that your title does not really fit us. There was a sense in which we all were Renaissance scientists – particularly Bernal himself but also Astbury and to a lesser degree Max Perutz and Fankuchen and me, wildly enthusiastic, trying to do far too many different things at the same time. We faced the data collection problem slowly and unwillingly – we could see that we could not reach our goals as far as proteins were concerned without very serious advances in computing and intensity measurements, which none of us were fitted to make; the war intervened and the advances were made for us so that the new generation could come in and carry us through".

Dorothy Wrinch differed from this group of high-minded scientists with wide-ranging interests and culture in one crucial respect: she had less caution, she was more willing to rush into print with speculations if, according to her conscience, these were legitimate.

9. A postscript upon numerological speculations

In the late 1930s, the data on the molecular weights of proteins which Svedberg had been measuring with the ultracentrifuge seemed to him to display conspicuous regularities (Table IX): all the molecular weights were, he thought, multiples of a ca. 17 500 unit. In the Royal Society Symposium ('A Discussion on the Protein Molecule') of November 17th, 1938, Svedberg reported his findings in these terms (Svedberg, 1939):

"A (. . .) result of ultracentrifugal and diffusion investigations is the finding of a rule of simple multiples for the molecular weights of the proteins (. . .) the masses of most proteins, even those of widely different origin, show similar regularities. If we choose 17 600 as the unit the majority of the proteins may be divided into eleven classes with molecular weights which are multiples of this unit by factors containing powers of 2 and 3".

TABLE IX

Molecular weights from ultracentrifugal determinations for various proteins (from Svedberg, 1939)

Protein	Molecular weight
Gliadin	26 000
Egg albumin	42 100
Lactoglobulin	39 800
Hemoglobin	68 000
Serum albumin	68 500
Serum globulin	158 000
Amandin	329 000
Thyroglobulin	639 000
Homarus hemocyanin	777 000
Octopus hemocyanin	2 780 000
Helix pomatia hemocyanin	6 650 000

I shall borrow from the same general presentation of the protein structure problem with which Svedberg opened the Royal Society Symposium his summary of the Bergmann and Niemann study (Svedberg, 1939):

"Following up an idea first expressed by Astbury (1934), Bergmann and Niemann (1937a, b) postulate a general law which expresses the constitution of the proteins as a function of the 'frequencies' of the various amino acids. Each amino-acid residue is supposed to occur at regular intervals in the polypeptide chain. The one with the highest frequency F_i, or lowest absolute number N_i, according to the expression $F_i.N_i = N_t$, where N_t is the total number of amino-acid residues, determines the minimum molecular weight. The frequency values F_i deduced by Bergmann and Niemann from the analytical data available are very nearly multiples of powers of 2 and 3, and so are the N_t. The molecular weights deduced by Bergmann and Niemann agree very well with those found by means of ultracentrifugal sedimentation. The classification of the proteins according to molecular weight attempted on the basis of ultracentrifugal data may be explained by the law postulated by Bergmann and Niemann".

Wrinch, as we know, took the Svedberg determinations very seriously; they appeared to corroborate her cyclol theory of protein structure. She also agreed completely with the Bergmann–Niemann 'numerological law'. We find her stating, during the same Royal Society Symposium, that the 'various amino acids occur in proteins in general, not in a random manner, but in powers of two and three' (Wrinch, 1939).

Ironically, her arch-opponent Linus Pauling agreed with her on this point: he had been convinced by his younger colleague Carl Niemann at CalTech of the validity of his numerological considerations. And we find indeed that the Pauling–Niemann article on 'The structure of proteins' (Pauling and Niemann, 1939) includes a full section on this theme:

"Considerable evidence has been accumulated", wrote Pauling and Niemann (1939), "suggesting strongly that the stoichiometry of the polypeptide framework of protein molecules can be interpreted in terms of a simple basic principle. This principle states that the number of each individual amino acid residue and the total number of all amino acid residues contained in a protein molecule can be expressed as the product of powers of the integers two and three. (. . .) The evidence regarding frequencies of residues involving powers of two and three leads to the conclusion that there are 288 residues in the molecules of some simple proteins. It is not to be expected that this number will be adhered to rigorously".

And a similar note of caution is struck by none other than The Svedberg himself (Svedberg, 1939):

"As pointed out by Astbury (1937), the frequencies calculated by Bergmann and Niemann would in many cases require that one and the same place in the polypeptide chain be occupied by two or more different amino acid residues. Furthermore, the scheme is not easily compatible with the general structure of proteins established by X-ray measurements. The analytical data available at present, especially those relating to the amino acids of low percentage, are rather uncertain. (. . .) As an example the deductions by Bergmann and Niemann concerning haemoglobin may be mentioned. Here from the cystein content, 0.5%, they calculate $N_i = 3$ for this amino acid and find the total number of residues in the molecule to be 576. Their molecular weight value is 69 000, in excellent agreement with ultracentrifugal and osmotic values. Now the ultracentrifugal investigations have shown that haemoglobin can split reversibly into half molecules and that the two halves are probably equal (Steinhardt, 1938). X-ray measurements indicate the same (Bernal and others, 1938; Crowfoot, 1938). Bergmann and Niemann ought therefore to have found a minimum weight of about 34 500 instead of 69 000 for haemoglobin".

But, by and large, the Bergmann–Niemann 'law' was quickly accepted, including by some geneticists (Goldschmidt, 1938) because, according to Fruton (1979) 'it appeared to suggest the operation of a mathematical principle that might link Mendel's laws to the structure of proteins, which were considered at that time to be the material basis of heredity'. Another important reason for the widespread acceptance of the Bergmann–Niemann distribution law has been pointed out by Karlson (1979): one finds repeatedly in the history of protein structure the notion that proteins resemble carbohydrates in composition (Braconnot, 1819) or in the arrangement of the component units. Karlson writes thus (1979):

"The basic idea was at that time – I think we should acknowledge this – to find repeating units within large macromolecules. You know we have the repeating unit of maltose in glycogen and starch; we have the repeating unit of cellobiose in cellulose (. . .) It was generally believed that the nucleic acids were made out of tetranucleotides in repeating units, and it was more or less tacitly assumed that a protein would also be built up as a form of repeating unit of amino acids. Therefore, the magic numbers of Bergmann and Niemann had something to do with this belief, I think, that there was a repeating unit in all macromolecules".

In any case, and even though it received widespread support, the Bergmann–Niemann scheme also drew some flak at the time when it was most popular. We find for instance A. Neuberger making these common-sense remarks at the November 1938 meeting of the Royal Society (Neuberger, 1939; emphasis added):

"The frequency hypothesis of Bergmann and Niemann states that the number (a) of amino-acid residues in a protein is expressed by the formula $a = 2^m 3^n$, where m and n are whole numbers. Now the average error in the determination of individual amino acids in a protein may be fairly taken to be 5–6%; assuming the possibility of an error of 6%, calculation indicates a very high degree of probability (approx. 80%) that a *purely random distribution of amino acids should give values in apparent accordance with the formula of the frequency hypothesis*".

To which should be added the humorous remarks of R.D. Hotchkiss (1979) and N.W. Pirie (1979) made during the New York Academy of Sciences meeting of May 31, June 1 and 2, 1978. First, Dr. Hotchkiss:

"While we are interested in burying ideas of numerology, or rather premature numerology, I would like to tell you something that I almost forgot. At an earlier New York Academy of Sciences meeting sometime in the 30s, Drs. Vickery and Cohn were reporting analytical data on purified crystalline proteins and attempting to work backward from the amino acid content to show that the data were quite compatible with the molecular weight, and at that particular time the molecular weights were alternatively the 35 000 of Svedberg or the 40 000 discovered by Pedersen. They were calculating from amino acid compositions as if there were let's say 19 molecules in one case or 21 for the other molecular weight, and they were quite happy with the data. It struck me that almost any random data might give the same fit if you have allowable numbers, somewhere between 9 and 30 units, so I took *The New York Times* for that particular day (I have it still) and out of the bond quotations I picked a region where a variety of bond prices were listed. I calculated the molecular weight of bonds from *The New York Times*. The fit of the protein data was something with an error of about 4.9, the fit of *The New York Times* bond quotation was about 2.3, less than half the error of the actual amino acid data".

Now, to Pirie who mentioned that

" . . . he did something similar (Pirie, 1940) and pointed out that the only long gaps in the earlier part of the sequence of numbers that can be expressed as $2^n \times 3^m$ are 19–23 and 37–47. The other gaps are small compared with the probable error in amino acid measurements. Therefore, the only numbers that would be excluded by the Bergmann–Niemann hypothesis are 22 and 42. With the 1939 level of analytical precision, the hypothesis seemed to me to have no observational basis".

REFERENCES

Astbury, W.T. (1934) Cold Spring Harbor Symp. Quant. Biol., 2, 15.
Astbury, W.T. (1937) Nature, 140, 968.
Ayling, Jean (1930) The Retreat from Parenthood, Kegan Paul, Trench, Trubner, London.
Bergmann, M. and Niemann, C. (1937a) J. Biol. Chem., 118, 301.
Bergmann, M. and Niemann, C. (1937b) Science, 86, 187.
Bernal, J.D. (1939) Nature, 143, 74.
Bernal, J.D. and Crowfoot, D. (1934) Nature, 133, 794.
Bernal, J.D., et al. (1938) Nature, 141, 521.
Bradshaw, G.F., Langley, P.W. and Simon, H.A. (1983) Science, 222, 971.
Brasseur, H. and Rassenfosse, A. de (1936) Bull. Soc. Roy. Sci. Liège, no. 5, 123.
Chandrasekhar, S. (1979) Physics Today, July 1979; from a lecture given at the International Symposium in Honor of Robert R. Wilson on 27 April 1979.
Conan Doyle, Sir Arthur (1891) The Strand Magazine, II (Adventures of Sherlock Holmes): A Scandal in Bohemia.
Crowfoot, D. (1935) Nature, 135, 591.
Crowfoot, D. (1938) Proc. Roy. Soc. (London), A164, 580.
Frank, F.C. (1936) Nature, 138, 242.
Fruton, J.S. (1979) Ann. N.Y. Acad. Sci., 325, 1.
Goldschmidt, R.B. (1938) Sci. Monthly, 46, 268.
Harker, D. (1936) J. Chem. Phys., 4, 381.
Haurowitz, T. (1938) Z. Physiol. Chem., 256, 28.
Haurowitz, T. and Astrup, T. (1939) Nature, 143, 118.
Hodgkin, D.C. (1950) Letter to Dorothy M. Wrinch of 22 June; The Dorothy Wrinch Papers, Sophia Smith Collection, Smith College, Northampton, MA 01060 (reproduced by permission).
Hodgkin, D.C. (1951) Letter to Dorothy M. Wrinch of March 4; The Dorothy Wrinch Papers, Sophia Smith Collection, Smith College, Northampton, MA 01060 (reproduced by permission).
Hodgkin, D.C. (1979) Ann. N.Y. Acad. Sci., 325, 33.
Hodgkin, D.C. (1984) Letter to Pierre Laszlo of April 2nd.
Hodgkin, D.C. and Riley, D.P. (1968) in Structural Chemistry and Molecular Biology (Rich, A. and Davidson, N., Eds.), Freeman, San Francisco, pp. 15–28.
Hotchkiss, R.D. (1979) Ann. N.Y. Acad. Sci., 325, 34.
Huggins, M.L. (1939) J. Am. Chem. Soc., 61, 755.
Karlson, P. (1979) Ann. N.Y. Acad. Sci., 325, 17.
Langmuir, I. (1939a) Proc. Phys. Soc., 51, 592; quoted by Fruton, J.S. (1979).
Langmuir, I. (1939b) Letter to Linus Pauling of 17th May; The Dorothy Wrinch Papers, Sophia Smith Collection, Smith College, Northampton, MA 01060 (reproduced by permission).

Langmuir, I. (without date, 1940s presumably) in General Electric Research and Development Center (Hall, R.N., Ed.), Schenectady, NY, report 68-C-035.

Langmuir, I., Schaefer, Wrinch, D.M. (1937) Science, 85, 76.

Lloyd, D.J. (1932) Biol. Rev., 7, 254.

Lloyd, D.J. and Marriott, (1937) Trans. Faraday Soc., 29, 1228.

Mirsky, E.A. and Pauling, L. (1936) Proc. Natl. Acad. Sci. USA, 22, 210.

Neuberger, A. (1939) Proc. Roy. Soc., B127, 25.

Newman, J.H. (1848) Letter to Mrs. William Froude of 27 June; quoted in The Oxford Dictionary of Quotations, Oxford University Press, 3rd ed., 1979.

Patterson, A.L. (1935) Z. Krist., 90, 517.

Pauling, L. (1956) Letter to William T. Scott of 14 February; The Dorothy Wrinch Papers, Sophia Smith Collection, Smith College, Northampton, MA 01060 (reproduced by permission).

Pauling, L. (1983) Letter to Pierre Laszlo of 13th June.

Pauling, L. and Niemann, C. (1939) J. Am. Chem. Soc., 61, 1860.

Pauling, L. and Sherman, J. (1933) J. Chem. Phys., 1, 606.

Pirie, N.W. (1940) Ann. Rep. Chem. Soc., 36, 351.

Pirie, N.W. (1979) Ann. N.Y. Acad. Sci., 325, 34.

Polanyi, M. (1935) Letter to Dorothy Wrinch dated 23rd May; The Dorothy Wrinch Papers, Sophia Smith Collection, Smith College, Northampton, MA 01060 (reproduced by permission).

Russell, B. (1968) The Autobiography of Bertrand Russell, vol. 2 (1914–1944), Little, Brown, Boston, pp. 132–133; 203.

Schneer, C.J. (1983) Am. Scientist, 71, 254.

Steinhardt, J. (1938) J. Biol. Chem., 123, 543.

Sturdivant, J.H. (1968) in Structural Chemistry and Molecular Biology (Rich, A. and Davidson, N., Eds.), Freeman, San Francisco, pp. 3–11.

Svedberg, T. (1934) Science, 79, 327.

Svedberg, T. (1939) Proc. Roy. Soc. London, B127, 1.

Svedberg, T. and Eriksson-Quemsel, (1935–36) Tabul. Biol. Per., 5, 351.

Wrinch, D.M. (1934) Nature, 134, 978.

Wrinch, D.M. (1936a) Nature, 137, 411.

Wrinch, D.M. (1936b) Nature, 138, 241.

Wrinch, D.M. (1936c) Nature, 138, 651.

Wrinch, D.M. (1937) Nature, 139, 972.

Wrinch, D.M. (1938a) Letter to Niels Bohr dated 16th December; The Dorothy Wrinch Papers, Sophia Smith Collection, Smith College, Northampton, MA 01060 (reproduced by permission).

Wrinch, D.M. (1938b) J. Am. Chem. Soc., 60, 2005.

Wrinch, D.M. (1938c) Science, 88, 148.

Wrinch, D.M. (1939) Proc. Roy. Soc. London, B127, 24.

Wrinch, D.M. (1943) Isis, XXXIV, 232.

Wrinch, D.M. (1943) Letter to Dorothy Crowfoot-Hodgkin dated September 29; The

Dorothy Wrinch Papers, Sophia Smith Collection, Smith College, Northampton, MA 01060 (reproduced by permission).

Wrinch, D.M. and Langmuir, I. (1938) J. Am. Chem. Soc., 60, 2247.

Wrinch, D.M. and Lloyd, F.J. (1936) Nature, 138, 758.

Chapter 14

A Note on Two Wise Men

1. The diversity among eminent scientists

In writing science history, one stands the risk of selectivity: heaping praise upon too few scientists. The temptation is to latch on to the great discoveries only.

True, this is how the reward system of science operates; by giving out prizes. It is good and fine to bestow awards on those who have richly deserved to be thus honored.

But one should be wary of replacing a continuous ribbon with a string of pearls! To take an extreme position, were one to consider solely the immense merits of Nobel laureates in the advancement of biochemistry this might qualify as good science journalism – but assuredly not as good scientific history. Not only such hagiography ignores the cumulative impact from the 'rank-and-file' or the consequences from introduction of a new technology.

But even for those who relish their history with heroic actors on a brightly lit stage, the lone handsome figure of the daring discoverer is insufficient. It does not carry enough real flesh and blood with it. In order to achieve some degree of verisimilitude, the floodlights must be turned on other members of the cast, too. The great discoverers – for science history is still heavily marked by the ethnocentric analogy with the great voyages of discovery, which brought colonial riches to Europe, as reflected in use of this term – had as companions (perhaps as equals in the eyes of the historian) those imaginative thinkers who having come up with new ideas, were also influential in their dissemination. I shall be concerned here with two such men, J.D. Bernal and D'Arcy Wentworth Thompson. Since reliable biographies are available (Goldsmith, 1979; and D'Arcy Thompson,

1958, respectively), I shall content myself with short summaries on
these two scientists.

2. Desmond Bernal (1901–1971): a mini-biography

John Desmond Bernal (Des, as he was known to his friends) was
born in Ireland. His father was a farmer, his mother (a Stanford
alumna) was a journalist. Very early on, Bernal was inflamed by the
two great passions of his life: revolt against social injustice, and
science.

"At the age of seven Des, having chanced to read a lecture by Faraday on the chemistry of the candle, tried his first scientific experiment, carefully copying the word
'hydrogen' (which he had never previously encountered) and the recipe of how to
make this from 'diluted sulphuric acid and granulated zinc'. His mother, who knew
even less about science than her son, was then persuaded to obtain these substances
from a chemist (. . .) However, when they were combined in a flask in the garden
according to Faraday's instructions, to Des's extreme disappointment absolutely
nothing happened. He decided to take one more look before bedtime and, as it was
already dark, he struck a match to see whether anything had resulted. There immediately followed a magnificent explosion from which, happily, he escaped 'absolutely
convinced of the truth of science'." (Wooton, 1980).

Bernal went as an undergraduate to Cambridge University
(1919–22) where he made numerous friends, met and married his
wife Eileen Sprague, and earned for himself the sobriquet of 'Sage'
from Dora Grey (Hodgkin, 1980) because of the Catholicity of his
interests and knowledge. This was the time also when he became an
atheist and a Marxist. Desmond Bernal and his wife joined the Communist Party and the Labour Party in 1923. Henceforth, Bernal was
to be a political activist.

As a young student in Cambridge, J.D. Bernal worked out a deri-
vation of the 230 space groups using the quaternions devised by Sir
William Rowan Hamilton a century earlier. This paper got Bernal
instant recognition for its scientific merit, with a college prize and
acceptance by the Cambridge Philosophic Society. Furthermore, on
the strength of it and of a recommendation from the lecturer in
Mineralogy, Arthur Hutchinson, who described Bernal as a 'shy,

diffident, retiring kind of creature, but something of a genius' (quoted by Hodgkin, 1980), Sir William Bragg had him appointed to a research position at the Royal Institution in London. This is how Bernal became an x-ray crystallographer.

I shall borrow from his obituary in *The Times* (1971) to summarize very briefly this period: it recalls that

"at the Royal Institution, to which he went in 1922, Bernal was allotted space to construct his own X-ray tube and other apparatus. He set out to apply space group theory rigidly to the deduction of the crystal structure of graphite, and in the process developed the method of interpreting rotation photographs by the use of the reciprocal lattice charts which bear his name. (. . .) From graphite he became deeply interested in the problem of the metallic state, and then in the wider subject of crystal chemistry under the influence of V.M. Goldschmidt".

In the fall of 1927, Bernal returned to Cambridge as the first lecturer in Structural Crystallography. Finding himself again in this intellectual hot-bed, Bernal embraced wholeheartedly his two passions of science and politics, forming a group of concerned scientists together with J.B.S. Haldane, Joseph and Dorothy Needham and C.H. Waddington among others. These were extremely productive years. Bernal let himself be guided by his wealth of interests, and started seeking 'the physical basis of life' (Norton, 1980). He continued his structure determinations by x-ray but he had many other interests. He was assuredly not a run-of-the-mill x-ray crystallographer! He showed that the Windaus-Wieland steroid structure was untenable and proposed the

"now accepted sterol skeleton. Other subjects he explored included vitamin B1, the sex hormones, liquid crystals, Richelle salt and various problems in coordination chemistry, the theory of muscular contraction, and the distribution of minerals on the earth's crust in relation to earthquake shocks.

In 1933 he took the first X-ray photographs of single crystals of a protein, pepsin, and followed this a year or two later, in collaboration with I. Fankuchen, with a remarkable series of experiments on tobacco mosaic virus. These defined both the size and the shape of the virus particle, and showed it to have a regular internal structure explored much later in his laboratory by R[osalind] Franklin and A[aron] Klug" (Obituary in the London *Times*, 1971).

In 1937, Bernal was appointed Professor at Birkbeck College in

London, to be effective at the beginning of 1938. During all these years, Bernal was a very active Marxist and communist. He traveled often to the Soviet Union. He helped to set up the British association of scientific workers, 'whose task he saw as the emancipation of science from its "historical role" both as "a support for the established order in a hierarchical society" and as "the instruments of industrialists" ' (Bernal, as quoted by Wooton, 1980).

Bernal had total faith in such an emancipated science, 'he deified (. . .) science. (. . .) In his interpretation, science included the application of logical reasoning to every human problem, chemical, mechanical, medical, economic or social' (Wooton, 1980). In 1929, Bernal had published his first book. This short essay entitled *The World, the Flesh and the Devil* (Bernal, 1929) has for a sub-title: 'An Enquiry into the Future of the Three Enemies of the Rational Soul'. It is both a remarkably percipient, if not downright prophetic, and a remarkably naughty little book. Bernal announces space travel and space stations. He makes plenty of other predictions, his book in Norton's terms is 'a sort of non-satirical *Brave New World*' (Norton, 1980). The aspects I find less palatable are when Bernal advocates not only total control of humanity by the scientists, but also a complete remodeling of man, body and soul, according to the most rational scientific blueprint. This ambition of an overall eugenics is a little unpleasant in its totalitarian undertone.

The typical 19th-century belief in progress often went with the naive expectation that the tools to achieve the desired improvement of the human lot would be provided by science. Often also the 19th-century thinkers failed to make the necessary distinctions between science and technology. And, since it was the age of triumphant physical science, they sincerely believed that rational thought as embodied socially in scientists and engineers would be a cure-all. Hence, it is not surprising that the underlying ambition was to understand and control natural phenomena, totally if possible. Bernal was such a 'committed systematiser', according to a felicitous description by his co-worker and friend (Hodgkin, 1980).

However, for a number of reasons such as unequal development of the various scientific fields of science at a given time, the non-linear advancement of science in a given field, this ideal of total understanding not only fails but can be warped unconsciously into the Procustean bed of a totalitarian ideology (Laszlo, 1977). Bernal was an eminent representative, even if he was well-meaning, and squarely on the side of the people against the oligarchy, of this intellectual trend.

His bad luck - and this is one of the ironies of intellectual history - was to entertain this faith in science and in rational thinking at the very time when other scientists pointed to some natural limits to rationality: Gödel's theorem (1931), the Heisenberg uncertainty principle (1927), the concept of measurement as analysed by Niels Bohr (1925) all partake of this recognition of an intrinsic limitation in the understanding of nature.

John Desmond Bernal had been carried away in a totally different direction. For him, science had been in the past the instrument of the ruling classes and, with the advent of communism in the U.S.S.R., had begun at long last to be applied to the emancipation of the masses. He was far from being alone in such beliefs, in the 1920s and the 1930s. And, because he had such diverse and extensive knowledge in science history, he pioneered Marxist historiography of science. We are much indebted to him for the *Social Function of Science* (Bernal, 1939) and *Science in History* (Bernal, 1954) - the former book fortunately almost entirely devoid of the political dogmatism which mars the latter.

To return to the nasty side of *The World, the Flesh and the Devil*, Bernal writes thus, in the conclusion to his clever and brilliantly written essay (Bernal, 1929):

"Thus the balance which is now against the splitting of mankind might well turn, almost imperceptibly, in the opposite direction. The whole question is one largely of numbers, and would become entirely so as soon as the quantity and quality of population were controlled by authority. From one point of view the scientists would emerge as a new species and leave humanity behind; from another, humanity - the humanity that counts - might seem to change en bloc, leaving behind in a relatively primitive state those too stupid or too stubborn to change. The latter view suggests another biological analogy: there may not be room for both types in the same world and the old mechanism of extinction will come into play. The better organized beings will be obliged in self-defence to reduce the numbers of the others, until they are no longer seriously inconvenienced by them".

Plate 5. Professor Brand and Dorothy Crowfoot. (Courtesy Prof. Crowfoot-Hodgkin.) (See pages 295 and 296.)

Perhaps, such morally debatable statements are congruent with actions by J.D. Bernal, much later, in the late 40s and early 50s when as the staunch Stalinist he had become he came to the support of the charlatan geneticist Trofim D. Lysenko, 'writing in 1949, for example, that the Lysenkoists surely had a point when they insisted that no one had isolated the gene' (Norton, 1980). Unfortunately, as remarks Goldschmidt (1980) this was 'a rather defeatist comment *at a time when the material basis of the operation of the gene was to come under increasing study in his own laboratory*' (emphasis added). Thus Bernal's 'scientific reputation was irretrievably damaged' (Wooton, 1980). This single episode serves to explain why Bernal has become a much-maligned scientist.

This was a pity because his laboratory at Birkbeck College was very active in the 1950s, in such diverse fields as x-ray crystallography of proteins and viruses, micro-focus x-ray tubes, electronic computers, cements and building materials: Professor Bernal was pushing pure science and applied science simultaneously.

His finest hour, I feel, was during World War II when he served as scientific adviser to Lord Mountbatten. There are some superb stories of their fruitful collaboration, especially on the preparation of D-day (the invasion of Normandy). After conceiving the idea of an artificial harbor,

"Bernal (. . .) turned his versatile mind to the geological problems involved in choosing the spot on the French coast at which the invasion force should land. First, he consulted the current copy of the *Guide Bleu* and noted its warning to tourists that at Arromanches the beach was sloping and muddy. Clearly that would not do; the ships would get stuck. Next question: how thick was the sand along the beaches? To answer it, Bernal tried to reconstruct the history of movements in the structure of beaches. In the course of his researches, he unearthed an 'Anglo-Norman' poem [*Roman de rue*] which described how William [the Conqueror] had crossed the sea on his way from Cherbourg to Ries on a ridge of rock at low tide. The continued existence of the ridge was then confirmed by aerial photograph" (Wooton, 1980).

3. D'Arcy Wentworth Thompson (1860–1948): a portrait

My task is made extremely easy here, because Professor Medawar has provided us in his little gem of a book on scientific research, *The*

Art of the Soluble, with a superb portrait of D'Arcy Wentworth Thompson (Medawar, 1967):

"D'Arcy Wentworth Thompson was an aristocrat of learning whose intellectual endowments are not likely ever again to be combined within one man. He was a classicist of sufficient distinction to have become President of the Classical Associations of England and Wales and of Scotland; a mathematician good enough to have had an entirely mathematical paper accepted for publication by the Royal Society; and a naturalist who held important chairs for sixty-four years, that is, for all but the length of time into which we must nowadays squeeze the whole of our lives from birth until professional retirement. He was a famous conversationalist and lecturer (the two are often thought to go together, but, seldom do), and the author of a work which, considered as literature, is the equal of anything of Pater's or Logan Pearsall Smith's in its complete mastery of the *bel canto* style. Add to all this that he was over six feet tall, with the build and carriage of a Viking and with the pride of bearing that comes from good looks known to be possessed".

D'Arcy Wentworth Thompson is remembered nowadays mostly for his classic book *On Growth and Form* which he first published in 1917 and of which he gave a second, enlarged and revised edition in 1942 (which, as we saw in the previous chapter, Dorothy M. Wrinch reviewed for *Isis*). *On Growth and Form*, as Professor Medawar rightly points out, is a literary masterpiece. Besides its literary value, it is also a long essay – the first edition was nearly 800 pages long, the second edition has more than 1100 pages – about what might be termed the 'mechanical correlates of biological concepts'. D'Arcy points at the design of a bone as an engineering solution to the forces and stresses it withstands; at the design of a wing as the optimum solution to a problem in aerodynamics; and so on. 'He believed (. . .) that the laws of the physical sciences apply to living organisms' (Medawar, 1967) and he was a steadfast *un*believer in Darwinian evolution. *On Growth and Form* is an extremely stimulating book with profound insights. It has been influential because it proved, with a wealth of minutely described examples, that many biological structures could be understood with the help of quite simple physical concepts. As Medawar writes (1967):

"We are mistaking the direction of the flow of thought when we speak of 'analysing' or 'reducing' a biological phenomenon to physics and chemistry. What we endeavor to do is the opposite: to assemble, integrate, or piece together our conception of the phenomenon from our particular knowledge of its constituent parts. It was D'Arcy's belief, as it is also the belief of almost every reputable modern biologist, that this act of integration is in fact possible".

Hence, the widespread influence of *On Growth and Form* during the 1920s and the 1930s on the biochemists whose work is chronicled in the present book. They rightly saw in D'Arcy Wentworth Thompson a pioneer, someone who had done at the macroscopic level what they were striving to emulate at the microscopic level, by establishing a correspondence between the structure of biomolecules and biological function.

To cite but a few names, Dorothy Wrinch was enormously influenced by D'Arcy, as indicated in the previous chapter. As for Linus Pauling, whose contributions are analysed in Chapter 15, he claims to have been relatively little influenced by D'Arcy (Pauling, 1983):

"When I began X-ray work on the structure of crystals in the fall of 1922, under the supervision of Dr. Dickinson, I was led by him to study the theory of the crystallographic space groups. I learned about symmetry in the course of this study and of my study of the known crystal structures. At some time in the 1920s, I think, I read the book by D'Arcy W. Thompson *On Growth and Form*. I remember that I was interested in it, but I doubt that it had very much influence on my work. I think that I can say that it was not a main factor behind my model building approach".

3. Reasons for the widespread influence of these two wise men

It is no accident, nor coincidence, if Bernal and D'Arcy meet in this part of our history: they had a number of traits in common, one of which was their lasting influence. Bernal was nicknamed 'Sage' and he was, for a while the '*guru* of science' (Goldschmidt, 1980). Medawar (1967) writes, likewise, that 'the influence of [D'Arcy Wentworth Thompson] *On Growth and Form* in this country and in America has been very great but it has been intangible and indirect'.

If we analyse the similarities between Bernal and D'Arcy, the following points are probably most important:

(*i*) Both were extremely gifted scientists, imaginatively endowed: Bernal was 'exceptionally fertile in thinking up hypotheses and equally ruthless in his search for evidence to substantiate or refute them' (Wooton, 1980). His power of rapid synthesis was legendary (obituary in the London *Times*, 1971). He could equally well argue a position and then its opposite (Admiral Mountbatten, quoted by Hodgkin, 1980).

(*ii*) Both these scientists were polymaths, whose extent of knowledge in very many diverse fields was enormous. D'Arcy Wentworth Thompson reminds one somewhat of Whewell or Fox Talbot, to take the examples of 19th-century scientists who were scholars in quite distinct disciplines. The anecdote of Bernal basing on a medieval romance a tactic for crossing the Channel on D-day is in the same ilk.

(*iii*) Both could fruitfully transfer what they knew in one discipline (say, maths or mechanics for D'Arcy; maths or physics for Bernal) and transfer it to another (biology). Hence, D'Arcy and Bernal proved by their work that quite a number of biological concepts could be understood with a textbook in elementary physics.

(*iv*) Both gave priority to static, structural concepts. This is easily grasped in Bernal's case. In D'Arcy's, I must again quote Medawar at length (Medawar, 1967; emphasis added):

"If D'Arcy was an anatomist, he was the first completely modern anatomist, in that his conception of structure was of molecular as well as of merely visible dimensions, his thought travelling without impediment across the dozen orders of magnitude that separate the two. *The advances that have occurred in modern biophysics and structural biochemistry are comprehended within D'Arcy's way of thinking.* We all now understand, though the idea was revolutionary in its day, that molecules themselves have shapes as well as sizes: some are long and thin, others broad and flat, some straight, some branched. We also know that crystalline structure is enjoyed by the huge molecules of proteins, nucleic acids and polysaccharides (. . .) and that the structure of the cell surface and of some of the 'organelles' enclosed by it must be interpreted in molecular terms. (. . .) Biology, and the chemistry and physics that go with it, have grown more rather than less anatomical in recent years, and the anatomy is indeed Thompsonian, *one which recognizes no frontier between biological and chemical form*".

(*iv*) Both D'Arcy Wentworth Thompson and Desmond Bernal

excelled in presentation of their ideas: here I would stress not only the beauty of their written English, but also praise their *iconic* minds. They were tremendous scientific illustrators. 'A picture is worth a thousand words', according to the Chinese proverb. By this token, *On Growth and Form* with an already lengthy text, has millions of words!

Bernal had also a very graphic mind. Often, he could get at the essence of a complex phenomenon simply by building a molecular model. He was a true pioneer in model-building. In the mid-30s, 'Bernal had produced a very versatile design of balls and spokes' (Hodgkin, 1984). For instance, he did a balls-and-sticks representation for a liquid, using distances between atoms and coordination numbers generated at random by the simple trick of, as he wrote (Bernal, 1960) 'doing the job in my office, where I was interrupted every five minutes or so. This enabled me to achieve almost perfect randomness, because by the time I got back to work I had forgotten what I had been doing last'. In the same work (Bernal, 1960), he placed spheres of plasticine (each standing for a molecule in a liquid) in a rubber bladder and exhausted the air so that the balls were pressed together tightly enough to fill all the space. By this neat little trick, Professor Bernal could recover compressed spheres deformed into irregular polyhedra, and show that five-sided faces predominated on these.

Doing the research for this book, as I was laying the groundwork for one chapter or for the next, I oscillated between two mutually incompatible attitudes. Sometimes, it seemed as if nature was revealing itself to the researchers and showing them totally unexpected properties: the monodispersity of proteins (Svedberg), the bewildering complexity of the conformation of a protein (Kendrew; Perutz) are examples of such surprises.

At the other times, it seemed as if the scientists knew in advance the answer, as with the polypeptide theory of Emil Fischer. The establishment by Linus Pauling of the basic structural elements in proteins, the α-helix and the β-pleated sheet, which is described in the next section, belongs to this second category. Desmond Bernal and D'Arcy Wentworth Thompson, without making an explicit, definite epochal advance, were nevertheless very much in the background of the elucidation of the molecular structure of proteins.

REFERENCES

Anonymous (1971) Times of London, September 16, obituary of John Desmond Bernal.

Bernal, J.D. (1929) The World, the Flesh and the Devil, Routledge, London (reprinted 1970, Jonathan Cape, London).

Bernal, J.D. (1939) The Social Function of Science, Routledge, London.

Bernal, J.D. (1954) Science in History, Watts, London.

Bernal, J.D. (1960) The Structure of Liquids, Scientific Am., August issue; reprinted in Supplementary Readings for Chemical Bond Approach (1961) Freeman, San Francisco.

D'Arcy Thompson, R. (1958) D'Arcy Thompson, Oxford University Press, Oxford.

Goldsmith, M. (1980) Sage: A Life of J.D. Bernal, Hutchinson, London.

Hodgkin, D.C. (1980) John Desmond Bernal, Biogr. Memoirs Fellows Roy. Soc. (London), 26, 17.

Hodgkin, D.C. (1984) Letter to Pierre Laszlo of April 2.

Laszlo, P. (1977) Intelligence Totale, Idéologies Totalitaires, ERIS, Bulletin du GERSULP, 4, 1-17.

Medawar, P.B. (1967) The Art of the Soluble, Methuen, London (reprinted 1969, Penguin Books, Harmondsworth, Middlesex).

Norton, B. (1980) Times Literary Suppl., May 23, 576.

Pauling, L. (1983) Letter to Pierre Laszlo of June 13.

Wooton, B. (1980) London Rev. Books, 19 June-2 July, 8.

Chapter 15

Molecular Order in Proteins : the Contributions from Linus Pauling

The purpose in having this as a companion chapter to a preceding one (Chapter 13) which dealt with the cyclol theory of protein structure by Dorothy Wrinch, is assuredly not to contrast a success story with the history of a pathetic failure. Not only would this be somewhat of a caricature, in historical as well as in psychological terms. It might also, by providing Dorothy Wrinch's story as a foil to the narrative of Linus Pauling's achievements, further a heroic view of the history of science – which, at best, would be a little simplistic.

No, my perspective in writing these two 'parallel lives' is closer to a chaotic view of history. Intellectual history, no less than the factual history of kingdoms and wars and great men, is a tale 'full of sound and fury', with great confusion and bewildering complexity. My goal, avowedly ambitious, is for this beautiful tale of the unraveling of protein structure not to be told idiotically as if it signified nothing, to continue elaborating on the famous quotation from *Macbeth* !

1. Pauling and bonding theory

Let us take the story, not from its apparent start in 1936 when Pauling brought out his first paper on the molecular structure of proteins (Mirsky and Pauling, 1936), but much earlier in the preceding decade, when Linus Pauling treated the nature of the chemical bond with a quantum-mechanical approach. The reason for back-tracking so much has to do with recognition that Pauling's genius has been very much that of *an importer of concepts*. He revolutionized biochemistry by helping to turn it into a molecular biology, because, probably more than any other single person, he injected his deep, detailed and precise knowledge of structural chemistry into gaining an understanding of the make-up of biopolymers. This he did in the 1940s and 1950s.

However, there had been a much earlier precedent for Pauling's role as a go-between, transferring key notions from one field to another. This was in the late 1920s, when Linus Pauling translated the brand-new notions of quantum mechanics into the language of structural chemistry – and this will probably remain his greatest achievement in the eyes of historians of science. This earlier episode can serve as a model for Pauling's role as an intermediary, as the ambassador of discipline A into discipline B. Furthermore, the simple structural ideas and rules he formulated in 1930 in the transfer from quantum mechanics to structural chemistry continued to be all-important when, 20 years later, he completed successfully another transfer, from structural chemistry to biochemistry: this was when he proposed as main components for protein structure, the α-helix and the β-pleated sheet.

This is easily substantiated from the text of Pauling's Nobel Lecture (Pauling, 1954). In consonance with the citation made to him when awarded the Nobel Prize ('for his research into the nature of the chemical bond and its application to the elucidation of the structure of complex substances'), Pauling dealt with chemical bonding in most of his lecture: seven out of a total of eight pages in the printed version (Pauling, 1954). He gave less than one page to the elucidation of the structure of proteins, even though this crowning achievement was quite recent work which he had completed during the period 1950–1953 and which, undoubtedly, made the Nobel Committee choose him for the 1954 chemistry prize. Another piece of evidence, in confirmation of the prime importance given by Pauling himself to his work on chemical bonding, is the choice of a single illustrative article to be reprinted in the *Festschrift* volume entitled *Structural Chemistry and Molecular Biology* (Rich and Davidson, 1968) which his former students, colleagues and friends offered him for his 65th birthday in 1966: this is the 1931 *JACS* article on the nature of the chemical bond (Pauling, 1931).

Perusal of this seminal article, written when Pauling was 30 – true, he had already about 25 papers on crystal structures plus 10 papers on quantum mechanics and 4 papers on the chemical bond under his belt – is extremely useful to our purpose. Pauling stresses in it (Pauling, 1931) that the new rules he formulates will supplement those of Gilbert N. Lewis. They will allow one to predict and estimate the relative strengths of bonds, valence angles, restricted internal rotation around bonds, and bond orders. His rules consist of using only one eigenfunction (atomic orbital, as we would now say) from each atom, in looking for maximum overlap, and in ranking atomic orbitals according to their increasing energy, in the sequence $s, p, d, f, \ldots$ The bonds are two-electron bonds with paired spins for the electrons. Once utilized for a given bond, an electron pair is unavailable for another bond.

Some of the properties Pauling deduces from these rules are that p electrons form stronger bonds than s electrons; that bonds formed by p electrons are at right angles to one another as exemplified in bent H_2O or pyramidal NH_3; that formation of sp^3 hybrids explains tetrahedral carbon; that likewise trigonal sp^2 hybrids are responsible for the geometry in benzene or graphite; and that heteronuclear diatomics can be described as a mixture of the covalent and the ionic structures.

Perusal of this 1931 paper also indicates the reason why Linus Pauling became quickly such an influential chemist: *he explains things*. Moreover, in opposition to a radical chemical innovator, who would have militantly overthrown earlier ideas in the name of quantum chemistry, Pauling makes himself co-terminous with earlier knowledge. There is total continuity between his work and the earlier Lewis structures for molecules, with the notion of valence, etc. Furthermore, Pauling tackled *well-established phenomena*, such as tetrahedral carbon, the bent water molecule, the linear hydrogen cyanide molecule, free rotation around single bonds, and so on. In doing so, his gift is in focussing on essentials, even if his treatment is only approximate (in our time, Roald Hoffmann exemplifies a similar approach). And all these qualities are capped by a superb teaching talent: Pauling is easily understandable, he skips entirely sophisticated mathematics, the rules he gives are simple and they are widely applicable, to atoms in general.

One additional feature is worthy of note in this same article on 'The Nature of the Chemical Bond': Pauling's lifelong fascination with the hydrogen bond. He describes it there (Pauling, 1931) with an electrostatic structure, so that for instance a hydrogen bond between two oxygen atoms is written : $O^{2-}\ H^+\ O^{2-}$.

Before we turn to Pauling's solution to the protein problem, one more remark regarding both his didactic genius and his skill in transferring notions from one field to another. This has to do with his characteristic and widespread use of quotations: whenever Pauling gave a lecture or wrote an article for popularization, his text swarms with literary quotations. For instance, in 1946, in his Harrison Howe lecture in Rochester, Pauling quoted Voltaire ('the secret of being a bore is to tell everything'), Robert Browning ('The long habit of living indisposeth us for dying'), and he concluded by citing Walt Whitman, also on Life and Death ('The two old, simple problems ever intertwined,/ Close home, elusive, present, baffled, grappled,/ By each successive age insoluble, passed on/ To ours today – and we pass on the same') (Pauling, 1946). Likewise, in his Friday Evening Discourse at the Royal Institution in London (February 27, 1948), Pauling quoted in succession Thomas Wright, Chaucer, Francis Bacon, and Lucretius (Pauling, 1948). That Pauling was a well-read, erudite scholar who liked to bridge the 'Two Cultures' is extremely relevant to his historical role: it is an important key to it.

2. The protein research problem for Linus Pauling

Twice, Linus Pauling had a go at the structure of proteins: in 1936, when he published his paper with Mirsky to draw attention to the importance of hydrogen bonds and to an explanation of protein denaturation in terms of a change in molecular conformation; and in the early 1950s, when he proposed with Corey the α-helix and the β-pleated sheet (Mirsky and Pauling, 1936; Pauling and Corey, 1950; 1951a–h; 1953; Pauling et al., 1951).

The end of the 1930s is a good time to take stock of the protein structural problem as it appeared then to Linus Pauling: because this period lies in between his two thrusts at a solution, and because we are fortunate to have in writing his formulation of the protein research problem in articles which he published in 1939 and in 1940.

(i) Pauling could build on earlier work, that of Astbury, who had used x-rays to indicate (nearly conclusively) the intervention of polypeptide chains in the extended structure for certain types of fibrous proteins. Dorothy Hodgkin has given an excellent summary of this early work of Astbury, which started in 1926 and continued through most of the 1930s (Hodgkin, 1979; emphasis added):

"After his Christmas lectures at the Royal Institution on 'Old Trades and New Knowledge' in 1924 (W.H. Bragg) planned a new lecture 'The Imperfect Crystallization of Common Things' in 1926. He asked W.T. Astbury to take some (X-ray) photographs for him of fibers like wool and hair (. . .). Astbury (as Bernal recorded) took to them from the beginning; they led to his first formal appointment as lecturer in Textile Physics at Leeds in 1928 *and became his life's work*. With H.J. Woods and A. Street, he discovered that wool gave two types of X-ray, α when unstretched, β when stretched (Astbury and Woods, 1930; Astbury and Street, 1931). He deduced that the α photograph, which gave only two features, an equatorial 'reflection' of spacing 10 Å and a meridional 'reflection' of spacing 5.1 Å, represented some type of folded peptide chain that *could be reversibly stretched to a straight chain*, giving the β photograph. The β photograph, which showed two equatorial reflections of 10 Å and 4.65 Å and one meridional one, 3.4 Å, clearly, to his eyes, was the same type as that of silk fibroin. As the years passed, he found the same types of photograph given by many proteins, keratin, myosin, epidermin, fibrinogen (. . .)".

Hence, Pauling could build on this prior knowledge, the distinction among fibrous proteins between those like silk fibrin and β-keratin which had an extended polypeptide chain, and those like α-keratin and other fibrous proteins which existed in a folded configuration (Pauling and Niemann, 1939). And Pauling accepted the Astbury work as essentially right, even if the analysis of the x-ray had been done with 'great ingenuity but without full accuracy' (Bernal, 1968), because of one feature – which, as we shall find repeatedly, was of overwhelming importance to Pauling – : continuity with the structures for simple organic molecules. The Astbury (1933) structure for β-keratin, Pauling pointed out (Pauling, 1940), was consistent with those obtained for glycine (Albrecht and Corey, 1939) and diketopiperazine (Corey, 1938).

(*ii*) To the notion of a polypeptide chain, as proposed by Emil Fischer (Chapter 12), which could exist in a fully extended configuration in fibrous proteins, as shown by Astbury, Pauling added his own finding of the planarity of the peptide bond due to the amide resonance. Here his view was in conflict with conventional chemical knowledge (Bernal, 1968):

"The elementary chemical view was that the configuration about the bond between the first carbon atom and the succeeding amide nitrogen was to be a simple, single bond capable of any orientation. Pauling, however, pointed out that there would be some resonance between the planar carboxy group and the carbon–nitrogen bond, resulting in a shortening of the carbon–nitrogen bond and the introduction of a partial resonance of 50 percent double-bond character that would prevent rotation about the bond".

In 1939, in his critique with Niemann of Dorothy M. Wrinch's cyclol theory, Pauling put the magnitude of this amide resonance at 21 kcal.mol^{-1} (Pauling and Niemann, 1939).

He felt strongly about its intervention, having worked on this problem during the past few years, as he indicated in his Nobel Lecture (Pauling, 1954):

"(. . .) the amide group, an important structural feature of proteins, can be described as resonating between two structures, one with the double bond between the carbon atom and the oxygen atom, and the other with the double bond between the carbon atom and the nitrogen atom (Pauling, 1932). (. . .) the structure with the double bond between carbon and oxygen should contribute somewhat more to the normal state of the amide group than the other structure; experience with other substances and acquaintance with the results of quantum mechanical calculations suggest the ratio 60% : 40% for the respective contributions of these structures. A 40% contribution of the structure with the double bond between the carbon atom and the nitrogen atom would confer upon this bond the property of planarity of the group of six atoms; the resistance to deformation from the planar configuration would be expected to be 40% as great as for a molecule such as ethylene, containing a pure double bond, and it can be calculated that rotation of one end by 3° relative to the other end would introduce a strain energy of 100 cal.mol^{-1}. The estimate of 40% double-bond character for the C–N bond is supported by the experimental value of the bond length, 1.32 Å, interpreted with the aid of the empirical relation between double-bond character and interatomic distance" (Pauling et al., 1935).

After his predictions had been borne out by the x-ray structure determinations for glycine (Albrecht and Corey, 1939) and for dike-topiperazine (Corey, 1938), Pauling was adamant about the planari-ty of the peptide bond.

(iii) Combining points (i) and (ii) with Occam's razor, a tool for which Linus Pauling had great affection - Pauling in his JACS arti-cle of 1940 on structure and properties of antibodies refers to it somewhat humorously (Pauling, 1940) as 'the rule of parsimony (the use of the minimum effort to achieve the result)' - he came up with his 'stack of pancakes' structure for globular proteins (Pauling, 1940). He arranged the polypeptide chain in its extended configura-tion, so that there would be a number of parallel peptide chains between which he drew hydrogen bonds. In Pauling's words (Pau-ling, 1940) :

" . . . a layer of finite size could be constructed by running a single polypeptide chain back and forth. A globular protein could then be made by building several such layers parallel to one another and in contact, like a stack of pancakes, the layers being held together by side-chain interactions as well as by the polypeptide chain itself. A pro-tein molecule with, for example, roughly the shape and size of a cube 40 Å on edge might contain four layers, each with about eight strings of about twelve residues each".

This structural hypothesis of 1940 is fully equivalent, in the episte-mological sense, with the structural hypotheses of 1950-51 (Pauling and Corey, 1950; 1951a-d). Of course, the former was abandoned, and the latter triumphed: but this is the perspective from hindsight! Let us note also that the hypothesis of 1940 is pregnant with the future β-pleated sheet of 1951 (Pauling and Corey, 1951c).

(iv) Another idea of Pauling in 1940 about the structure of pro-teins which did not stand very well the test of time was his notion of 'proline interference' as an explanation for the polymorphism of immunoglobulins (Pauling, 1940). The presence of proline (or hydroxyproline), because it was incompatible with an extended chain, would cause the 180° turns demanded by the 'stack of pan-cakes' structure. Since his theory of immunoglobulins (see below) assumed a multiplicity of terminal conformations with similar ener-gies, between which the antigen could pick the particular conforma-tion uniquely suited for specific binding, Pauling thought again that

structure breaking by proline would provide the necessary polymorphism (Pauling, 1940; italicized by myself):

" . . . *the end parts of the globulin polypeptide chains contain a very large proportion
(perhaps one-third or one-half) of proline and hydroxyproline residues and other
residues which prevent the assumption of a stable layer configuration*".

And Pauling went on (Pauling, 1940) to give the example of gelatin,
in which about one third of the residues are proline or hydroxyproline, as a protein molecule 'not characterized by a single well-defined
molecular configuration':

(v) Indeed, the master idea for Pauling in the late 1930s and early
1940s was that (Pauling and Niemann, 1939) 'a native protein with
specific properties must possess a definite configuration, involving
the coiling of the polypeptide chain or chains in a rather well-
defined way'. (The same idea, although in less concise a formulation,
had been already expressed three years before by Pauling and Mirsky (1936).) Notice the insistence: biological specificity arises from
a well-defined geometry of its molecular correlate, the protein. This
is an extremely important notion, originated by Pauling, to which I
shall return below.

(vi) But in 1939 and 1940, when for instance he criticized forcefully the Wrinch theory of protein structure, Pauling was conscious of
expressing, not only his own private view, but a more general consensus. He shared his notions about the structure of proteins with a
whole group of people which included assuredly J.D. Bernal and D.
Crowfoot-Hodgkin, W.T. Astbury, R.B. Corey, C. Niemann, etc.
Thus, the modern reader finds, tucked into a footnote to their article
(Pauling and Niemann, 1939) this sense of expressing the dominant
paradigm:

"We believe that our views regarding the structure of protein molecules are essentially the same as those of many other investigators interested in this problem".

Lest it be thought that I am making too much out of this single
sentence, the same point is reiterated twice in this article (Pauling
and Niemann, 1939): a sub-heading for the last section reads 'The

Present State of the Protein Problem' and the article itself, even though it is predominantly a devastating critique of the Wrinch cyclol theory, is entitled (somewhat misleadingly) *The Structure of Proteins* !

(*vii*) The title of the 1939 article by Pauling and Niemann was self-assertive; yet, it did not propose a detailed, fully worked-out geometrical structure for proteins. Why not? Or, to put the same question in a slightly different formulation, what prevented Pauling, after he had correctly imputed hydrogen bonds as the secondary forces responsible for holding the native conformation of a protein with Mirsky in 1936, from experimenting with molecular models and discovering the α-helix and β-pleated sheet structural elements much earlier? Why a 14-year delay between Mirsky and Pauling (1936) and Pauling and Corey (1950)? Basically, there is nothing in the Pauling and Corey papers of 1950–51 which Pauling could not have achieved in the late 1930s. How can one explain the fact that he did not rush immediately to such conclusions?

Part of the answer is prudence. Linus Pauling was a cautious man. He had just witnessed Dorothy Wrinch deluding herself with the cyclol theory, because she refused to see some internal contradictions in her proposals. And Pauling, perhaps because of a positivistic bias, surely because of his scientific upbringing as an x-ray crystallographer, starting in 1922, would not proceed with the protein problem till more data would become available. Thus we find, in the 1939 Pauling and Niemann paper, the notion that prior knowledge of the x-ray structure of amino acids is required (Pauling and Niemann, 1939; emphasis added):

"A protein molecule, containing hundreds of amino acid residues, is immensely more complicated than a molecule of an amino acid or of diketopiperazine. Yet despite attacks by numerous investigators no complete structure determination for any amino acid had been made until within the last year, when Albrecht and Corey succeeded, by use of the Patterson method, in accurately locating the atoms in crystalline glycine (Albrecht and Corey, 1939). The only other crystal with a close structural relation to proteins for which a complete structure determination has been made is diketopiperazine (Corey, 1938). *The investigation of the structure of crystals of relatively simple substances related to proteins is being continued in these Laboratories*".

This notion of the x-ray structure of amino acids as a pre-requisite for an understanding of protein structure was part of the paradigm mentioned earlier (point (*vi*) above). We find for instance J.D. Bernal prefacing a paper on the crystal structures of about 15 natural substances with these words (Bernal, 1931; emphasis added):

"a knowledge of the crystal structure of the amino acids is essential for the interpretation of the X-ray photographs of animal materials : silk fibroin, keratin, collagen, *proteins*, etc., which have been studied by this method for the first time in the last few years".

I shall quote Bernal again to point out an apparently stringent limitation, which may also have been responsible for delaying Pauling's discovery of the α-helix by many years. Linus Pauling was a trained x-ray crystallographer, and the notion of non-integral repeats along a helical structure as embodied in the α-helix was not only quite novel when he introduced it in 1950 but also a challenging, not to say a heretical one (Bernal, 1968; emphasis added):

"These (Pauling, 1950) ideas straight away very much simplified the problem of protein structure *but they required very large modifications of the basic ideas of crystallography.* These ideas had contained the restriction that a helix was possible in crystals only with a helicity of 2-, 3-, 4-, and 6-fold symmetry, the screw axes of elementary space-group theory. It was not a new idea, by any means, that the peptide chain could be helical, but this limitation appeared much too stringent to account for the variety of protein structures. In fact, there was no real reason why the crystallographic limitation of symmetry should apply to the internal structure of a molecule. It only strictly applied to relations of separate molecules in the same cell. *The stroke of genius on the part of Pauling was to abandon the idea of integral repeats along a helix and to substitute a helix of peptides with an irrational and, therefore, not exactly repeating structure".*

(*viii*) Besides these reasons – caution, the need to know more about amino acids, the implausibility of helical structures – I submit that during the late 1930s Linus Pauling did not care that much about the structure of proteins. Rather, he put the emphasis on their biological significance. From 1935 to 1945, Pauling published 10 papers on proteins, eight of which deal with oxygen binding and the magnetic properties and structure of hemoglobin. During the same period, he also published 21 studies on antigen-antibody precipitation, most of which are quite specialized and rather technical. Out of a total of 31 papers dealing with proteins in general, *only five addressed themselves to the general problem of the structure of proteins.* What Pauling was mostly interested in were the molecular correlates of biological complementarity. He stated this research program in his paper with Delbrück entitled *The Nature of the Intermolecular Forces Operative in Biological Processes* (Pauling and Delbrück, 1940). They wrote:

"It is our opinion that the processes of synthesis and folding of highly complex molecules in the living cell involve, in addition to covalent-bond formation, only the intermolecular interactions of van der Waals attraction and repulsion, electrostatic interactions, hydrogen-bond formation, etc., which are now rather well understood. These interactions are such as to give stability to a system of two molecules with *complementary* structures in juxtaposition, rather than of two molecules with necessarily identical structures; we accordingly feel that complementariness should be given primary consideration in the discussion of the specific attraction between molecules and the enzymatic synthesis of molecules".

When Pauling and Delbrück wrote this, Pauling had already been working for four years to 'understand and interpret serological phenomena in terms of *molecular* structure and *molecular* interactions' (Pauling, 1940; emphasis added). By making the simplest, most parsimonious hypothesis, that of the bivalence of antibodies – using an obvious *simile* with the structure of small molecules – Pauling could visualize the three-dimensional network formed in the antibody-antigen precipitation. And, as I have indicated in point (*iv*) his first thoughts to account *in molecular terms* not only for the specificity in the recognition of a given antigen by the corresponding immunoglobulin but also for the apparent identity of all the immunoglobulins, were to postulate identical chain sequences with a vast number of conformations that could be assumed by the end parts.

In the late 1930s and early 1940s, to explain biological specificity and its determinant mode of complementarity was Linus Pauling's over-riding concern. Following the above-quoted paper outlining *A Theory of the Structure and Process of Formation of Antibodies* (Pauling, 1940), Pauling worked out his ideas in a series of approximately 30 papers during the period 1940-1950. And when I put to him the central question of this book in the following form:

"In our century, from hindsight, the penetration of molecular concepts in biochemistry appears to have been rather slow. It was resisted mostly from a vitalist standpoint. For instance, Sumner and Northrop (1946), Staudinger (1953) and yourself (1954) all received the Nobel Prize for work performed mostly in the 1930s. Would you agree with such an analysis? Would you help me in identifying the main factors for such a resistance to accepting molecular order as a basic feature of biology?"

Pauling answered by stressing the importance of his ideas and work on molecular complementarity as an explanation of biological specificity (Pauling, 1983; LP makes a small error in the title of his 1940 *JACS* article which I have quoted above because he is working in Palo Alto from memory and not from his Big Sur house, where he has his early papers):

"I think that the most important factor in the resistance to molecular concepts in biochemistry in the period around fifty years ago was that no molecular explanation of biological specificity had been developed. I had thought a great deal about this problem during the 1930s, and had finally published my ideas about detailed molecular complementariness in structure of interacting molecules in 1940, in a paper with the title *The structure of antibodies and the nature of serological reactions*. Also in 1940 Max Delbrück and I published our note stating that Pascual Jordan was wrong in contending that the special quantum mechanical interaction of identical molecules would permit a molecule such as a gene to determine the synthesis of another molecule identical with itself. We stated that it is detailed complementariness in structure that permits such interactions to take place, and that the gene consisted of two strands, mutually complementary, each of which, after unravelling, could serve as the template for the synthesis of a new strand identical not to itself but to the other one. I amplified this in perhaps a dozen papers during the next eight years, in addition to the large numbers describing the results that we had obtained in our immunochemical investigations. I think, however, that this work, published almost entirely in chemical journals, was ignored by biologists, so that the idea of molecular complementariness, which was the basis for the discovery of the double helix, was delayed until the discovery of the double helix of DNA. I note that in modern books on biology the history of the concept of detailed molecular complementariness as the basis for life is almost entirely ignored".

The reader will recall that we have already dealt in Chapter 6 with complementarity as an essential component of Pauling's 'scientific style'.

3. Nature of the secondary bonds

The first systematic study of the hydrogen-bonding phenomenon, as we term it nowadays, dates back to 1920 (Latimer and Rodebush, 1920). When Pauling, in the 1930s, puzzled about the nature of the forces holding the conformation of a protein molecule in a well-defined arrangement, both his learning and his taste favored hydrogen bonds. I would even go further and say that Pauling could not help favoring hydrogen bonds as responsible for holding the polypeptide chain into a well-defined coil, which is the substance of his famous article with Mirsky (Mirsky and Pauling, 1936).

The hydrogen bond was an old favorite of Linus Pauling. At 27 he had proposed an electrostatic model for this phenomenon, exemplified when a hydrogen atom is wedged in between two oxygens, a situation which he represented as $O^{-2} H^+ O^{2-}$ (Pauling, 1928) using a point-charge model. Thus, from very early on Linus Pauling thought he understood the essentials of hydrogen bonding, a pervasive phenomenon whose presence he encountered often in his structural work.

Here is what Pauling wrote recently about his early interest in hydrogen bonding (Pauling, 1983):

"I had become interested in the hydrogen bond at the beginning of my research career. Latimer and Rodebush had written their important paper on the hydrogen bond in 1920, and I probably read it at that time or a year or two after its publication. It soon became clear to me that hydrogen bonding was an important factor in determining the structure of inorganic crystals. I remember mentioning the hydrogen bond in my 1928 paper on quantum mechanics and molecular structure published in *Proceedings of the National Academy of Sciences*. Sometime around 1930 I published the structure of ammonium hydrogen difluoride on the basis of the idea that the deformation from the structures of the corresponding sodium and potassium compounds results from the formation of hydrogen bonds. My student and I published many structures of minerals and other inorganic compounds in which hydrogen-bond formation plays an important part. I had also formulated structures of some organic gas molecules, such as the dimer of formic acid, by invoking hydrogen-bond formation. It was clear to me that polypeptide chains would form hydrogen bonds. Mirsky, during the period that he spent in Pasadena, gave me a large number of facts about the properties of proteins, and I was able to develop a comprehensive theory of protein structure, as you know".

Pauling was not alone in stressing the importance and ubiquity of hydrogen bonding. The year following publication of Pauling's theoretical work on the hydrogen-bond interaction (Pauling, 1928) appeared the first x-ray determination of the structure of ordinary ice (ice Ih) (Barnes, 1929): in full agreement with Pauling's predictions, it showed bent water molecules were arranged in a tetrahedral array so that each H_2O had four oxygen atoms as nearest neighbors at an oxygen–oxygen distance of 2.76 Å. A few years after Pauling gave his elementary electrostatic description in the course of his investigation into the nature of the chemical bond (Pauling, 1928), this description was greatly refined by Bernal and Fowler (1933).

Their collaboration started during enforced inactivity in an airport terminal, to while away the time (Hodgkin, 1980). This is how Edsall summarizes the Hodgkin account of this epochal encounter (Edsall, 1983) :

"In September 1932 Bernal and Fowler, with other colleagues from East and West, had been involved in theoretical discussions for physicists and chemists at the School of Physical Chemistry in Moscow. The British group planned to return home by air and arrived at the airport at 4 a.m. as directed. There was a thick fog; no flights were possible, and in the new airport there was no place to eat or even sit. They walked up and down talking, and Fowler asked Bernal about water and the relation between its structure and its behavior. Bernal poured forth the story of water as it was then known – he knew nearly everything that was known. The fog did not lift for 12 hours; their talk continued, and Fowler said to Bernal 'you must write it up' ".

Bernal and Fowler's contribution was to refine somewhat the Pauling first-order analysis by explicit inclusion of dispersion and repulsive forces. They studied hydrogen bonding in water, using a point-charge model for the water molecule. This was complemented by an estimate of the dispersion forces from a modified London formula, and they evaluated the 'repulsive energy from the condition that the net force on a molecule must vanish' (Eisenberg and Kauzmann, 1969). They arrived thus at a figure of 7.1 kcal.mol^{-1} for the energy of a hydrogen bond. They postulated that the geometric arrangement of molecules in liquid water was related to that in ice. In Edsall's words, however (Edsall, 1983) 'the idea, obviously due to Bernal, that liquid water could be described as a mixture of regions resembling crystals of ice, tridymite and quartz (the proportions varying with temperature) did not prove fruitful and was abandoned'. It is significant, in my opinion, that in the early 1930s both Pauling at CalTech and Bernal in Cambridge would share these two interests, of the nature and properties of hydrogen bonds, and of protein structure. In 1935, Pauling went on to suggest that, among the ice polymorphs, ordinary ice is characterized – because of the hydrogenbond network – by a proton-disordered structure (Pauling, 1935): 'at each oxygen site there are six possible hydrogen-bonded orientations for the water molecule, and these occur statistically, with equal probability' (Kamb, 1968).

Thus, it was natural that Linus Pauling should think of the hydrogen bond to stabilize a well-defined conformation of the polypeptide chain in globular proteins (Mirsky and Pauling, 1936). Furthermore, such a concept was consistent with the methodological precept he seems to have given himself very early on: not to seek other factors in the construction of large, complex molecules than those already operative in much simpler structures. Another factor in the publication of their ideas about protein structure by Mirsky and Pauling in 1936 was the prior appearance in 1931 of a paper by a Chinese scientist, who had proposed that biological specificity was expressed in a well-defined geometric arrangement of the polypeptide chain of a protein (Wu, 1931). It is interesting to note that in his early papers on protein structure, Pauling repeatedly gives credit to Wu for pointing to the relationship between the well-defined properties of native proteins and a definite configuration for the polypeptide chain (see for instance Pauling and Niemann, 1939; Pauling, 1940).

Not long after the appearance of the predictions by Mirsky and Pauling (1936) regarding the molecular structure of the proteins, appeared the first x-ray structural determination from crystals of an amino acid: Albrecht and Corey published in 1939 their structure of α-glycine (Albrecht and Corey, 1939). This was an extremely important epistemological step: one will recall the extreme reluctance of Linus Pauling to speculate about the details of protein structure in the absence of hard data on the structure of the component amino acids.

This structure of α-glycine, more specifically, was important on two counts. It confirmed Pauling's expectation, based on resonance theory, of the planarity of the amide group. However, and whereas he had always worked on the assumption of the linearity of hydrogen-bonded systems – three atoms $O_1 \ldots H-O_2$ linked by a hydrogen bond are in a line – , the crystal structure of α-glycine was consistent with the coexistence of linear *and* bifurcated hydrogen bonds (Donohue, 1968):

"[Albrecht and Corey (1939)] found that the $-NH_3^+$ group has four near oxygen neighbors, at 2.76, 2.88, 2.93, and 3.05 Å. Although they had no direct evidence for the location of the hydrogen atoms, they also found that it was possible to orient a tetrahedrally disposed $-NH_3^+$ group such that one hydrogen atom was very nearly directed towards the oxygen atom at 2.76 Å, the second hydrogen atom was very nearly directed towards the oxygen atom at 2.88 Å, and the third hydrogen atom was 'sharing its bond-forming capacity nearly equally between two nearest oxygen atoms (at 2.93 and 3.05 Å) in the adjacent layer' ".

Before we move on to the model-building which in the early 1950s suggested to Pauling and Corey (1950; 1951) the α-helix and the β-pleated sheet as important structural components of proteins, I wish to raise a minor issue: why did Pauling emphasize specifically or exclusively hydrogen bonds for the weak forces stabilizing a protein conformation rather than, e.g., hydrophobic attractions between the side chains? We now know that the interior of globular proteins consists predominantly of the side chains from hydrophobic residues; and that such hydrophobic contacts or 'forces' play an important role in maintenance of the native structure for a protein. Furthermore, such hydrophobic interactions *had been recognized in the late 1930s* [this a full 20 years before Kauzmann's classical paper on hydrophobic interactions (Kauzmann, 1959)]: we find, for instance Dr. J.F. Danielli, in the Royal Society Symposium of 17 November 1938, in London (Danielli, 1939; emphasis added), interpreting his experiments on protein monomolecular films in such terms:

"Similar forces involving a free energy change of the same order of magnitude will operate on the hydrocarbon residues of the protein in solution, tending *to roll the polypeptide chain into a form where the hydrocarbon residues are in close contact and the polar groups are directed towards the aqueous phase*".

Dorothy Wrinch also mentioned the possibility of such hydrophobic interactions, to give another example (Wrinch, 1939). She considered them, in addition to other factors, as a source of stabilization for (her cyclol) protein structure; and she wrote (Wrinch, 1939) of 'the coalescence of the hydrophobic groups in the interior of the cage'. Hydrophobic interactions were pushed aside by Linus Pauling in the late 30s, probably on two counts: (*i*) geometrically speaking, they involved many more atoms than did hydrogen bonds; and therefore, they were less simple to envisage on a molecular model; (*ii*) and their energy was assuredly lower, by at least a factor of two, than the energy of a typical hydrogen bond. Indeed, the latter was implicit in the argument opposed to Langmuir and Wrinch who had mentioned hydrophobic interactions as a possibility for stabilizing cyclol structures (Langmuir and Wrinch, 1939): in his article with Niemann, Pauling wrote:

"The stabilizing effect of the coalescence of the hydrophobic groups has been estimated (Langmuir and Wrinch, 1939) to be of about 2 kcal.mol^{-1} per CH_2 group, and to amount to a total for the insulin molecule about 600 kcal.mol^{-1}. (. . .) The maximum of 600 kcal.mol^{-1} from this source is still negligibly small compared with the total energy difference to be overcome [between a polypeptide and a cyclol structure] . . . " (Pauling and Niemann, 1939).

4. Model building

I have devoted an earlier section (Chapter 4, section 1) to 'Linus Pauling, as the Pioneer Molecular Architect'. He was indeed one of the pioneers in the use of molecular models in the 1920s and 1930s.

By the 1940s, all the scientists (especially the x-ray crystallographers) working on the protein structure problem toyed with models, hoping to come up with a reasonable structure matching the experimental evidence. These are, for instance, Hodgkin and Riley's recollections on insulin-related model-building (Hodgkin and Riley, 1968):

"There was a period in the early 1940's, before the real breakthrough in protein X-ray analysis, when we all turned to model building. By comparison with other proteins, the molecule was so small that it seemed that, at least for insulin, we might be able to

278   MOLECULAR ORDER IN PROTEINS

devise a model that would fit the diffraction effects. The first problem we considered
was the form of chain folding present. Partly led on by the trigonal symmetry of the
crystal, we first built a model of a spiral peptide chain with a three-fold screw axis of
symmetry, which we soon discovered had been built before us, both by M.L. Huggins
(1943) and by H.S. Taylor (1941). We then built trial structures in which short
lengths of such chains were placed parallel with the insulin three-fold axis in positions suggested by Bernal's observations on the insulin Patterson projections and the
correlation of parts of the distribution with particles in close packing. Luckily we had
with us Dr. Käthe Schiff, who destroyed this model practically at birth. She convinced us that the rather even distribution of observed intensities in insulin excluded
any structure based predominantly on chains parallel to the c-axis and also any chain
structures having a regular repeat along the chain axis".

There were many other attempts at model-building, besides those of
Wrinch (preceding chapter), Hodgkin and Riley (1968), Astbury
(1939), Huggins (1942; 1943), Taylor (1941), etc.
    Pauling's orgy of model-building, which was to lead to the estab-
lishment of the α-helix and β-pleated sheet as important components of the protein structure, happened in 1948 while he was
bedridden (probably with the flu) in Oxford, England (Hodgkin,
1979; emphasis added):

"The alpha helix was just a stroke of genius on the part of Linus Pauling himself.
(. . .) I know because I was there. He was having a cold or flu in Oxford the year he
was there and was rolling bits of paper around his fingers, which took him away from
the crystallographic repeats and allowed the helix to go its own way".

Hodgkin's account is complemented by Bernal's, who gives pictures
(Figs. 3 and 4 in Bernal, 1968) of the unrolled planar networks of
parallel polypeptide chains with hydrogen bonds joining them
which Pauling was twisting around his fingers, and which appeared
later in print (Pauling et al., 1951):

"(Pauling) built models by drawing planar extended sheets of peptides on the correct
scale and then rolled them up somewhat askew to form tubes, joining the edges so as
to obtain a perfect fit of the atoms in the peptide chain. Thus he arrived at two
helices, both nonintegral [abandoning the unnecessary restriction of integral repeats
along a helix], the α or 3.7 residue helix and the γ or 5.1 residue helix. In the α helix he
found that each planar amide group could be hydrogen-bonded to the third amide
group beyond it all along the helix, and that in the γ helix each is bonded to the fifth
amide group beyond it" (Bernal, 1968).

But Linus Pauling, ever the cautious, exact and deliberate scientist, did not rush into print as soon as (1948) he had these exciting models into his hands. He went back to CalTech, in Pasadena, where he knew he could find the best structural data gathered over the years mostly by Dr. R.B. Corey from the x-ray pattern of simple amino acids and peptides.

This had been his avowed strategy from the very start. Here again we have J.D. Bernal's recollections (Bernal, 1968; emphasis added):

"By 1940 it was clear that a successful attack on the complete protein structure could be made, but there were still many difficulties. Two modes of attack suggested themselves: the first was a straightforward X-ray crystallographic study of crystalline protein, using all the techniques of an advanced crystal analysis. Computers were not available for this until much later, in the mid-1950s. The second was a model building method *based on an exact knowledge of the structure of amino acids and smaller peptides* themselves and an attempt to build up the protein a priori and then check the structure by X-ray methods. I remember very well discussing the problem with Pauling just *before the war. He was in favor of the second method*, which I thought indirect and liable to take a very long time. Nevertheless, it was Pauling's ideas that were to have a decisive effect on the result. The series of structures of the amino acids came from R.B. Corey and his colleagues at the Gates and Crellin Laboratories at the California Institute of Technology. These workers were able to establish far more accurately than I had been able to do (Bernal, 1931) the structure of these molecules. The key to the configuration of the polypeptide structure lay in the X-ray analysis of the di- and tripeptides. Here Pauling's chemical intuition enabled him to introduce the decisive formulations that were to help to fix the configuration".

And indeed, Pauling and Corey's first report of 'Two hydrogen-bonded spiral [sic] configurations of the polypeptide chain' (Pauling and Corey, 1950) opens with the following sentences:

"During the past fifteen years we have been carrying on a program of determination of the detailed atomic arrangement of crystals of amino acids, peptides, and other simple substances related to proteins, in order to obtain structural information that would permit the precise prediction of reasonable configurations of proteins. We have now used the information to construct two hydrogen-bonded spiral configurations of the polypeptide chain, with the residues all equivalent, except for variation in the side chain".

Noteworthy in this first paragraph are the allusion to the elucidation of the structure of α-glycine (Corey, 1938), the (assuredly Pauling-inspired) notion of continuity in the principles of molecular architecture from simple amino acids to protein, and the need for accurate structural information on simpler systems prior to any 'precise prediction' about protein conformation. A further methodological constraint, which perhaps originated at the time of the Pauling and Niemann (1939) critique of Wrinch's cyclol theory, was the requirement of equivalence between *all* the amino acid residues. This very point is forcefully reiterated in the next sentence (Pauling and Corey, 1959; emphasis added):

"We have attempted to find all configurations for which the residues have the *interatomic distances and bond angles found in the simpler substances and are equivalent*, and for which also *each* CO group and NH group is involved in the formation of a hydrogen bond".

Interestingly, this letter to the editor of *JACS* (Pauling and Corey, 1950) carries a single reference pointing out that if Huggins (1943) had described a somewhat similar three-residue 'spiral', his earlier proposal differed from theirs in *essential structural details* (emphasis added).

I do not want to give the impression that Linus Pauling was a finicky and picky individual, on the contrary: his considerable successes as a scientist have been due to this emphasis on establishing conclusions on the most accurate data available. Leaps of intuition were useful in formulating hypotheses, but these had to be firmly grounded on a wealth of structural facts.

Such deliberation on Pauling's part could have cost him the priority in establishing the nature of the ordered structural parts of proteins, as it was later to lose him the credit for the discovery of the double helix ! Indeed, he was beaten into print by a few months in 1950 when Sir Lawrence Bragg, J.C. Kendrew and M.F. Perutz published their ideas on 'Polypeptide chain configurations in crystalline proteins'. Their article in the *Proceedings of the Royal Society* (Bragg et al., 1950) was submitted on March 31, 1950. Pauling and Corey's communication to *JACS* was submitted on October 16, 1950.

However, Bragg, Kendrew and Perutz failed to discover the α-helix. Their article is a detailed and systematic survey of plausible structures for a folded polypeptide chain. It is based, like Pauling and Corey's work, on the postulate that (Bragg et al., 1950) 'those structures in which all NH and CO groups are hydrogen-bonded (are regarded) as inherently the more probable, because their free energy is presumably lower'. It posits an analogy between the nature of the chains in globular proteins and the regular folded α-configuration found by Astbury in fibrous proteins such as hair and wool and characterized by the presence of three residues per repeat of 5.1 Å. Indeed, when Bragg, Kendrew and Perutz examined closely the Patterson projection from their x-ray diffraction pictures of hemoglobin and myoglobin, they found chains to be present as the 'relatively few units which stand out from the rest of the structure because of their greater density' (Bragg et al., 1950). Hemoglobin showed 'a somewhat irregular but clearly marked rod of density through the origin and parallel to the a axis (. . .) (which showed) a periodicity with a repeat of rather over 5 Å'. Bragg, Kendrew and Perutz then proceeded to build molecular models, using bond lengths and bond angles as obtained by Corey from his work on simple amino acids and peptides, and did a systematic survey of what they thought were all the possible structures for a folded polypeptide chain. They came to the following conclusion (Bragg et al., 1950):

"We have preferred structures in which the repeat (or pseudo-repeat) contains three amino-acid residues (. . .) We do not feel, however, that we can entirely exclude the possibility that the repeat contains four residues . . . ".

Clearly, they were prevented from discovering the α-helix by a kind of a mental block, the unnecessary assumption of an integral number of residues per repeat. And Pauling's leap forward – which he could perform because not only was he an x-ray crystallographer like the trio from the Cavendish laboratory in Cambridge, England, but he had also unique chemical expertise – was 'the negative notion of the nonintegral character of the helical twist' (Bernal, 1968).

5. The confirmation from Max Perutz

From then on, everything went very rapidly. In their 1950 communication to *JACS*, Pauling and Corey had reported the existence of two helical structures: one with 3.7 residues per turn and a unit translation d per residue of 1.47 Å [this is the α-helix, whose accepted parameters are 3.61 residues per turn, $d = 1.50$ Å and a pitch $p = 5.41$ Å (Dickerson and Geis, 1969)]; the other (γ-helix) with 5.1 residues per turn and a unit translation of 0.96 Å. This short note also stated the probable existence of the α-helix in α-keratin and some other fibrous proteins, and of the γ-helix in supercontracted keratin (Pauling and Corey, 1950). Shortly afterwards, Pauling and Corey came out with a whole flurry of papers (Pauling and Corey, 1951a–h; Pauling et al., 1951), of which one proposed the β-pleated sheet as another ordered structure present in proteins (Pauling and Corey, 1951c), while the others detailed the helical structures (Pauling et al., 1951; Pauling and Corey, 1951a) and showed their pertinence to the structure of feather rachis keratin (Pauling and Corey, 1951e), of hair and muscle proteins (Pauling and Corey, 1951f), of collagen (Pauling and Corey, 1951g). Pauling and Corey also made some speculations as to the 'polypeptide-chain configuration in hemoglobin' (Pauling and Corey, 1951h).

Then came a crucial confrontation. Pauling's structures for proteins, the α-helix and the β-pleated sheet, which (it will be recalled) had been devised on the basis of *chemical* principles were presented to the crystallographers at the International Union of Crystallography conference in Stockholm (1951). In Bernal's words 'Pauling's chemical hypothesis (. . .) was so comprehensible that it won immediate acceptance' (Bernal, 1968). And, 'literally within hours after reading the first report of the CalTech work, M.F. Perutz of Cambridge had proved the existence of the α-helix in hair keratin and other proteins' (Dickerson and Geis, 1969). Bernal writes that 'Perutz (. . .) had spent years vainly trying to solve the crystal structure of hemoglobin by orthodox methods' (Bernal, 1968). Now, Perutz seized upon the α-helix proposed by Pauling and Corey and found 'a previously overlooked but critical reflection [1.5 Å]' (Dickerson and Geis, 1969) which proved the presence of α-helices in the structure of hemoglobin.

Perutz wrote, in his communication to *Nature* (Perutz, 1951) of having 'found a new reflexion from planes perpendicular to the fibre axis at a spacing of 1.50 Å which corresponds to the repeat of the amino-acid residues along the chain'. He compared various structures, as follows (Table X) and concluded 'that the only structure to give a 1.5 Å reflexion is that of Pauling, Corey and Branson (1951)' (Perutz, 1951).

Table X

The only structure to give a 1.5 Å reflexion is the α-helix (Perutz, 1951)

Structure proposed by	Repeat of pattern	Screw axis	Spacing of first 0k0 reflexion (Å)
Ambrose, Elliott and Temple (1949)	5.0–5.6	2	2.5–2.8
Bragg, Kendrew and Perutz (1950)	5.4	2	1.8
Huggins (1943)	5.2	3	1.7
Bragg, Kendrew and Perutz (1950)	5.6	4	1.4
Astbury and Bell (1941)	10.2	2	5.1; 1.7
Pauling, Corey and Branson (1951)	1.5	–	1.5

One can still feel Perutz's exhilaration from the last sentence of his letter to *Nature* (Perutz, 1951):

"The discovery of the 1.5 Å reflexion shows that even relatively disordered substances like hair may contain an atomic pattern of such high intrinsic regularity that it gives rise to X-ray diffraction effects at spacings where they had never before been suspected".

Max Perutz's self-denial is admirable. He could have stuck obdurately to the structures proposed in the 1950 article he had co-authored with Bragg and Kendrew (Bragg et al., 1950). Instead, he rushed to the lab and made x-ray photographs by oscillating 'the specimens about a direction normal to the (keratin or poly-γ-benzyl-L-glutamate) fibre axis' so that he could find this 1.5 Å reflexion (Perutz, 1951).

6. By way of commentary

Thus in 1951 Pauling was proven right. And by that time everyone (but she) knew that Dorothy Wrinch was hopelessly wrong. But one should not rush to the conclusion that Pauling's ideas about protein structure had *always* been *absolutely* right. He made mistakes too.

He let his name appear in the 1939 *JACS* article with Niemann with all the numerology in it. He made too much of 'proline interference' (Pauling, 1940). He conceived of globular proteins as 'stacks of pancakes' (Pauling, 1940).

Even the (essentially correct) ideas of Pauling and Corey in 1950–51 had their flaws: Pauling had again this incorrect idea of a compact arrangement in globular proteins of α-helical rods, somewhat similar to the earlier 'stacks of pancakes':

"Pauling made the (. . .) simplifying but incorrect assumption that the structure of the globular proteins consisted of rods of polypeptides arranged parallel to each other in different kinds of order. Crick (1952; 1953) was able to show that this was incompatible with the intensities of the X-ray reflections, which ought to be, on this hypothesis, much stronger than those observed" (Bernal, 1968).

Clearly, this was a case of over-enthusiasm on Pauling's part, once he had discovered the α-helix and it was psychologically to be expected that he should seek it in every single globular protein even when it was absent. As Bernal percipiently wrote (Bernal, 1968).

"the only thing that was wrong in Pauling's hypothesis, but carefully not stated, was the implication that the α helix was an important structural feature of *all* globular proteins. If it had been stated as *some* globular proteins, it would have been correct as well as illuminating".

But this is not the place to stretch this epistemological point: even if the naive philosophy of the scientists themselves gives such overwhelming importance to 'being right' and to 'not being wrong' – the latter is far from being synonymous with the former – the philosopher of science singles out other features in the conceptual fabric of the scientific endeavor.

To close, I wish to stress again the determinant of Pauling's success in elucidating the structure of proteins, in that he was an importer of chemical concepts into crystallography. This view is shared by many and has been cogently expressed by Bernal (1968):

"Pauling's work in molecular biology can thus be seen to be absolutely fundamental in nature and we ought to inquire how it came itself to be based on leading ideas that Pauling held from the very start. We can distinguish two of these ideas. The first is his intuitive hold of quantum chemistry, which enabled him to think in terms of quantized energy levels from the outset. As hinted before, in his later years Pauling could afford to be very free with his quantum theory.

The second leading idea is Pauling's appreciation of quantitative geometrical linking as found in all kinds of substances from metals to nucleic acids. He knew his atoms and their various states and binding conditions so well that he was prepared to break with what are after all only conventions – such as the regularities of classical crystallography – if they could not be fitted into these regularities".

Thus, innovation is often rooted in transference of existing concepts from one field into another.

REFERENCES

Albrecht, G.A. and Corey, R.B. (1939) J. Am. Chem. Soc., 61, 1087.

Ambrose, E.J., Elliott, A. and Temple, R.B. (1949) Nature, 163, 859.

Astbury, W.T. (1933) Trans. Faraday Soc., 29, 193.

Astbury, W.T. (1939) Annu. Rev. Biochem., 8, 113.

Astbury, W.T. and Bell, F.O. (1941) Nature, 147, 696.

Astbury, W.T. and Street, A. (1931) Phil. Trans. Roy. Soc., 230, 75.

Astbury, W.T. and Woods, H.J. (1930) Nature, 126, 913.

Barnes, W.H. (1929) Proc. Roy. Soc. London, A125, 670.

Bragg, Sir L., Kendrew, J.C. and Perutz, M.F. (1950) Proc. Roy. Soc. London, A203, 321.

Bernal, J.D. (1931a) Z. Krist., 48, 363.

Bernal, J.D. (1931b) Z. Krist., 69, 467.

Bernal, J.D. (1968) in Structural Chemistry and Molecular Biology (Rich, A. and Davidson, N., Eds.), Freeman, San Francisco, p. 370.

Bernal, J.D. and Fowler, R.H. (1933) J. Chem. Phys., 1, 515.

Corey, R.B. (1938) J. Am. Chem. Soc., 60, 1598.

Crick, F.H.C. (1952) Acta Cryst., 5, 381.

Crick, F.H.C. (1953) Acta Cryst., 6, 600.

Danielli, J.F. (1939) Proc. Roy. Soc. London, B127, 34.

Dickerson, R.E. and Geis, I. (1969) The Structure and Action of Proteins, Harper and Row, New York.

Edsall, J.T. (1983) TIBS, 30, 1.

Eisenberg, D. and Kauzmann, W. (1969) The Structure and Properties of Water, Oxford University Press, p. 144.

Hodgkin, D.C. (1979) Ann. N.Y. Acad. Sci., 325, 121.

Hodgkin, D.C. (1980) John Desmond Bernal, Biogr. Memoirs Fellows Roy. Soc. London, 26, 17.

Hodgkin, D.C. and Riley, D.P. (1968) in Structural Chemistry and Molecular Biology (Rich, A. and Davidson, N., Eds.), Freeman, San Francisco, pp. 15-28.

Huggins, M.L. (1942) Annu. Rev. Biochem., 11, 27.

Huggins, M.L. (1943) Chem. Rev., 32, 195, 211.

Kamb, B. (1968) in Structural Chemistry and Molecular Biology (Rich, A. and Davidson, N., Eds.), Freeman, San Francisco, p. 507.

Kauzmann, W. (1959) Adv. Protein Chem., 14, 1-63.

Langmuir, I. and Wrinch, D.M. (1939) Nature, 143, 49.

Latimer, W.M. and Rodebush, W.H. (1920) J. Am. Chem. Soc., 42, 1419.

Mirsky, A.E. and Pauling, L. (1936) Proc. Natl. Acad. Sci. USA, 22, 439.

Pauling, L. (1928) Proc. Natl. Acad. Sci. USA, 14, 359.

Pauling, L. (1931) J. Am. Chem. Soc., 53, 1367.

Pauling, L. (1932) Proc. Natl. Acad. Sci. USA, 18, 293.

Pauling, L. (1935) J. Am. Chem. Soc., 57, 2680.

Pauling, L. (1940) J. Am. Chem. Soc., 62, 2643.

Pauling, L. (1946) Chem. Eng. News, 24, 1064.

Pauling, L. (1948) Nature, 161, 707.

Pauling, L.C. (1954) Nobel Lecture in Les Prix Nobel en 1954 (Liljestrand, M.G., Ed.), Kungl. Boktryckeriet P.A. Norstedt and Söner, Stockholm, 1955, pp. 91-99; (1955) Angew. Chem., 67, 241; (1956) Science, 123, 255.

Pauling, L. (1983) Letter of June 13 to Pierre Laszlo.

Pauling, L. and Corey, R.B. (1950) J. Am. Chem. Soc., 72, 5349.

Pauling, L. and Corey, R.B. (1951a) Proc. Natl. Acad. Sci. USA, 37, 235.

Pauling, L. and Corey, R.B. (1951b) Proc. Natl. Acad. Sci. USA, 37, 241.

Pauling, L. and Corey, R.B. (1951c) Proc. Natl. Acad. Sci. USA, 37, 251.

Pauling, L. and Corey, R.B. (1951d) Nature, 168, 550.

Pauling, L. and Corey, R.B. (1951e) Proc. Natl. Acad. Sci. USA, 37, 256.

Pauling, L. and Corey, R.B. (1951f) Proc. Natl. Acad. Sci. USA, 37, 261.

Pauling, L. and Corey, R.B. (1951g) Proc. Natl. Acad. Sci. USA, 37, 272.

Pauling, L. and Corey, R.B. (1951h) Proc. Natl. Acad. Sci. USA, 37, 282.

Pauling, L. and Corey, R.B. (1953) Proc. Roy. Soc. London, B141, 21.

Pauling, L. and Delbrück, M. (1940) Science, 92, 77.

Pauling, L. and Niemann, C. (1939) J. Am. Chem. Soc., 61, 1860.

Pauling, L., Brockway, L.O. and Beach, J.Y. (1935) J. Am. Chem. Soc., 57, 2705.

Pauling, L., Corey, R.B. and Branson, H.R. (1951) Proc. Natl. Acad. Sci. USA, 37, 205.

Perutz, M.F. (1951) Nature, 167, 1053.

Rich, A. and Davidson, N. (Eds.) (1968) Structural Chemistry and Molecular Biology, Freeman, San Francisco.

Taylor, H.S. (1941) Proc. Am. Phil. Soc., 85, 1.

Wrinch, D.M. (1939) Nature, 143, 482.

Wu, H. (1931) Chinese J. Physiol., 5, 321.

Chapter 16

Sequence Analysis

During the 1940s and early 1950s, culminating with publication by Sanger and Thompson of the complete sequence for insulin in 1953, determination of the arrangement of amino acids in proteins ran parallel to the other advances in the understanding of protein structure. In this chapter, I shall focus mainly on the insulin story because of its historical significance and because of its impact on the field. I shall first set the stage, as it were, for this classic achievement of Sanger, describing the tools he could utilize in the mid-1940s, and indicating the background to his own work. Then, I shall briefly go through the main stages in the deduction by Sanger and his co-workers of the insulin sequence.

1. Introduction

Interestingly, Sanger opened his Nobel lecture (1958) with a statement provocative for the historian:

"In 1943 the basic principles of protein chemistry were firmly established. It was known that all proteins were built up from amino acid residues bound together by peptide bonds to form long polypeptide chains. Twenty different amino acids are found in most mammalian proteins, and by analytical procedures it was possible to say with reasonable accuracy how many residues of each one were present in a given protein" (emphasis added).

This affirmation of the polypeptide theory as firmly established by 1943 should be contrasted with Sanger's own less sanguine and more accurate assessment when he wrote these other sentences, which I have already quoted in this book (Sanger, 1952):

"As an initial working hypothesis it will be assumed that the peptide theory is valid, in other words, that a protein molecule is built up only of chains of α-amino (and α-imino) acids bound together by peptide bonds between their α-amino and α-carboxyl groups. While this peptide theory is almost certainly valid, it should be remembered that it is still a hypothesis and has not been definitely proved. Probably the best evidence in support of it is that since its enunciation in 1902 no facts have been found to contradict it".

Indeed, if we refer back to Sanger's published views at the time of the start of his pioneering work on the insulin structure, we find that he was much more cautious than the opening fanfare praise of the polypeptide theory at the outset of his 1958 Nobel lecture would have us believe. Note for instance the guarded language when he wrote as follows (Sanger, 1945):

"In some proteins, and particularly in insulin, the number of free amino groups (Van Slyke) is far in excess of that which can now be ascribed to lysine, which suggests that the protein must contain residues of certain amino-acids *which are condensed in such a way* that their α-amino groups remain free" (emphasis added).

While the first point (the a posteriori optimism) can be ascribed to the triumphant context of the Nobel award and to the necessity for Sanger to summarize the field briefly before narrating his personal contribution, this leaves the second point to be answered: why did he pick 1943 as his reference year for a date when the polypeptide theory was already firmly established? Since this date coincides with the start of his own investigations on the insulin structure, this must mean only one thing, viz. that by 1943 Sanger had acquired the personal conviction that the polypeptide theory of Emil Fischer was right in every respect, and that time was ripe for the elucidation of the sequence of a protein. Asked by the author of this book about this very point, Sanger's research director in 1943 confirmed it:

"I am fairly certain that Sanger at that time, like myself, felt that the polypeptide theory of Fischer was right in every respect" (Neuberger, 1984b).

And Sanger himself wrote as follows:

"This seems to imply that I started my work with the grand idea that I was going to determine the complete sequence of a protein. This was certainly not true, and I think is very rarely the way that research is carried out. What happened was that, after completing my Ph.D. with A. Neuberger, A.C. Chibnall arrived as the new Professor of Biochemistry in Cambridge and I joined his group. He had been working mainly on amino acid analysis, and particularly of insulin. This work is described in his Bakerian lecture (*Proceedings of the Royal Society*, **B131**, 136). They had also determined the free amino groups of insulin by the Van Slyke method and had shown that there were more free amino groups than could be accounted for by the lysine content. He suggested that I follow this up and try to develop a method for general identification of N-terminal residues. The work developed successfully after several failures. Throughout the work on insulin it was usually a question of determining the next experiment based on what one had already found, but it was considerably later that I decided to try and complete the whole sequence" (Sanger, 1985).

With respect to the first proposition, even if Sanger shared in the consensus of the biochemists in 1943 that proteins consisted of polypeptide chains, there would still be much interest in finding out experimentally how the amino acid beads were strung along the chain: randomly, or according to a scheme? As he wrote (Sanger, 1958):

"The most widely discussed theory was (in 1943) that of Bergmann and Niemann, who suggested that the amino acids were arranged in a periodic fashion, the residues of one type of amino acid occurring at regular intervals along the chain. At the other extreme there were those who suggested that a pure protein was not a chemical individual in the classical sense but consisted of a random mixture of similar individuals".

With respect to the second proposition, we must also admire Sanger's foresight: in 1943, there was no good, universal tool for the separation and characterization of amino acids and peptides from protein hydrolysates. One could be hopeful, however, because Martin and Synge (1941) had proposed a little earlier their partition method on silica gel. By contrast and by way of calibration of Sanger's courage and daring, in 1943 Synge published a review of

"the partial hydrolysis products that had been obtained from proteins. In retrospect, it is remarkable how few peptide sequences had been rigorously defined - approximately half a dozen dipeptides but only two tripeptides, one each from hydrolysates of gelatine and casein. Indeed, among naturally occurring peptides, only the structures of carnosine, anserine, and glutathione were known" (Smith, 1979).

Sanger had the vision to initiate his determination of the insulin sequence at a time when the tools he would need were unavailable, and he had the luck to time his research effort just when these very tools were in the works: the early and mid-1940s saw development, in succession, of partition chromatography on silica gel (Martin and Synge, 1941) and of frontal analysis and displacement on active carbon (Tiselius, 1940), of partition chromatography on paper (Consden et al., 1944), and of chromatography on granular starch (Synge, 1944); all these techniques were devised for separation of amino acids from complex mixtures.

"Fred Sanger was remarkably good at picking out a problem which suited his particular ability, and he was not deterred by difficulties and disappointments. His persistence was perhaps the quality which impressed me most whilst he was a Ph.D. student" (Neuberger, 1984b).

The methodology needed was, first, a good method to separate the amino acids from protein hydrolysates; and second, a procedure for individual identification of each resulting fraction. Chromatography was the answer to the first question (and continues to be nowadays: ion-exchange chromatography on columns of sulfonated polystyrene is one of the standard methods for separation of amino acids from protein hydrolysates). 'In 1938, when Martin and Synge entered the field, the isolation of a single amino acid from the acid hydrolysate of a protein could take as long as two months' (Gordon, 1979). They were able to build on the conceptual foundations earlier laid by Tiselius (1940), who made the important distinction between three chromatographic types, frontal analysis, displacement development, and elution development. Synge (1939), Martin and Synge (1941) made a dual advance: theoretical, by introducing the notion of partitioning between a stationary phase and a mobile phase; practical, by building a countercurrent apparatus in which acetylamino acids partitioned themselves between an aqueous and an organic solvent phase.

Initially, detection was performed with a pH indicator introduced in the stationary aqueous phase. Sensitivity was the main problem. Hence, Martin and Synge set about exploring a variety of substances all known to give colored reaction products with amino acids, till they hit upon ninhydrin. The readers will recall that α-amino acids give an intense blue color when heated with this simple bicyclic molecule, while imino acids such as proline give a yellow color. The technique – which more than 40 years later is still in widespread use – is so sensitive that it can detect a microgram of an amino acid or of a peptide: Martin and Synge were in business!

They had built upon the already well-known ninhydrin reaction: but their main contribution was elaboration of the partition chromatography methods, which proved to be an important breakthrough for the peptide sequencing to be described. To return to detection using a pH indicator:

"In fact, this detection system was not used for silica gel chromatograms (Martin and Synge, 1941) because very soon after the great sensitivity of ninhydrin on filter paper as a color reagent for amino acids had been demonstrated, Martin made the essential intellectual jump and suggested that the filter paper itself might be a suitable chromatographic matrix (Consden, Gordon and Martin, 1944)" (Gordon, 1979).

The comment has been made that this epochal discovery of paper chromatography could (or should) have been made much earlier:

"(Martin) remarked that he thought paper chromatography might have been discovered anytime after 1800 or thereabouts, but in fact of course it was not" (Edsall, 1979).

This comment of J.T. Edsall during a *Symposium on the Origins of Modern Biochemistry: a Retrospect on Proteins*, held by the New York Academy of Sciences on May 31, June 1-2, 1978, was answered as follows (Gordon, 1979):

"You probably know, Prof. Edsall, that in the dyestuffs industry in the 19th century in Germany, paper chromatograms of a fair degree of sophistication were in fact being used. *However, the scientists did not pay any attention.* Then there was the oil industry in this country in which chromatography was used quite a lot. *And still the scientists did not pay much attention"* (emphasis added).

Another answer has the merit of simplicity:

"When, during World War II, the budgets for biochemical research in Great Britain diminished markedly, the scientists developed and exploited paper chromatography (. . .) a very powerful but remarkably inexpensive technique" (Stetten, 1980).

Going back to the ninhydrin technique, not only can it detect the presence of an amino acid: quantitatively, the concentration of the amino acid is proportional to the optical absorbance of its solution after it has been heated with ninhydrin – a later development, for which Moore and Stein are responsible. Finally, the identity of the amino acid is ascertained by its elution volume, i.e. by the volume of a buffer solution required to pull the amino acid away from the chromatographic column, if on silica gel; or by the position of the ninhydrin-colored spots on the filter paper.

There were parallel developments in the chemical characterization of amino acids. For instance, Erwin Brand in his laboratory at Columbia was engaged during the late 1930s in a systematic effort to devise photometric methods for the determination of various amino acids. He devised methods, e.g. for the photometric determination of cystine, cysteine by reaction with phosphotungstic acid (Kassell and Brand, 1938) and of tryptophan and tyrosine by the Millon reaction (Brand and Kassell, 1938). In this manner, they could show for example the presence of eight tyrosine and of two tryptophan residues per mole of denatured ovalbumin (Brand and Kassell, 1938).

2. First determination of the sequence of a simple peptide

Martin and Synge first applied their recently developed paper chromatographic method to the determination of the sequence of an antibiotic peptide, gramicidin S. The acid hydrolysates contained dipeptides, which were characterized, after separation by paper chromatography, by comparison with authentic synthetic samples. From these studies, gramicidin S appeared to possess the sequence -Val-Ornithyl-Leu-Phe-Pro- and thus was shown to be a cyclic pentapeptide (Consden et al., 1947). In fact, we now know it to be a repeating decapeptide. 'Prior to 1949, this was the longest peptide sequence known!' (Smith, 1979).

3. Complete amino acid analysis of β-lactoglobulin

A landmark paper was the report by Brand and his co-workers on the complete amino acid composition of β-lactoglobulin (Brand et al., 1945). This is a most impressive piece of work, bringing to bear on the problem a wealth of methods, many of which had been devised in Brand's own laboratory, for determining the amounts of the various amino acids. Isotope dilution methods, bioassays, spectrophotometric determinations, enzymatic methods were all exploited in finely tuned conjunction.

The empirical formula thus reported for β-lactoglobulin was Gly_8–Ala_{29}– Val_{21}– Leu_{50}– Ile_{27}– Pro_{15}– Phe_9– Cys_8– Met_9– Trp_4– Arg_7–His_4– Lys_{33}– Asp_{36}– Glu_{24}– Gln_{32}– Ser_{20}– Thr_{21}– Tyr_9 $(H_2O)_4$. This article is also noteworthy because, conceptually, it is entirely within the framework of polypeptide theory: true, with due caution, it states it as a working hypothesis ('... it is assumed that β-lactoglobulin is a chemical compound in which the constituent amino acids are present in integral molar quantities (...) it is assumed that the amino acids (RNH_2COOH) are present in the molecule in peptide linkage') (Brand et al., 1945). But at the end of his article, Brand makes this statement in the Discussion (which he repeats almost verbatim afterwards in the summary):

"We can conclude with some certainty that in this protein and perhaps in many others, the constituent amino acids are primarily linked by typical peptide bonds; other linkages, such as atypical peptide bonds (as in glutathione), esters, anhydrides, or imides, if they are present at all, can be present only in small numbers" (Brand et al., 1945; emphasis added).

It is most impressive to realize that as early as 40 years ago, a protein with M_r of 42 000 totalling 370 residues with 366 peptide bonds (Brand et al., 1945) could have its detailed amino acid composition determined with good accuracy: 'this was the first time that such an analysis was reported and it had a major impact on the general thinking of the field' (Smith, 1979). Brand knew that his contribution had historical significance; his article opened with the following paragraph which stresses it, perhaps a trifle clumsily:

"From the point of view of organic chemistry, the establishment of the empirical formula of proteins (cf. Mulder in 1838), in terms of constituent amino acids, is fundamental to an understanding of their constitution and to the development of theories of protein structure".

4. The sequence of insulin

From the preceding paragraphs, it should be clear that Sanger displayed a mixture of daring and caution in attacking the (at the time) awesome task of determining the amino acid *sequence* of a protein. The required methodology was not yet fully available, and he would have to devise himself some of the necessary tools. Conversely, he chose insulin as the target molecule because there was already a substantial amount of hard information available on this important protein molecule.

Two groups, that of Brand* at Columbia, and another at Cambridge, England, were determining the accurate amino acid composition of insulin by methods similar to those which had been successfully applied to the β-lactoglobulin molecule. In 1946, the results came out (Table XI) (Tristan, 1946; Macpherson, 1946; Rees, 1946; Brand, 1946).

Another reason for the choice of insulin as the target was an important piece of information, the identity of one of the terminal residues. In 1935, Jensen and Evans had treated insulin with phenylisocyanate and they had isolated the phenylhydantoin of phenyl-

---

*See Plate 5 on page 254.

TABLE XI

Amino acid composition of insulin

Amino acid	Cambridge workers	New York workers (Brand)
Gly	6.9	7.4
Ala	6.8	–
Val	8.0	9.1
Leu	12.0	12.3
Ile	2.6	2.7
Pro	2.7	3.1
Phe	5.9	5.7
Tyr	8.7	8.2
Glu	15.0	16.4
Asp	--	6.1
Ser	6.0	6.6
Thr	2.1	3.2
Cys/2	12.5	11.0
Arg	2.1	2.4
His	3.8	4.1
Lys	2.1	2.1
Amide	11.6	15.1

alanine from the hydrolysate: hence, Phe was one of the N-terminal residues (Jensen and Evans, 1935). This was brought to Sanger's attention by his Ph.D. mentor, Neuberger (Neuberger, 1984a).

This may have dictated Sanger's decision to attack the sequence determination *from the N-terminus*. According to his 1945 article, the background to his work was the suggestions

(i) "that the free amino groups in proteins may be the ε-amino group of lysine (as) first suggested by Skraup and Kaas (1906)" and (ii) "that the free amino groups of proteins over and above the ε-amino groups of lysine are due to terminal residues of polypeptide chains, and that the number of these groups must therefore give a measure of the number of polypeptide chains in the protein (Chibnall, 1942)" (Sanger, 1945).

Frederick Sanger thus set about identifying the N-terminal residues. According to a reliable witness of Sanger's efforts "the reason why he concentrated on the N-terminus was most probably the fact that reagents were available to combine with free amino groups such as carboxylic acid chlorides or sulphonyl chlorides. We indeed started with the latter reagents, i.e. toluene sulphonyl chlorides, but the use of fluorodinitrobenzene which was in many ways more suitable was started by Fred Sanger after I left Cambridge" (Neuberger, 1984b).

In the dinitrophenyl (DNP) method (Sanger, 1945) 1-fluoro-2,4-dinitrobenzene reacts with the uncharged α-NH$_2$ group to form a yellow-colored derivative of the peptide (or protein). This reaction is performed under such mild conditions (room temperature; a bicarbonate aqueous solution) that hydrolysis of peptide bonds within the peptide (or protein) is avoided. This statement was shown, a number of years afterwards, not to be 100% fool-proof: with certain proteins having Asp or Asn as the N-terminal residue, the DNP amino acid is split off during the treatment, thus liberating as the apparent end-group what is in fact the second residue (Weygand and Junk, 1951; Thompson, 1953). It is remarkable that 'the bond between DNP and the terminal amino group is stable under conditions that hydrolyse peptide bonds. Hydrolysis of the DNP-peptide in 6 N HCl yields a DNP-amino acid' (Stryer, 1981) which is then identified by its chromatographic properties. For this purpose,

"the partition chromatographic method of Gordon et al. (1943) was employed to separate the DNP-derivatives, using a stationary aqueous phase adsorbed on the silica gel and a moving organic phase. It was, of course, unnecessary to use an indicator in the aqueous phase as the derivatives themselves are colored" (Sanger, 1945).

Three colored products were found, identified as DNP-Phe, DNP-Gly and DNP-Lys. Quantitatively, there were two of each of these residues per insulin molecule. Thus, both the ϵ-amino groups of lysine are free in the insulin molecule. Since there are four free α-amino groups, two glycine and two phenylalanine residues

"the results suggest that the insulin sub-molecule of molecular weight 12 000 is made up of four open peptide chains, two of these having terminal glycyl-residues and the other two terminal phenylalanyl-residues respectively" (Sanger, 1945).

In his words

"it seemed probable that the chains of insulin were joined together by the disulfide bridges of cystine residues. Insulin is relatively rich in cystine and this was the only type of cross-linkage that was definitely known to occur in proteins" (Sanger, 1958).

The next step, therefore, became the cleavage of the insulin molecule into two polypeptide chains by splitting the disulfide bridges. Sanger (1947; 1949) used oxidation with performic acid – a method introduced by Toennies and Homiller (1942) – to convert cystine residues ($-CH_2S-$) into cysteic acid residues ($-CH_2-SO_3H$) thus breaking the cross-links. 'Performic acid also reacts with methionine and tryptophan residues, the two amino acids *which fortunately were absent from insulin*' (Sanger, 1958; emphasis added). Fractionation of the oxidized insulin (Sanger, 1947) was effected by precipitation methods (Sanger, 1949a). Two fractions were obtained.

"One (fraction A) contained glycine and the other (fraction B) phenylalanine N-terminal residues. Fraction A was acidic and had a simpler composition than insulin, in that the six amino acids: lysine, arginine, histidine, phenylalanine, threonine, and proline, were absent from it. It thus had no basic amino acids, which were found only in fraction B. (. . .) Fraction A contained about 20 residues per chain, four of these being cysteic acid and fraction B had 30 residues, two of which were cysteic acid" (Sanger, 1958).

For the next step, an entirely logical one ['Fred Sanger is a highly methodical and logical person' (Neuberger, 1984b)], Sanger returned to his DNP method:

"The method can be adapted in determining the sequence of amino acids which occupy positions near to the terminal amino acid residues. For this purpose the protein is submitted to partial hydrolysis leading to the liberation of a series of N-2:4 dinitrophenyl (DNP) peptides. These differ from the other products of partial hydrolysis in that they are acids which can be extracted from acid solution by organic solvents and can thus be obtained relatively free from other unsubstituted peptides (Sanger, 1949b).

By an analysis of these peptides it was possible to determine the N-terminal sequence to four or five residues along the chain" (Sanger, 1958).

In this manner, Sanger found

"that both the terminal phenylalanyl residues are present in the form of the tetrapeptide sequence phenylalanyl-valyl-aspartyl-glutamic acid, both the lysyl residues which are in the same chains are present in the sequence threonyl-prolyl-lysyl-alanine and both the terminal glycyl residues are present in the pentapeptide sequence glycyl-isoleucyl-valyl-glutamyl-glutamic acid" (Sanger, 1949b).

His other conclusion was 'that the insulin molecule is built up of two pairs of very similar, if not identical, polypeptide chains' (Sanger, 1949b): this last conclusion was predicated upon the M_r of 12 000 which Sanger believed to be that of insulin. 'The other alternative, which was later shown to be the case, was that the actual molecular weight was 6000' (Sanger, 1958).

So far, since its inception in 1943, Sanger had carried out all this work on the insulin structure single-handedly: quite a feat! However, it had taken him already half a decade to determine the sequences of two short sub-structures, a tetrapeptide (from fraction B) and a pentapeptide (from fraction A). At this pace, determination of the full sequence of insulin, with its 51 residues, could have taken another half-century.

Fortunately, Sanger could now draw upon the paper chromatography technique and follow the example of Consden, Gordon, Martin and Synge who had applied it successfully to the determination of the sequence of gramicidin S (Gordon et al., 1947).

Another lucky event came Sanger's way:

"At this point (1949) I was joined by Dr. *Hans Tuppy who came to work in Cambridge for a year.* Although we did not seriously envisage the possibility of being able to determine the whole sequence of one of the chains *within a year*, it was considered worth while to investigate the small peptides from an acid hydrolysate using essentially the methods that had been applied to 'gramicidin-S' " (Sanger, 1958; emphasis added).

It turned out that Sanger and Tuppy worked so efficiently as a team that, by the end of the year, they had almost identified the whole 30 residues sequence in fraction B! The acid hydrolysates from fraction B were, after gross separations had been effected by ionophoresis, ion-exchange chromatography, and adsorption on charcoal, subjected to fractionation by two-dimensional paper chromatography. As an illustration (Sanger, 1958) these are the results which allowed identification of two sequences containing cysteic acid residues as Leu- $CySO_3H$- Gly and Leu- Val- $CySO_3H$- Gly (Sanger and Tuppy, 1951a):

TABLE XII

Cysteic acid peptides identified in a partial acid hydrolysate of fraction B (Sanger, 1958)

$CySO_3H$-Gly	$CySO_3H$-Gly
Val-$CySO_3H$	
	Leu-$CySO_3H$
Val-($CySO_3H$,Gly)	
	Leu-($CySO_3H$,Gly)
Leu-(Val, $CySO_3H$)	
Leu-(Val, $CySO_3H$,Gly)	
Leu-Val - $CySO_3H$-Gly	Leu-$CySO_3H$-Gly

However, the identification of four of the 30 residues in fraction B still eluded Sanger! He decided, as a last resort, to try using proteolytic enzymes:

"Initally we had refrained from using them since it was considered that they might bring about re-arrangement of the peptide bonds by transpeptidation or actual reversal of hydrolysis. (. . .) Proteolytic enzymes are much more specific than is acid since only a few of the peptide bonds are susceptible" (Sanger, 1958).

By determining the sequences of various peptides produced by the action on the phenylalanyl chain (fraction B) of pepsin, trypsin and chymotrypsin, the complete sequence of this phenylalanyl chain was revealed (Sanger and Tuppy, 1951b) as: Phe- Val- Asp- Glu- His- Leu - Cys- Gly- Ser- His- Leu- Val- Glu- Ala- Leu- Tyr- Leu- Tyr- Leu- Val- Cys- Gly- Glu- Arg- Gly- Phe- Phe- Tyr- Thr- Pro- Lys- Ala. (This had been a most productive post-doctoral year for Dr. Tuppy!) Sanger then attacked fraction A with similar methods and found its sequence to be (Sanger and Thompson, 1953): Gly- Ile- Val- Glu- Glu- Cys- Cys- Ala- Ser- Val- Cys- Ser- Leu- Tyr- Glu- Leu- Glu- Asp- Tyr- Cys- Asp.

"Having determined the structure of the two chains of insulin the only remaining problem was to find how the disulfide bridges were arranged. About this time it was shown by Harfenist and Craig (1952) that the molecular weight of insulin was of the order of 6000, so that it consisted of two chains containing three disulfide bridges, and not of four chains as we had originally thought" (Sanger, 1958).

The difficulty came from a disulfide interchange reaction: after careful study, Sanger found that it was possible to prevent it by adding thioglycolic acid in small amounts to the acidic solution of insulin being hydrolysed. In this manner he could establish the position of the three disulfide bridges, which brought to its close, 12 years after its inception, this major venture of modern biochemistry (Sanger et al., 1955a; Ryle and Sanger, 1955; Ryle et al., 1955; Brown et al., 1955; Harris et al., 1956).

Success had resulted from a combination of hard work and clear-headed, logical thinking; but there had been also a continual succession of strokes of luck (or timely decisions on Dr. Sanger's part), in chronological order: the Phe terminus already identified by Jensen and Evans (1935); the polypeptide theory seemed to be alright (1943); the invention of paper chromatography (1944); the DNP group was stable under 6 N HCl treatment (1945); the absence in the insulin molecule of Met and Trp residues which allowed the use of HCO_3H for cleavage of the disulfide bridges (1947); and the coming of Dr. Tuppy on the scene, when someone like him was most needed (1949).

"The impact of Sanger's achievement was immediate. First of all, in addition to the relatively few laboratories already studying the sequences of proteins, many more were encouraged to work in this field. Secondly, although the three-dimensional structure of insulin remained unknown for some years, there was now hope that this problem could be approached with the aid of the knowledge of the amino acid sequence. Indeed, the structures of a number of other proteins were solved by X-ray analysis before that of insulin. Third, the success with insulin firmly established that the primary linkage between amino acid residues in proteins is the peptide bond. Further, proteins could contain one or more than one type of peptide chain. Fourth, the sequence in each chain was completely unique containing no repetitive sequences or patterns. Finally, Sanger and his associates used both partial acid hydrolysis and enzymic hydrolysis to obtain the peptides used in ascertaining the structure of insulin. An important aspect of these studies was the proof that proteolytic enzymes acted at sites in the longer peptide chains that could be predicted from the earlier knowledge of the action of these enzymes on synthetic substrates. A subsequent result is the now general practice of using the chains of insulin or other proteins of known sequence to determine the specificity of proteolytic enzymes of unknown specificity, a method first used to determine the specificity of subtilisin by Tuppy (1953)" (Smith, 1979).

This statement by Smith summarizes admirably and extremely concisely the impact of the Sanger breakthrough. The historian has, however, to elaborate upon it a little further and to bring in a few complementary facts.

Frederick Sanger received the Nobel prize for chemistry in 1958 for his unraveling of the insulin sequence. I wish to stress three features of this reward:

(*i*) it came very soon after completion of the work: one might say that it was almost immediate, since the insulin complete sequence was published in the *Biochemical Journal* in 1955 and the Nobel Committee must have reached its decision in the early months of 1958;

(*ii*) a second unusual feature is the promise or prophecy made by Tiselius in his Presentation speech:

"(. . .) very likely *you are more apt to look ahead.* It was Alfred Nobel's intention that his prizes should not only be considered as awards for achievements done but that they should also serve as *encouragement for future work.* We are confident that you are a worthy recipient of the Nobel award *also in this sense*" (Tiselius, 1958; emphasis added).

Indeed, as is well known, Sanger went on to earn himself a second Nobel prize two decades later.

(*iii*) needless to say, the *scientific* significance of the work was the first determination of the full amino acid sequence of *a* protein; the *public* significance of the work – which surely was in the mind of the members of the Nobel Committee and which made their task much easier – was that it concerned *insulin*, with its considerable importance as an hormone and treatment for diabetes, a disease very much in the public eye.

Which brings in the characteristics of the work, and here I should like to mention half a dozen additional points:

(*iv*) Sanger's success was that of craftsmanship. It could be contrasted with the failure of the ambitious task force which had been assembled, at the same time and on both sides of the Atlantic, in Britain and in the US, with major industrial laboratories cooperating, to work on the chemical synthesis of penicillin. By contrast, it is noteworthy that the first protein sequence was determined almost single-handedly and in a British university laboratory:

(*v*) the work squarely belongs in the category of scientific research as puzzle-solving; and perhaps, since its inception is contemporary of World War II, the analogy with code-breaking by cryptographers, especially with the British success in 'breaking' the German Enigma machine (Jones, 1978) should receive at least passing mention even though Sanger could not have known of it at the time he started unraveling the sequence of insulin;

(*vi*) in order to read the insulin message, Sanger opted for a methodology by recurrence, chopping off one residue at a time starting from the (known) phenylalanine terminus of the molecule: one might *term* it a Cartesian strategy!

This statement oversimplifies things quite a bit. As Sanger wrote after reading my script:

"My method was not really a question of 'chopping off one residue at a time' but was a much more random method of partial hydrolysis. The real method of 'recurrence' was Edman's, which is the major method used today. It is interesting that this method and the DNP one were described at about the same time. The main reason why the DNP method was used more at first was that the reagents were coloured and thus helped greatly with the chromatography. When efficient methods of fraction collection and identification of the phenylthiohydantoins were developed the Edman procedure superseded the DNP method" (Sanger, 1985).

(*vii*) His tactic for doing so was the use of *organic chemical* techniques, by preparing derivatives of individual amino acids which could be characterized easily, the DNP compounds; this tactic, which places him in the intellectual lineage of Emil Fischer, can be contrasted with e.g. that of Brand who, at the same time, was bringing to bear on related problems (the amino acid composition of β-lactoglobulin) the whole repertory of existing *biological* techniques.

Sanger used organic chemical techniques very much of his own initiative, since

"undergraduate teaching in biochemistry at Cambridge was not to any great extent based on organic chemistry and Fred Sanger's knowledge of ordinary chemical techniques was not very extensive when he started on his Ph.D. work with me. I believe it is fair to say that he learned most of the chemical techniques which he used later while he was engaged on his Ph.D. work. There was some chemistry done in the biochemistry department at the time, first of all by Pirie, who left the department about 1940, and by Bell who was a carbohydrate chemist, but most of the staff were somewhat remote from organic chemistry. When A.C. Chibnall succeeded Sir Frederick Gowland Hopkins as professor of biochemistry in 1943, the emphasis of the department shifted of course to some extent to protein chemistry, but this happened after Sanger had obtained his Ph.D." (Neuberger, 1984b),

which Sanger himself further clarifies in the following manner:

"I think the main point of Neuberger's remarks is that I *had been* exposed to organic chemistry, which I learned from him while doing my Ph.D. and when I joined Chibnall's group. Neuberger and Chibnall had much more influence on my work than any other members of the biochemistry laboratory" (Sanger, 1985).

(*viii*) The DNP method invented by Sanger is a *labeling* technique, of the type which has proven immensely valuable to biochemistry as a whole;

(*ix*) it may be asked why Sanger decided to work from the N-rather than from the C-terminus (see point *vi* above); in any case, his invention of the DNP method slanted the whole field of protein sequencing towards reading the sequence starting from the N-terminus, for a good many years;

(x) indeed, there have been enduring features from the Sanger insulin saga (1943-1955): even in automatic sequenators, the protein continues to be chopped off into smaller peptides whose N-termini are determined by the DNP method or by the Edman method (Edman 1950; 1953);

(xi) of interest also is indeed the *conservatism* of biochemical techniques: ironically, the Sanger technique of acid hydrolysis with 6 N HCl - which also continues to enjoy widespread use - is essentially identical to the experimental protocol by which Proust (1819) and Braconnot (1820) had first isolated leucine and glycine (see Chapter 12, section 1)!

(xii) I wish to elaborate further upon a point briefly mentioned in the quoted paragraph (Smith, 1979): ' . . . the sequence in each (protein) chain was completely unique containing no repetitive sequences or patterns'. This is extremely important. Sanger himself, in his Nobel Lecture (Sanger, 1958) stressed that, prior to the determination of the insulin sequence, biochemists ranked themselves between two extremes, those believing that proteins had a highly organized pattern, perhaps such as Bergmann and Niemann had proposed, and those who thought that the arrangement of the amino acids might be at random. The former were the majority during the period 1943-1955. However, this statement should not be construed to imply widespread support for the Bergmann and Niemann theory. According to the testimony of Neuberger:

"We [Neuberger and Sanger] both felt convinced that without any experimental evidence however *the peptide sequence in the protein was regular and orderly*, without however accepting the Bergmann and Niemann hypothesis which I had already attacked before the war" (Neuberger, 1984; emphasis added).

When Sanger reported that the latter was the case, this finding created intense shock. For instance, the extremely influential mathematician John von Neumann, who had much interaction with biologists, exclaimed:

"I shudder at the thought that highly efficient purposive organizational elements, like the protein, should originate in a random process" (von Neumann, 1955).

I believe that this highly articulate sentence by von Neumann, which has pronounced echoes in it of Einstein's *Gott würfelt nicht*, also deserves some elaboration. When von Neumann writes of 'highly efficient purposive organizational elements', à propos of proteins, he has clearly in mind biosystems based upon proteins such as enzymes, with their admirable selectivity, efficiency, and rapidity of action. Besides, there is a marked teleological flavor to his sentence: the 'purposive' epithet indicates that, if a protein molecule fulfills a given and extremely well-defined function, then, in some way, its own architecture should embody the function. If we try to make more explicit what von Neumann found intellectually scandalous (and it was intellectually scandalous at the time, since this finding *appeared* to run against the grain of all the progressive development of a molecular biochemistry since the beginning of the century), it is that biological evolution could seize upon any odd arrangement of amino acids into a chain and not only make do with it but, in some mysterious way, elevate it to a degree of efficiency in function which looked almost magical.

REFERENCES

Brand, E. (1946) Ann. N.Y. Acad. Sci., 47, 187.

Brand, E. and Kassell, B. (1938) J. Biol. Chem., 131, 489.

Brand, E., Saidel, L.J., Goldwater, W.H., Kassell, B. and Ryan, F.J. (1945) J. Am. Chem. Soc., 67, 1524.

Brown, H., Sanger, F. and Kitai, R. (1955) Biochem. J., 60, 556.

Chibnall, A.C. (1942) Proc. Roy. Soc. B, 131, 136.

Consden, R., Gordon, A.H. and Martin, A.J.P. (1944) Biochem. J., 38, 224.

Consden, R., Gordon, A.H., Martin, A.J.P. and Synge, R.L.M. (1947) Biochem. J., 41, 596.

Edman, P. (1950) Acta Chem. Scand., 4, 283.

Edman, P. (1953) Acta Chem. Scand., 7, 700.

Edsall, J.T. (1979) Ann. N.Y. Acad. Sci., 325, 104.

Gordon, A.H. (1979) Ann. N.Y. Acad. Sci., 325, 95.

Gordon, A.H., Martin, A.J.P. and Synge, R.L.M. (1943) Biochem. J., 37, 79.

Harfenist, E.J. and Craig, L.C. (1952) J. Am. Chem. Soc., 74, 3087.

Harris, J.I., Sanger, F. and Naughton, M.A. (1956) Arch. Biochem. Biophys., 65, 427.

Jensen, H. and Evans, E.A. (1935) J. Biol. Chem., 108, 1.

Jones, R.V. (1978) Most Secret War. British Scientific Intelligence 1939-1945, Hamish Hamilton, London.

Kassell, B. and Brand, E. (1938) J. Biol. Chem., 125, 115.

Macpherson, H.T. (1946) Biochem. J., 40, 470.

Martin, A.J.P. and Synge, R.L.M. (1941) Biochem. J., 35, 1358.

Neuberger, A. (1984a) Letter of August 2 to Pierre Laszlo.

Neuberger, A. (1984b) Letter of November 7 to Pierre Laszlo.

Rees, M.W. (1946) Biochem. J., 40, 632.

Ryle, A.P. and Sanger, F. (1955) Biochem. J., 60, 535.

Ryle, A.P., Sanger, F., Smith, L.F. and Kitai, R. (1955) Biochem. J., 60, 541.

Sanger, F. (1945) Biochem. J., 39, 507.

Sanger, F. (1947) Nature, 160, 433.

Sanger, F. (1949a) Biochem. J., 44, 126.

Sanger, F. (1949b) Biochem. J., 45, 563.

Sanger, F. (1952) Adv. Protein Chem., 7, 1.

Sanger, F. (1958) in Nobel Lecture (Chemistry) 1942-1963, Elsevier, Amsterdam, p. 544.

Sanger, F. (1959) Science, 129, 1340.

Sanger, F. (1985) Letter of February 27th to Pierre Laszlo.

Sanger, F. and Thompson, E.O.P. (1953) Biochem. J., 53, 353, 366.

Sanger, F. and Tuppy, H. (1951a) Biochem. J., 49, 463.

Sanger, F. and Tuppy, H. (1951b) Biochem. J., 49, 481.

Sanger, F., Thompson, E.O.P. and Kitai, R. (1955) Biochem. J., 59, 509.

Skraup, Z.H. and Kaas, K. (1906) Liebigs Ann., 351, 379.

Smith, E.L. (1979) Ann. N.Y. Acad. Sci., 325, 107.

Stetten, Jr. D.W. (1980) Persp. Biol. Med., 357.

Stryer, L. (1981) Biochemistry, Freeman, San Francisco, p. 23.

Synge, R.L.M. (1939) Biochem. J., 33, 1913.

Synge, R.L.M. (1944) Biochem. J., 38, 285.

Thompson, E.O.P. (1953) Biochim. Biophys. Acta, 10, 633.

Tiselius, A. (1940) Ark. Kemi. Mineral. Geol., 14B, 22.

Tiselius, A. (1958) in Nobel Lecture (Chemistry) 1942–1963, Elsevier, Amsterdam, p. 541.

Toennies, G. and Homiller, R.P. (1942) J. Am. Chem. Soc., 64, 3054.

Tristram, G.R. (1946) Biochem. J., 40, 721.

Tuppy, H. (1953) Monatsh. Chem., 84, 996.

von Neumann, J. (1955) Letter to George Gamov, July 25, quoted by Heims, S.J. 1980; John von Neumann and Norbert Wiener. From Mathematics to the Technologies of Life and Death, MIT Press, Cambridge, p. 154.

Weygand, F. and Junk, R. (1951) Naturwissenschaften, 38, 433.

Chapter 17

The Merrifield Automated Supported Peptide Synthesis

On May 26, 1959, Bruce Merrifield, a biochemist in his late thirties wrote in his notebook :

"There is a need for a rapid, quantitative, automatic method for the synthesis of long chain peptides. A possible approach may be the use of chromatographic columns where the peptide is attached to the polymeric packing and added to by an activated amino acid, followed by removal of the protecting group with repetition of the process until the desired peptide is built up. Finally the peptide must be removed from the supporting medium" (Henahan, 1971).

This was singularly prescient. After Merrifield got the go-ahead from his research supervisor at the Rockefeller Institute, Dr. D.W. Woolley, and started carrying out the idea he had thus spelled out, only a little over two years were needed to demonstrate its practicability. He was able to report in 1962, to the Federation of American Societies for Experimental Biology conference in Atlantic City, that he had synthesized a dipeptide from alanine and leucine, and a tetrapeptide Leu-Ala-Gly-Val, with close to 100% yield in the coupling steps, each of which took of the order of four hours (Merrifield, 1962). By another two years Merrifield had synthesized, by his solid-phase method, the nonapeptide bradykinin, with the activity (smooth muscle contraction, etc.) of the natural material (Merrifield, 1964). Yet another five years, and Merrifield, who in the meantime had built a machine to automate the multi-step process, had achieved the syntheses of several peptides, including a decapeptide from tobacco mosaic virus and the cyclic antibiotic valinomycin, the hormones angiotensin and oxytocin; and synthesis of proteins as well, such as insulin and ribonuclease (Henahan, 1969).

[311]

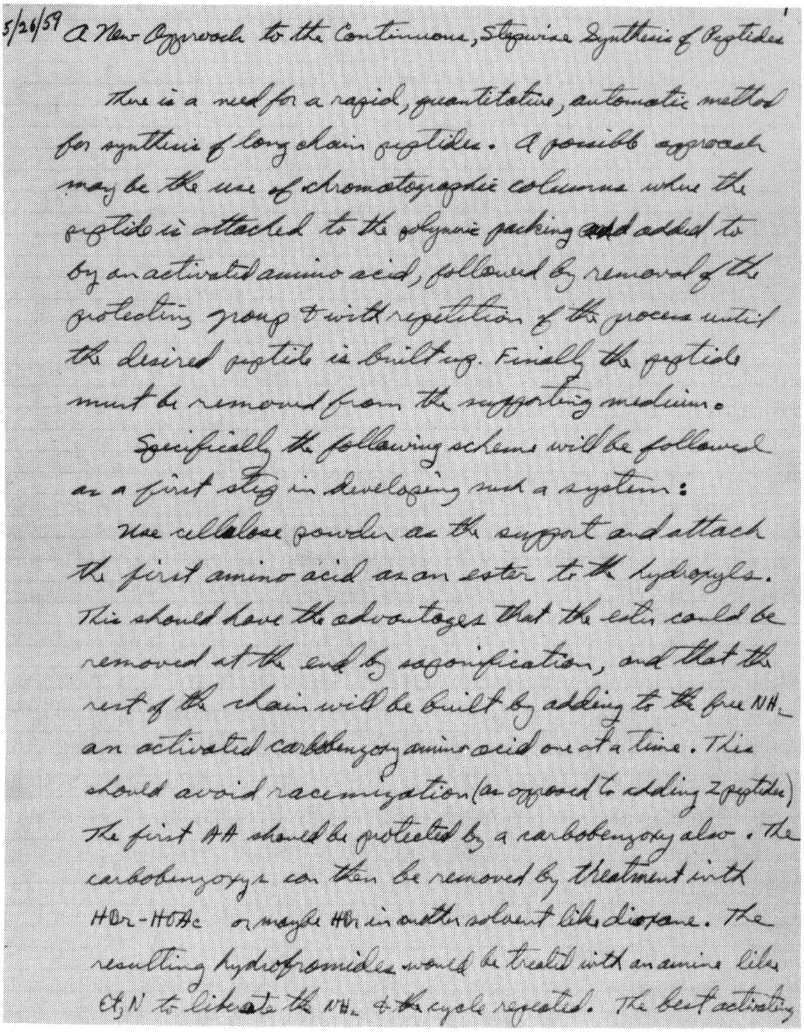

5/26/59 A New Approach to the Continuous, Stepwise Synthesis of Peptides

There is a need for a rapid, quantitative, automatic method for synthesis of long chain peptides. A possible approach may be the use of chromatographic columns where the peptide is attached to the polymeric packing and added to by an activated amino acid, followed by removal of the protecting group & with repetition of the process until the desired peptide is built up. Finally the peptide must be removed from the supporting medium.

Specifically the following scheme will be followed as a first step in developing such a system:

Use cellulose powder as the support and attach the first amino acid as an ester to the hydroxyls. This should have the advantages that the ester could be removed at the end by saponification, and that the rest of the chain will be built by adding to the free NH₂ an activated carbobenzoxy amino acid one at a time. This should avoid racemization (as opposed to adding 2 peptides) The first AA should be protected by a carbobenzoxy also. The carbobenzoxy can then be removed by treatment with HBr-HOAc or maybe HBr in another solvent like dioxane. The resulting hydrobromides would be treated with an amine like Et₃N to liberate the NH₂ & the cycle repeated. The best activating

Plate 6. The 5/26/59 entry on page 1 of Dr. Merrifield's notebook. (Courtesy R.B. Merrifield.)

After having thus sketched this success story, impressive in its rapid unfolding, I want to go over its main aspects: inception; implementation; and mechanization; which I shall do in this (chronological) order.

1. Inception

Dr. Merrifield was surely frustrated by the poor yields of polypeptides with the classical methods of peptide chemistry. This gave him the idea of solid-phase peptide synthesis. Here is the story, as Merrifield confided it 10 years afterwards to a scientific journalist :

"(Merrifield) synthesized a number of larger polypeptides (20 to 40 amino acids of random sequence) that seemed to act like catalytic enzymes in the hydrolysis of certain nitrophenyl esters. The project soon became bogged down because it took so long to synthesize the peptides – several months in the case of the simpler pentapeptides. It would have taken even longer to synthesize larger molecules with well-defined amino acid sequences. At that point (1959) Dr. Merrifield told himself there must be a better and quicker way to make polypeptides.
    One of the most frustrating aspects of conventional peptide synthesis was the fact that the product had to be removed from the reaction vessel and purified each time a new amino acid was added to the growing peptide chain. The result was a case of decreasing returns, in which more of the desired product was lost with each purification step" (Henahan, 1969).

Creativity is an obscure and mysterious process. That Merrifield seized upon the idea to affix one of the ends of the desired peptide product onto a solid support may have originated with the term "*stationary* phase" as used in chromatography. Chromatographic methods for the separation of the peptides he was preparing were indeed and by necessity extremely familiar to Merrifield. His initial idea, as consigned to his notebook in May 1959 was to

"use cellulose powder as the support and attach the first amino acid as an ester to the hydroxyls. This should have the advantages that the ester could be removed at the end by saponification, and that the rest of the chain will be built by adding to the free NH_2 an activated carbobenzoxy amino acid one at a time" (Henahan, 1971).

In this manner,

"the anchored amino acid would stay in the reactor while unneeded reagents and solvent were removed by filtration and thorough washing" (Henahan, 1969).

When Merrifield carried out this scheme, everything worked nicely as planned, and he could form a dipeptide on the cellulose support. However, removal of the product from the support proved to be impossible: treatment by strong acid turned everything to tar! Since Merrifield was so close to his goal, he experimented with various other porous solid supports, especially with resins of various sorts such as methacrylates. He found that tiny beads of polystyrene co-polymerized with about 2% divinylbenzene worked nicely. In order to covalently link amino acid No. 1 to the support, he functionalized the polymer by chloromethylation of 10–15% of the benzene rings of the polystyrene, by the reaction :

$$ R\ C_6H_5 + H_3C\text{-}O\text{-}CH_2Cl \xrightarrow[\text{cat.}]{SnCl_4} R\ \text{-}C_6H_4\text{-}\ CH_2Cl + CH_3OH $$

2. Implementation

Implementation of the solid-phase peptide synthesis thus started with 'attachment of the protected C-terminal amino acid to the insoluble solid particle (,) chloromethylated copolystyrene–2% divinyl benzene' (Merrifield, 1964). He then protected the amino end of amino acid No. 1 with a benzyloxycarbonyl group (as he had planned initially in 1959), made the triethylamine salt with the carboxylate terminus and reacted it with the chloromethyl anchor on the resin (Merrifield, 1963). Before proceeding to the key step of coupling amino acid No. 2 (leucine) to amino acid No. 1 (alanine, the reader will recall) using N,N'-dicyclohexylcarbodiimide (Sheehan and Hess, 1955) as the coupling reagent, Merrifield washed away solvents and unreacted reagents carefully, and removed the protecting group with dilute HBr. 'The end result was a dipeptide consisting of leucine and alanine, which could be easily removed from the polymer support with anhydrous HBr in a solution of trifluoroacetic acid' (Henahan, 1971).

Shortly thereafter, Merrifield made a slight change to yet improve his procedure, replacing the carbobenzoxy with the *t*-butyloxycarbonyl (*t*-BOC) as a protecting group. 'The *t*-BOC group could be removed with such mild treatment with acid (1 N HCl in acetic acid, 30 min, 25°C) that loss of peptide from the resin was almost completely avoided' (Merrifield, 1964). Thus the general scheme could be written (Merrifield, 1964):

$$(H_3C)_3COC\overset{O}{\overset{\|}{N}}\overset{R_1}{\overset{H}{C}}H-COO^- + ClCH_2C_6H_4\textcircled{P} \rightarrow$$

$$(H_3C)_3COC\overset{O}{\overset{\|}{N}}\overset{R_1}{\overset{H}{C}}H-COOCH_2C_6H_4\textcircled{P} \underset{(C_2H_5)_3N}{\overset{HCl-HOAc}{\longrightarrow}}$$

$$(H_3C)_2 = CH_2 + CO_2 + H_2N\overset{R_1}{\overset{|}{C}}HCOOCH_2C_6H_4\textcircled{P}\overset{(H_3C)_3COC\overset{O}{\overset{\|}{N}}\overset{R_2}{\overset{H}{C}}H-COO^-}{\underset{diimide}{\longrightarrow}}$$

$$(H_3C)_3COC\overset{O}{\overset{\|}{N}}\overset{R_2}{\overset{H}{C}}HC\overset{O}{\overset{\|}{N}}\overset{R_1}{\overset{H}{C}}HCOOCH_2C_6H_4\textcircled{P}\overset{HBr-CF_3COOH}{\longrightarrow}$$

$$(H_3C)_2C=CH_2 + H_2N\overset{R_2}{\overset{|}{C}}HC\overset{O}{\overset{\|}{N}}\overset{R_1}{\overset{H}{C}}HCOOH + BrCH_2C_6H_4\textcircled{P}$$

And the reactivity of the diimide as the coupling reagent was high enough to allow 'the reaction time to be shortened to 2 hours so that a complete cycle, i.e. the lengthening of the peptide chain by one amino acid residue, could be completed in four hours' (Merrifield, 1964).

Then came an important test for the method: could it be applied to the synthesis of a 'larger and more complicated peptide' (Merrifield, 1964)? He chose the nonapeptide bradykinin because of its physiological importance, and because it contained three residues (Arg, Ser, and Pro) which might be tricky to introduce with the solid-phase peptide synthesis. It took Merrifield about a year to

Plate 7. Dr. Bruce Merrifield, ca. 1963, in front of the first manually operated
solid-phase reaction vessel and shaker. (Courtesy R.B. Merrifield.)

work out all the procedural details, at the end of which he had the
satisfaction to report:

"At the rate of two residues per day the lengthening of the chain to this extent (the
protected nonapeptide) required only four days, and the time and labor needed for the
usual purification of intermediates was largely eliminated since the peptide was *not*
removed from the reaction vessel until the entire sequence was completed. (. . .)
Eight days were required to go from the *t*-BOC-nitro-L-arginyl-resin to the chromato-
graphically purified bradykinin. The peptide isolated from the column was homogeneous, and indistinguishable from authentic bradykinin by paper electrophoresis,
paper chromatography and counter-current distribution (. . .) The preparation was
found to possess the full (biological) activity of bradykinin" (Merrifield, 1964).

By contrast, 'it took Rockefeller (Institute's) Dr. John M. Stewart
12 months to make three bradykinin analogs by the older technique.

With the solid-phase method, he found he could make about 50 in the same period'. (Henahan, 1971). *The author of this book had reached this point in the manuscript when, by a lucky coincidence, the award of the 1984 Nobel Prize in chemistry to Dr. R.B. Merrifield was announced!*

3. Mechanization

The next part of the story became the mechanization and automation of the whole procedure for polypeptide synthesis: this was an obvious development, once a molecule as large as bradykinin had been synthesized. For this purpose, the machine to be built should have a supply of the various amino acids to be added one at a time in a reactor, reservoirs for the various solvents (and reagents), and a program to dictate the various steps in the precise sequence needed. As Merrifield said:

"Originally, the idea was sort of helter-skelter, until John Stewart, who was also a ham radio man, cooperated on the electrical end of things. I functioned as plumber, and between the two of us we put something together in the basement at home" (see Henahan, 1971).

A working model had been put together by 1965, with a programming drum activating as many as 30 different microswitches, to open and close valves in between the reagent and solvent containers, on one end, and the reactor, on the other (each valve had 12 orifices controlled by the general program). The initial performance of the Merrifield machine was quite promising: it coupled six new amino acids every 24 hours. Using it, Merrifield and his co-workers synthesized a variety of peptides. The insulin synthesis, once automated, took less than three weeks (Marglin and Merrifield, 1966).

The crowning achievement for the new mechanical method (Merrifield, 1965; 1967; Marshall and Merrifield, 1965; Marglin and Merrifield, 1966; Merrifield et al., 1966) though, was the synthesis of ribonuclease (Gutte and Merrifield, 1969). A protected linear polypeptide of 124 amino acid residues with the sequence of ribonuclease

A was synthesized: this required '369 chemical reactions and 11 931 steps of the automated peptide synthesis machine without any intermediate isolation steps'. (Gutte and Merrifield, 1969). The quality of the result, while not optimal, was respectable, with an overall yield of 17% implying that 'about 1.4% of the peptide chain was lost from the resin at each cycle of the synthesis' (Gutte and Merrifield, 1969), and with a specific activity 13% of that of the native enzyme. In later syntheses, the two Rockefeller researchers increased the activity to 78% by employing a battery of refined purification techniques (Gutte and Merrifield, 1971). By way of parallel and contrast, the same issue of *JACS* which carried the announcement by Gutte and Merrifield (1969) published, back to back, a communication reporting the synthesis by a 26-man team (!) from Merck, Sharp and Dohme of the part of RNase known as the S-protein (Hirschmann et al., 1969). Marglin and Merrifield (1970) wrote, 'as a goal, this synthesis of RNase was consistent with the traditions of organic chemistry for the total synthesis of a natural product'. I shall also borrow their conclusion from a useful review by Marglin and Merrifield (1970):

"Our ability to synthesize peptides is now well established. The synthesis of glucagon, secretin, and calcitonin (. . .) and of several other peptides of this general size leaves little doubt that additional peptides in this range, containing almost any sequence of amino acids, could be prepared in an acceptable yield and purity. We certainly will see the synthesis of many new analogs of these important peptide hormones, and products with increased activity and stability and fewer pharmacological side effects can be expected. The field has been moving into the area of synthetic proteins and the future efforts of many peptide chemists will turn in that direction. The early experiments on ribonuclease have shown that it is possible to synthesize a protein that possesses true enzymic activity. The difficulties are clearly much greater than with the smaller molecules, and problems of yield and especially of purity become of paramount importance. But improvements in both chemical methods of synthesis and physical methods of purification and analysis can be expected".

But I wish to single out, for its historical relevance the very last sentence in this fine review:

"The important information about enzymes that can be obtained from X-ray crystallography and from chemical modifications of native enzymes *can now be supple-*

mented by data obtained from chemical synthesis" (Marglin and Merrifield, 1970; emphasis added).

Emil Fischer could rest happily in his grave: the tradition he had started was well alive!

REFERENCES

Gutte, B. and Merrifield, R.B. (1969) J. Am. Chem. Soc., 91, 501.

Gutte, B. and Merrifield, R.B. (1971) J. Biol. Chem., 246, 1922.

Henahan, J.F. (1971) Chem. Eng. News, August 2, p.22 gives a photostatic reproduction of the whole corresponding page from Merrifield's notebook;see also Plate 6.

Hirschmann, R., Nutt, R.F., Veber, D.F., Vitali, R.A., Varga, S.L., Jacob, T.A., Holly, F.W. and Denkewalter, R.G. (1969) J. Am. Chem. Soc., 91, 507; and preceding communications.

Marglin, A. and Merrifield, R.B. (1966) J. Am. Chem. Soc., 88, 5051.

Marglin, A. and Merrifield, R.B. (1970) Annu. Rev. Biochem., 39, 841.

Marshall, G.R. and Merrifield, R.B. (1965) Biochemistry, 4, 2394.

Merrifield, R.B. (1962) Fed. Proc., 21, 412.

Merrifield, R.B. (1963) J. Am. Chem. Soc., 85, 2149.

Merrifield, R.B. (1964) Biochemistry, 3, 1385.

Merrifield, R.B. (1965) Science, 150, 178.

Merrifield, R.B. (1967) Rec. Prog. Hormone Res., 23, 41.

Merrifield, R.B., Steward, J.M. and Jernberg, N. (1966) Anal. Chem., 38, 1905.

Sheehan, J.C. and Hess, G.P. (1955) J. Am. Chem. Soc., 77, 1067.

PART III

The Hemoglobin Story

(Chapters 18–26)

The red color of blood

assumes great importance in all cultures. It is a poetic commonplace.
Oscar Wilde wrote thus:

"He did not wear his scarlet coat,
  For blood and wine are red,
And blood and wine were on his hands
  When they found him with the dead" (Wilde, 1898).

Robert Browning came with this splendid line:
"Sunset ran, one glorious blood-red, reeking into Cadiz bay".

Many writers have also referred to the moral stain from blood-shed-
ding, Daniel Defoe for instance:
"Nature has left this tincture in the blood,
That all men would be tyrants if they could".

Coming back to the intense fascination which the color of blood
evokes, here is a line from Emily Dickinson:
"Sang from the Heart, Sire,
Dipped my Beak in,
If the Tune drip too much
Have a tint too Red

Pardon the Cochineal –
Suffer the Vermilion –
Death is the Wealth
Of the Poorest Bird.

(. . .)" (Dickinson, ca. 1865).

This oxymoron figure of speech (' . . . the wealth of the poorest . . . ')
on which I end my quote from this four stanzas poem leads on to a
religious conclusion. In between, Emily Dickinson builds on many
analogies, of the poet and of the bird, of the cardinal (bird) and the
cardinal (high priest). She likens sincerity in writing to a bird song
which, in turn, is compared to a ballad, to a liturgy, to a choral. Even
in the most mundane reading, Emily Dickinson has three words to
denote the color of blood, *red, cochineal,* and *vermilion.*

Chapter 18

Historical Notes about Hemoglobin and its Various Names

This vermilion color of the blood was ascribed by Lavoisier to oxy-gen binding (Lavoisier, 1777). Lavoisier had earlier the same year, in the Easter public session of the French Academy, reported the results of his October 1776 experiments with Trudaine de Monti-gny. These experiments dealt with the mechanism of respiration. They showed (in modern terminology) that oxygen was the active gas, and that carbon dioxide was evolved in the lungs. Later, the same year 1777, in his *Mémoire sur la Combustion en Général*, Lavoisier pointed to the analogy between respiration and the com-bustion of carbon, with heat evolution. As Henry Guerlac puts it,

"a persuasive argument seemed to him to be that, just as the calcination of mercury and lead results in red powders so the absorption of oxygen gives a bright red color to arterial blood" (Guerlac, 1975).

Lavoisier went on, using the calorimeter, to determine in collabora-tion with Laplace the heat evolved by a guinea pig exhaling a given quantity of carbon dioxide (Guerlac, 1975).

To determine the nature of the red substance present in the blood, the difficulties for the chemists at the beginning of the 19th century we now know were awesome. Hemoglobin associates an inorganic element, iron, with a porphyrin together with a quite large protein. This is a tough structure to crack. Yet, by the 1830s, chemists had established the presence of iron; and they were starting to distin-guish, in an uncertain and groping manner, true, the heme compo-nent from the globulin proteic part.

Another difficulty was conceptual, i.e. to identify hemoglobin as the respiratory pigment. Again, the ingenuity and intuition of our forebears are remarkable. After the initial stroke of genius by Lavoisier came an extremely lucid statement by Liebig, in the 1840s. Thereafter, the connection became gradually firmer and firmer. To the extent that in 1864, when hemoglobin was christened by Hoppe-Seyler, its dual composition (heme plus protein) was secure, and its respiratory function was also well established. Hoppe-Seyler could draw also upon another feature – to which he had contributed but which reflected mostly the work of Stokes – the absorption spectrum of hemoglobin which had become available.

Yet another daunting task was the determination of the elemental composition and molecular weight for such a large molecule as hemoglobin. These became known (for the monomer rather than for the tetramer, but with quite good accuracy), as early as 1886! A present-day chemist can only be impressed by the precocity of all these determinations, with the techniques available.

Hence, it is not surprising from hindsight if the complexity of hemoglobin's constitution (we shall deal further on with its detailed molecular architecture) took almost a century to unravel (if one takes as an arbitrary starting point the studies by Lavoisier just prior to the French Revolution).

One may note in passing that the historical disturbances, the French Revolution which stopped at an early stage Lavoisier's explorations in physiological chemistry, the Napoleonic wars afterwards, probably delayed progress in the hemoglobin field, perhaps by as much as a whole generation (25 years).

The chemical and biochemical terminology merely reflects the underlying scientific knowledge. Accordingly, it took a long time also for the terminology relative to hemoglobin and its constituent parts to be settled and to coincide with that in use nowadays.

After this rapid summary, may we return to the actual history. Prévost and Dumas made a physicochemical study of blood in the early 1820s (Prévost and Dumas, 1821; 1823). They reported that 'the coloring matter in blood appears to be made of an animal substance combined with iron peroxide' (Prévost and Dumas, 1823). Berzelius quantified this finding, he determined as half of a percent the amount of iron peroxide present in this red-colored substance of blood which M.E. Chevreul (1786–1889) named hematosine in 1827 (Chevreul, 1827). Such a small quantity of iron was difficult to determine, to ascertain its role was even more difficult:

"If one does not know which difference exists between the actual composition of hematosine in the arterial blood and in the venous blood, one is even further from knowledge of the role played by iron peroxide in the hematosine of which it is always a component" (Thénard, 1836).

At the time Louis Jacques Thénard (1777–1857) wrote his influential book, hematosine was known also under the names of zoohematine and hematochroïne. It had been isolated in two states, one water-soluble and the other water-insoluble. Some authors (Lecanu) held that hematosine was not the pure coloring ingredient of blood. Rather, Lecanu held that hematosine was a complex between this colored substance and albumin (Thénard, 1836; Lecanu, 1838). We shall come back to this point.

Work on hemoglobin proceeded, the fluctuations in designation reflecting the uncertainties about its chemistry and physiological function. In the late 1830s–early 1840s, the respiratory pigment of blood became known also as *cruorin*: one finds, for instance, the statement that 'the solution of cruorin is reddened less strongly by exposure to air' (Baly, 1840). The existing terminology in the year 1845 includes, besides the names already mentioned (hematosine, zoohematine, hematochroïne, cruorin), the *haematin* and *haematoglobulin*. Haematin is listed among the various aspects which Mulder's 'protein' could assume:

"Protein and its various modifications – gelatin, bilin, and the products of its metamorphosis – haematin, urea, uric acid, etc." (Day, 1845a).

Haematoglobulin is a substance which 'according to Berzelius, (...) contains 100 parts of globulin and 5.3 of haematin' (Day, 1845b).

But the term 'cruorin' (or 'cruorine') seemingly had the edge, scarlet cruorine belonging to 'oxidized blood' while purple cruorine characterized 'deoxidized blood' (Stokes, 1864). In between Chevreul's 'hematosine' of 1827 and Stokes's 'cruorine' of 1864, this self-same

326

326 THE HEMOGLOBIN STORY

red substance of blood ascended to the conceptual status of a respiratory pigment.

During this whole period, from 1827 till the pivotal year 1864, the shifting terminology reflects considerable uncertainty as to the basic location of the respiratory phenomenon. Gustav Magnus concluded from his experiments that the change in blood color during respiration was due to chemical changes undergone by the blood in the lungs, a re-statement of the fundamental intuition of Lavoisier (Magnus, 1837).

Lecanu showed in the following year that Chevreul's hematosine contained iron, and he was convinced this was not an albuminoid substance:

"hematosine is a solid, odorless, tasteless, with a metallic sparkle and of a reddish black color which evokes the aspect of the red silver of mineralogists" (Lecanu, 1838).

Lecanu describes there only the heme; indeed Lecanu also reports that he made a precipitate from this hematosine and albumin (Debru, 1983a). Likewise, the haematin which Berzelius had studied had been heme separated from the proteic part; and indeed Berzelius, who shared Mulder's postulate of a single protein, denied during the same period of the late 1830s that haematin belonged with the animal substances made from protein (Berzelius, 1839). And like Lecanu, Berzelius reported that haematin formed a complex with 'globulin' (see above; Day, 1845b).

A decisive conceptual breakthrough was made, just a few years afterwards, by Justus von Liebig (1803-1873):

"(. . .) the blood corpuscles which have become dark red in carbonic acid return to a bright red at the contact of oxygen. (. . .) the blood corpuscles contain an iron combination. (. . .) The iron combination in the blood corpuscles behaves as an oxygen combination, since it is decomposed by hydrogen sulfide in the very same manner as iron oxides or other analogous iron combinations. (. . .) iron combinations, more than those from any other metal, have remarkable properties. The combinations from iron protoxide have the property to abstract oxygen from other oxygen-containing compounds; the combinations of iron oxide give up their oxygen with the greatest ease under other conditions. (. . .)" (Liebig, 1842).

I wrote above of 1864 being the pivotal year, because it saw both the pioneering spectroscopic experiments of Stokes and the naming of hemoglobin by Hoppe-Seyler. Sir George Gabriel Stokes (1819–1903) was an Irish mathematician and physicist. There is some irony, as writes Debru (1983b), that 'it was a highly reputed mathematician and physicist, not a physiologist, who was responsible for setting the bases of the chemical physiology of hemoglobin'.

Stokes had already extended spectroscopy towards shorter wavelengths. He recorded the spectra of numerous organic substances in solution, including alkaloids and glucosides (Stokes, 1862). 'Stokes in 1863 made the decisive step for the chemical physiology of hemoglobin' (Debru, 1983c). He reported two widely different spectra which (as earlier noted) he named 'cruorine'. These spectra differ in turn from those of the heme ('hematin'), which has also two different spectra depending upon its oxidation state. From these observations, Stokes concluded that cruorin forms a weak chemical combination with oxygen in the lungs (Stokes, 1863–1864).

Felix Hoppe-Seyler (1825–1895) had done studies somewhat similar to those of Stokes, although they were more limited in scope and proved to be partly wrong in their conclusions. He had obtained in 1862 the absorption spectrum of the colored substance of the blood in the visible. However, he did not find the absorption spectrum of this substance to depend upon its oxygenation state. But he found that the hemoglobin spectrum was invariant under a number of chemical conditions, in the presence of various gases and other reagents. If there were changes, these were reversible; furthermore, there was a basic identity of this hemoglobin, from the spectroscopic viewpoint, for at least all the vertebrates (Hoppe, 1862).

Hoppe-Seyler had earlier (1857) studied competitive binding of oxygen and carbon monoxide to hemoglobin. At the same time, Claude Bernard had interpreted the same phenomenon as being caused by the destruction, wrought by carbon monoxide, of the combination of oxygen with the red blood cells (Bernard, 1857). His own studies convinced Hoppe-Seyler that hemoglobin was an oxygen carrier (Debru, 1983d). And when he observed a unique characteristic spectrum for this compound, which a number of investigators (Funke, Lehmann, Teichmann, etc.) had already obtained in crystalline form by 1852–1853, Hoppe-Seyler felt justified to make a series of firm assertions. He named this substance 'hemoglobin', which consisted for him of the union of globulin, a protein, and hematin, an iron-containing non proteic part (Hoppe-Seyler, 1864).

It must have been the right time: the new name caught on immediately and stuck. A few examples of its use by contemporaries follow:

"The specific gravity of haemoglobin may by calculation be approximately estimated as 1.2 to 1.3" (Anon., 1869).
"Haemoglobin is the only ferruginous constituent of the blood corpuscles" (Watts, 1869–1872),

or, restating the Hoppe-Seyler definition:

"Called haemoglobin from its readily breaking up into globulin and haematin" (Huxley, 1872).

A further step in the breakdown of hemoglobin into its constituent parts was J.L.W. Thudichum's purification of 'iron-free hematin', which he called 'cruentine' and which we now name porphyrin (Thudichum, 1867).

Hemoglobin was for quite a while, during nearly 40 years (from 1852 to 1890, when Franz Hofmeister crystallized ovalbumin) the only protein available in crystalline form. Thus, it was isolated in crystalline form from many animal species. In 1877, a French translation of one of Hoppe-Seyler's books indicates that crystalline hemoglobin can be obtained from the blood of dogs, cats, hedgehogs, marmots, guinea pigs, rats, geese, horses, etc., and that it is less easily crystallized from human blood. Hoppe-Seyler recommends isolation of hemoglobin by a salting-out procedure, using an aqueous sodium chloride solution, followed by crystallization with ether, and subsequently by addition of cold ethanol:

"By stirring blood (red) cells from the rat, the guinea pig, the squirrel, or the dog with ether, a crystalline precipitate of oxyhemoglobin forms so rapidly that part of this substance remains on the filter when filtering the liquor" (Hoppe-Seyler, 1877).

Hoppe-Seyler goes on to make the important observation that the crystals of oxyhemoglobin differ in shape depending upon the animal species. Reichert and Brown, in the early 1900s, much enlarged upon these initial observations:

"Oxyhemoglobin crystals", writes Hoppe-Seyler, "are generally microscopic and they seldom are more than 3 mm; their shape depends on the nature of the blood and thus upon the animal species which served to their preparation. Those made from turkey-cock blood are the only to present a regular shape; these are rather large cubes rarely modified at their corners by octahedral facets. Those provided by squirrel blood have a tabular six-sided shape and belong to the hexagonal system. The tetrahedra and octahedra coming from guinea pig and rat blood appear to be representatives of the rhombic system. Finally crystals of dog blood, generally made of four-sided prisms, as well as those from goose blood, are rhombohedra or belong to the monoclinic system" (Hoppe-Seyler, 1877).

In 1909, appeared the massive compilation by Reichert and Brown. They reported the crystal systems and crystalline modifications for all these hemoglobins, which they tried to relate to the phylogeny, i.e. to an evolutionary scheme for these various species (Reichert and Brown, 1909).

The history of hemoglobin as a molecule continues with the first determination of its M_r by Zinoffsky in 1886, using quantitative analysis (Zinoffsky, 1886; see Chapter 1, section 4): he reported a value of 16 700 and the elemental formula $C_{712}H_{1130}N_{214}S_2FeO_{245}$.

I have referred already (Chapter 1, section 4) to the determination by Adair in 1925 of the M_r of hemoglobin using osmotic pressure: he reported a value of 66 700; this value was in nice agreement with the M_r of 68 000 determined in 1926 by Svedberg and Fahraeus with the ultracentrifuge.

REFERENCES

Anonymous (1869) Syd. Soc. Biennial Retrospect., 3.

Baly, (1840) tr. Muller's Physiol. (ed. 2), I, 133.

Bernard, Cl. (1857) Leçons sur les effets des substances toxiques et médicamenteuses, Paris; see also Bernard, Cl. (1859) Leçons sur les propriétés physiologiques et les altérations pathologiques des liquides de l'organisme, Paris.

Berzelius, J.J. (1839) Traité de Chimie, Bruxelles, III, 528.

Browning, R. Home Thoughts, from the Sea, in Oxford Dictionary of Quotations, 3rd ed., Oxford University Press, Oxford, 1979.

Chevreul, M.E. (1827) Dict. Sci. Nat., 47, 187.

Day, G.E. (1845) tr. Simon's Anim. Chem., (a) I, 5; (b) I, 43.

Debru, Cl. (1983) L'esprit des Protéines. Histoire et Philosophie Biochimique, Hermann, Paris, (a) p. 138; (b) p. 144; (c) p. 143; (d) p. 141.

DeFoe, D., The Kentish Petition, addendum 1.11, in Oxford Dictionary of Quotations, 3rd ed., Oxford University Press, Oxford, 1979.

Dickinson, E. (ca. 1865) in The Complete Poems of Emily Dickinson (Johnson, T.H., Ed.), Little, Brown, Boston, 1960, p. 482.

Guerlac, H. (1975) Antoine-Laurent Lavoisier, Chemist and Revolutionary, Ch. 12, Scribner, New York.

Hoppe, F. (1862) Arch. Pathol. Anat. Klin. Med. (Virchow), 23, 449.

Hoppe-Seyler, F. (1864) Arch. Pathol. Anat. Physiol. (Virchow), 29, 233.

Hoppe-Seyler, F. (1877) Traité d'Analyse Chimique Appliquée à la Physiologie et à la Pathologie. Guide Pratique pour les Recherches Cliniques. Schlagdenhauffen, transl., Paris, pp. 295-297 (my translation).

Huxley, T.H. (1872) Physiology, iii, 65.

Lavoisier, A.L. (1777) Mem. Acad. Roy. Sci., 192.

Lecanu, L.R. (1838) Ann. Chim. Phys., 67, 59.

Liebig, J. von (1842) Die organische Chemie in ihrer Anwendung auf Physiologie und Pathologie, Braunschweig, pp. 274-276; quoted in French translation by Debru (1983a) (my translation into English).

Magnus, G. (1837) Ann. Sci. Nat. (Zool.), 2nd series, 8, 79.

Prévost, J.L. and Dumas, J.B. (1821) Ann. Chim. Phys., 18, 286.

Prévost, J.L. and Dumas, J.B. (1823) Ann. Chim. Phys., 23, 55.

Reichert, E.T. and Brown, A.P. (1909) The Differenciation and Specificity of Corresponding Proteins and Other Vital Substances in Relation to Biological Classification and Organic Evolution: the Crystallography of Hemoglobins, The Carnegie Institution, Washington, DC.

Stokes, G.G. (1862) Phil. Trans. Roy. Soc. London, 152, II, 599.

Stokes, G.G. (1863-1864) Proc. Roy. Soc. London, 13, 357.

Stokes, G.G. (1864) Phil. Mag., 27, 388.

Thénard, L.J. (1836) Traité de Chimie Elémentaire Théorique et Pratique, Suivi d'un Essai sur la Philosophie Chimique et d'un Précis sur l'Analyse, Hauman, Bruxelles, vol. 2, p. 253 (my translation).

Thudinium, J.L.W. (1867) Rept. Med. Off. Privy Council X (Appendix 7), H.M.S.O., London, pp. 152; 200; 227.

Watts, (1869-1872) Dict. Chem., VI, 353.

Wilde, O. (1898) The Ballad of Reading Gaol, I, i.

Zinoffsky, O. (1886) Z. Physiol. Chem., 10, 33.

Chapter 19

Cooperativity of Dioxygen Binding: the Bohr Effect

When Zinoffsky proposed in 1886 his elemental formula and M_r for hemoglobin (Zinoffsky, 1886), this was not an altogether new finding. In fact, he was merely refining upon the earlier determinations by Hüfner who had reported a formula $C_{550}H_{852}N_{149}S_2FeO_{149}$ and an M_r, correspondingly, of 12 042 (Hüfner, 1883). Hüfner's main contribution is conceptual:

"Hüfner is (. . .) led to describe the dissociation of oxyhemoglobin in terms of chemical equilibrium, when he observes that this dissociation is never complete. It tends like all dissociation phenomena towards an equilibrium state. By being the first to introduce this theoretical concept, Hüfner made a fundamental clarification in the study of these phenomena, (he) laid the first theoretical basis for this study" (Debru, 1983a).

Yet, Hüfner had the misconceived notion that hemoglobin contained a single iron atom (see his formula above), which ruled out accurate accounting by mass action law of the oxygenation curve for hemoglobin. Nevertheless,

"by describing oxygen absorption by blood as a diffusion equilibrium followed by a chemical equilibrium (Hüfner, 1890), Hüfner is able to derive an expression which makes it possible to compare the experimental and the theoretical curves for hemoglobin oxygenation" (Debru, 1983a).

Gustav Hüfner succeeded Felix Hoppe-Seyler in the chair of physiological chemistry at Tübingen, which he held during the period 1872-1908. Like his predecessor, Hüfner lectured on organic chemistry. For him, physiological chemistry was not cut off from organic and physical chemistry (Kohler, 1982). This open-mindedness was fortunate. It gave Hüfner the impetus to make his case, and to hold his ground against rival claims. In this case, his chief opponent (Debru, 1983b) was the Danish experimental physiologist Christian Bohr.

Christian Bohr's two main claims to fame are siring his illustrious son Niels Bohr, the physicist; and discovering the effect which now bears his name. His intellectual background was rather similar to Hüfner's: by making too much of the opposition between these two scientists (Debru, 1983b), one might overlook their joint origin from the same school of thought. Christian Bohr had trained, in part, in the laboratory of Karl Ludwig. Karl Ludwig had himself been steeped in the new physicochemical paradigm at the Berlin School of Johannes Müller and Hermann Helmholtz, where he had trained prior to setting up an Institute for physiological chemistry in Leipzig (Kohler, 1982). After returning to Copenhagen, where he was appointed to the chair of physiology, Christian Bohr set about studying the physiology of blood and respiration. He did it in phenomenological manner.

His first innovation was a clean experimental technique for studying exchange of gases by the blood.

"Bohr derived arterial blood into an enclosure, where he could place it in contact with a well-determined volume of gases with known partial pressures. He repeated this procedure until an equilibrium was obtained between the pressure of gases dissolved in the circulating blood and the pressure of gases in the closed compartment, where it was being measured. In this manner, by comparing the pressures of the dissolved gases to the alveolar pressures, he was led to the idea of pulmonary respiratory processes occurring against the gas pressure gradient; hence, these processes can be explained only by an active absorption or excretion process, physical diffusion in itself is unable to account for these observations (Bohr, 1887; 1888). This idea will have a durable impact upon the history of physiology" (Debru, 1983c).

A few years later, Christian Bohr went on to determine competitive binding of dioxygen and carbon dioxide by hemoglobin (Bohr, 1892a). He recorded the corresponding dissociation curves for a wide range of partial pressures. These dissociation curves, he found, were rectangular hyperbolas, tending asymptotically towards a saturation limit when the partial pressure of O_2 or CO_2 keeps increasing. Bohr inferred from these experiments that CO_2 binding interfered with O_2 binding, and that the former came about from the proteic part of hemoglobin:

"binding of carbonic acid with the colorless part of the molecule produces (. . .) a change in the colored part which decreases oxygen absorption by the latter" (Bohr, 1892b, quoted by Debru, 1983d).

The bone of contention between Christian Bohr and Gustav Hüfner was coexistence of various hemoglobins in the blood of a given animal species, these respiratory pigments differing markedly in their affinity for oxygen. This finding was reported by the Danish scientist (Bohr, 1892a, b), it was violently challenged and denied by his German rival who maintained that the physicochemical properties were consistent with a single hemoglobin for each animal species.

In subsequent years, Bohr, who had shown in 1892 that CO_2 pressure could assist oxygen release by blood in the tissues (Debru, 1983e), attempted (rather unsuccessfully) to give an experimental answer to the converse effect, the influence of O_2 partial pressure on absorption and release of carbon dioxide by the blood (Bohr et al., 1903-1904).

In doing so, Bohr and his co-workers noticed that the family of dissociation curves they obtained for oxyhemoglobin in the presence of carbon dioxide at various partial pressures differed from the rectangular hyperbolas they had obtained earlier. These were now sigmoidal curves (Bohr et al., 1903-1904). Bohr was hard put to explain this feature. He came to the conclusion that it implied the presence of two iron atoms per hemoglobin molecule, a view consonant with contemporary ideas about the chemical make-up of hemoglobin, but again strongly adverse to the simple chemical ideas of Gustav Hüfner. As noted by Debru (1983f),

"In addition, Bohr, who had been unable to determine with certainty the maximum oxygen binding capacity of hemoglobin, was confronted by the problem of the great variability of this capacity according not only to the animal species, but even from one individual to another. Yet he was intent upon the hypothesis of a stoichiometric combination of oxygen with the heme iron. Bohr was thus forced to postulate the existence of several hemoglobins having differing maximum binding capabilities, he was forced to admit that hemoglobin did not consist of a single chemical species (Bohr, in Nagel, 1905-1909)".

But Bohr's chief contribution, that for which he is mostly remembered, was establishing the cross-relationship between the oxygen affinity of hemoglobin and the partial pressure of carbon dioxide. This is the phenomenon known as the Bohr effect. As Debru (1983g) writes perceptively:

"Hemoglobin was then (in the early 1900's) what it would be for a long time to come, raising an intense puzzlement: this is the unique case where a physiological investigation leads to an explanation, molecular in type. This is because the behavior shown by the dissociation curves cannot be distinguished in actuality from the molecular complexity it reveals. It cannot be interpreted without an additional hypothesis on molecular complexity and on the various resulting phenomena".

The first attempts at explanation, just like those of Christian Bohr himself, came back to the question of the M_r of hemoglobin. Bohr leaned toward a dimer. Waymouth Reid proposed, on the basis of osmometry, a trimer (Waymouth Reid, 1905). Gustav Hüfner conversely, using also osmometric measurements, maintained that oxyhemoglobin had an M_r of 16 700 and consisted of a single dioxygen molecule added to a single hemoglobin molecule (Hüfner and Gansser, 1907).

An important step forward was taken by Joseph Barcroft (1872-1947). He was the first to control carefully the ionic strength of the aqueous solutions of hemoglobin on which gas absorption isotherms were measured. He found the very shape of the dissociation curve of oxyhemoglobin to depend upon the presence of electrolytes (Barcroft, 1914): a dialyzed solution gives a rectangular hyperbola, whereas an undialyzed solution leads to a Bohr sigmoidal curve (Barcroft and Camis, 1909-1910; Barcroft and Roberts, 1909-1910).

Another key contribution came from the laboratory of John Scott Haldane (1860-1936), a British scientist like Barcroft. J.S. Haldane initially worked at Oxford where he taught chemical physiology, at first in an informal manner, since 1887. He resigned his readership in 1913 when he was passed over in favor of Sherrington for a chair in physiology (Kohler, 1982b).

Haldane, with C.G. Douglas and his son J.B.S. Haldane, studied in 1912 the dissociation curves for carbon monoxyhemoglobin; he showed that these had also a sigmoidal shape and could be shifted more or less depending upon the partial pressure of carbon dioxide. They interpreted their results by an equilibrium of the type $HbO_2 + Hb_2O_4 = Hb_3O_6$ (Douglas et al., 1912). The same J.S. Haldane group was able to make in 1914 the important generalization that dioxygen or carbon monoxide will shift, Bohr-effect like, the sigmoidal curve of carbon dioxide binding by hemoglobin. Thus, they reached the key-physiological interpretation of these effects: oxygenation of blood in the lungs helps releasing carbon dioxide, whereas conversely release of oxygen by the blood assists in the tissue binding of carbon dioxide (Christiansen et al., 1914).

As Henderson scathingly writes, when he alludes to the work of the Danish physiologist Christian Bohr and to the protracted realization by Haldane of the reciprocal relationship between the partial pressures of dioxygen and carbon dioxide, as mediated by hemoglobin:

"To those who have not themselves experienced that state of bewilderment which is the usual condition of the investigator, it must seem strange that the physiologists who were studying the respiratory function of the blood should not have drawn from the discovery of the variation of oxygen saturation with carbon dioxide pressure the conclusion that, since carbonic acid influences the oxygen equilibrium in blood, oxygen must influence the carbonic acid equilibrium. (. . .) Yet, so little are physiologists accustomed to mathematics, and such is the natural inertia of the mind, that this conclusion escaped us all and it remained for Christiansen, Douglas and Haldane (1914) to discover by experiment that the carbon dioxide dissociation curves of oxygenated and of reduced bloods are different" (Henderson, 1928).

This little gem of a text, lucid, modest and self-deprecatory toward the end, witty and perceptive, puts it all in a nutshell!

In the meantime, the British physiologist Archibald Vivian Hill (1886-1977) started investigating, in parallel to his study on the nature of muscular contraction, the possible effects of aggregation (of oligomerization, as we now know it to be the case) on the dissociation curve of oxyhemoglobin. This was the contribution of a young scientist close to the start of his career. A.V. Hill - who was thus working within the then dominant colloidal paradigm - assumed that he was dealing with a chemical equilibrium of the type $Hb_n + nO_2 = Hb_n(O_2)_n$. He further assumed that the oxygen binding sites could be totally saturated (Hill, 1910). A few years later, Hill proposed the equation:

$$K = \frac{[Hb]}{[HbO_2]} \cdot P_{O_2}^n$$

to account quantitatively for the sigmoidal oxygen dissociation curve first observed by Bohr. This is a phenomenological approach. The two parameters, K and n, have a clearly defined meaning, the former being an apparent mean equilibrium constant for binding of dioxygen at n equivalent sites, and the latter being a minimum aggregation number (or number of binding sites, on the basis of the postulated equilibrium). Hill further showed that a plot of ln $y/(1-y)$, where the saturation fraction $y = Kx^n/(1 + Kx^n)$, against ln x, where x is the partial pressure of oxygen, the so-called Hill plot, provides the value of this phenomenological Hill coefficient n (Hill, 1913).

G.C. Adair showed about 10 years later that the Hill coefficient n is the lower limit to the number of equivalent subunits. He first established by osmometry the M_r of hemoglobin as 67 000 (Adair, 1925a), which corresponded to a tetrameric structure Hb_4 (Adair, (1925b). Adair suggested that ligand binding to hemoglobin occurs in four distinct successive steps:

$Hb_4 \quad + X = Hb_4X$ (equil. constant K_1)
$Hb_4X \quad + X = Hb_4X_2$ (equil. constant K_2)
$Hb_4X_2 + X = Hb_4X_3$ (equil. constant K_3)
$Hb_4X_3 + X = Hb_4X_4$ (equil. constant K_4)

The fractional degree of saturation y is then given by an equation of the form:

$$4\,y = \frac{K_1X + 2\,K_1K_2X^2 + 3\,K_1K_2K_3X^3 + 4\,K_1K_2K_3K_4X^4}{1 + K_1X + K_1K_2X^2 + K_1K_2K_3X^3 + K_1K_2K_3K_4X^4}$$

(Adair, 1925c).

Another, related thread in the story was consideration of the acid–base equilibria in which the hemoglobin molecule is involved. In 1867, it was found that transport of carbonic acid occurred to a much greater extent in red blood cells than in the serum, and that hemoglobin behaves as a weak acid, capable of decomposing sodium bicarbonate (Zuntz, 1867; Preyer, 1867). Barcroft and Orbeli showed at a much later stage, after the Bohr effect had been well established, that lactic acid has a similar effect as carbonic acid on the dissociation of oxyhemoglobin, with the important physiological interpretation that lactic acid produced by muscular work will produce a higher release of oxygen from oxyhemoglobin in the muscular tissue (Barcroft and Orbeli, 1910–1911). Quite a number of investigators studied the acid–base behavior of hemoglobin in the presence of various ligands. An interesting finding was that

"the buffer value of hemoglobin in blood is almost ten times as great as that of the serum proteins (. . .) it is not only greater than that of the serum proteins but also exceptional among proteins in general. It is to be regarded as a well-marked adaptation" (Henderson, 1928).

Lawrence J. Henderson (1878–1942) had an engaging personality. His career was marked by personal and institutional rivalries. His unfortunate tale sheds a crude, unfavorable light upon the shortsightedness of university administrators (even those at Harvard!).

Henderson came from Boston-Yankee intelligentsia. He became a chemist, and taught physical chemistry as an instructor in the Department of Chemistry at Harvard College. Henderson had returned to Harvard after spending two years in Hofmeister's Strasbourg laboratory.

Then, in 1905, an emergency arose at Harvard Medical School: Edward S. Wood, in charge of physiological chemistry, was dying of cancer. Lawrence J. Henderson was therefore transferred part-time from the Department of Chemistry to the Harvard Medical School, where he assisted Carl Alsberg, appointed to replace Wood as head of the Department of Chemistry (of the Medical School). The third man in the makeshift troika, quickly improvised because of Wood's terminal illness, was Otto Folin, who was appointed research associate professor of physiological chemistry. As Kohler writes:

340   THE HEMOGLOBIN STORY

"On being appointed to the medical school, (Henderson) took up physicochemical
studies of electrolyte balance in blood, which he thought more appropriate to a medical context. But his enthusiasm for kinetics earned him the nickname of 'little k' "
(Kohler, 1982c).

We get thus the correct impression: Henderson's enthusiasm for
pure chemistry did not endear him to the medical students. By contrast, Otto Folin thrived both as a medical teacher, and as an effective administrator. Furthermore, Folin, whose research philosophy
was of the old medicinal chemistry type, attempted at first to get
Henderson into the medical fold. When this did not work, he tried
and succeeded in ousting Henderson.

"In 1909, Folin was appointed Hamilton Kuhn Professor of Biological Chemistry.
Henderson was deeply disappointed; he did not resign but stayed away, in part to let
Folin organize the department in his own way and in part because Folin's way was just
not his. (. . .) Henderson spent more and more of his time at Harvard College, where
he taught biological chemistry and the history of science and ultimately became
involved in general physiology, industrial physiology, and social theory" (Kohler,
1982c).

As Kohler writes elsewhere in his book:

"The romantic intellectual, Henderson delighted in new ideas, as long as they were
avant-garde; a brilliant teacher and thinker, Henderson was an indifferent administrator" (Kohler, 1982d).

During the early 1920s Henderson described the effect of a ligand L_B
upon the uptake by hemoglobin of another ligand L_A by having the
corresponding equilibrium constant K_A depend upon the concentration $[L_B]$ (Henderson, 1920). This was an important physicochemical generalization. Debru terms it a mathematical generalization
(Debru, 1983h) and pits it against the contemporary physiological
generalization by J.S. Haldane (Haldane, 1908; 1917; 1922).

Henderson also clarified convincingly the dependence of the oxyhemoglobin–carbonic anhydride interaction upon another chemical
equilibrium (or series of equilibria), that between aqueous carbonate, bicarbonate, and carbon dioxide. Henderson treats the 'blood as
a physicochemical system' (Henderson, 1921). This approach, in the
spirit of Jacques Loeb, met with considerable success.

It is a textbook example of the fruitfulness of a simple-minded 'physical' approach, in which a quite complex biological system is *modeled* with a well-understood mechanism entailing only few parameters. In this case, as he explains in progressive, didactic manner in his 1928 book, Henderson starts with an aqueous mixture of 30 mM base, 50 g.l^{-1} of serum proteins and he calculates with these data the carbon dioxide dissociation curve. He then considers an aqueous solution of hemoglobin (147 g.l^{-1}), of base (43 mM), and again builds the carbon dioxide dissociation curve. Finally, he combines both sets of conditions, and obtains a resultant curve, which mimics quite accurately that for oxygenated blood:

"the acid-base equilibrium of the proteins of blood, especially that of hemoglobin, determines the character of the carbon dioxide dissociation curves, at least for the case of normal blood, and makes possible the transport of carbonic acid" (Henderson, 1928).

Henderson ascertains that the Hill equilibrium constant K_{O_2} is 'approximately a linear function of the partial pressure of carbon dioxide' (Henderson, 1928). Henderson analyzes in detail acidification of the blood upon oxygenation, first advanced by J.S. Haldane as a working hypothesis (Christiansen et al., 1914). The Harvard physical chemist thus reaches a general conclusion, which he phrases in physiological terms:

"The fall in concentration of carbon dioxide in the lung is accompanied by a rise in the capacity of the blood to absorb oxygen, and the rise in concentration of carbonic acid in the tissues is accompanied by a fall in the capacity to absorb oxygen" (Henderson, 1928).

With the work of Henderson, the history of hemoglobin as oxygen carrier in the blood has finally, at long last, reached the stage where explanations of biological function are offered, convincingly, quantitatively, in physicochemical and molecular language. This was the very time, in the early 1930s when Linus Pauling started to follow his life-long interest in the hemoglobin molecule. As he writes himself:

"My first paper in this field dealt with the oxygen equilibrium of hemoglobin and its structural interpretation. Instead of four separate equilibrium constants for the addi-

Plate 8. Lawrence J. Henderson. (Courtesy Harvard University.)

tion of four successive dioxygen molecules to the hemoglobin molecule [the Hill–Adair approach], with their corresponding four values to the standard free energy of combination, I simplified the problem to a two-parameter one. I assumed that the free energy of addition of dioxygen to each of the four heme groups would be the same and that an additional contribution to the free energy of the molecule would be made by each interacting pair of hemes with attached dioxygens. I found that the experimentally obtained oxygen equilibrium curve could be approximated reasonably well by either one of two simple assumptions: that the four hemes be at the corners of a regular tetrahedron or at the corners of a square. I suggested, however, that the square arrangement might be closer to the truth, largely on the basis of symmetry arguments (Pauling, 1935)" (Pauling, 1980–1981).

In the next chapter we shall continue to describe the contributions of Linus Pauling to the understanding of hemoglobin as a molecular object.

REFERENCES

REFERENCES

Adair, G.S. (1925a) Proc. Roy. Soc. London, 108 A, 627.
Adair, G.S. (1925b) Proc. Roy. Soc. London, 109 A, 292.
Adair, G.S. (1925c) J. Biol. Chem., 63, 493.
Barcroft, J. (1914) The Respiratory Function of the Blood, Cambridge University Press, Cambridge, 1st ed., ch. IV.
Barcroft, J. and Camis, M. (1909-1910) J. Physiol., 39, 118.
Barcroft, J. and Orbeli, L. (1910-1911) J. Physiol., 41, 355.
Barcroft, J. and Roberts, F. (1909-1910) J. Physiol., 39, 143.
Bohr, Ch. (1887) Zbl. Physiol., 1, 293.
Bohr, Ch. (1888) Zbl. Physiol., 2, 437.
Bohr, Ch. (1892a) Skand. Arch. Physiol., 3, 76.
Bohr, Ch. (1892b) Skand. Arch. Physiol., 3, 47.
Bohr, Ch., in Nagel, W. (1905-1909) Handbuch der Physiologie der Menschen, I, 94.
Bohr, Ch., Hasselbalch, K. and Krogh, A. (1903-1904) Zbl. Physiol., 17, 661.
Christiansen, J., Douglas, C.G. and Haldane, J.S. (1914) J. Physiol., 48, 244.
Debru, Cl. (1983) L'esprit des Protéines. Histoire et Philosophie Biochimiques, Hermann, Paris, pp. (a) 149; (b) 147; (c) 148-149; (d) 150; (e) 151; (f) 152-153; (g) 154; (h) 159 (my translation).
Douglas, C.G., Haldane, J.S. and Haldane, J.B.S. (1912) J. Physiol., 44, 275.
Haldane, J.S. (1908) Nature, 78, 553.
Haldane, J.S. (1917) Organism and Environment as Illustrated by the Physiology of Breathing, Yale University Press, New Haven, CT.
Haldane, J.S. (1922) Respiration, Yale University Press, New Haven, CT.
Henderson, L.J. (1921) J. Biol. Chem., 46, 401.
Henderson, L.J. (1928) Blood. A Study in General Physiology, ch. IV, Yale University Press, New Haven, CT.
Hill, A.V. (1910) J. Physiol., 40, iv-vii (Proc. Physiol. Soc.).
Hill, A.V. (1913) Biochem. J., 17, 471.
Hüfner, G. (1883) Z. Physiol. Chem., 8, 358.
Hüfner, G. (1890) Arch. Physiol. (Du Bois-Reymond) Physiol. Abt., 1.
Hüfner, G. and Gansser, E. (1907) Arch. Anat. Physiol. (Engelmann) Physiol. Abt., 209.
Kohler, R.E. (1982) From Medical Chemistry to Biochemistry. The Making of a Biomedical Discipline, Cambridge University Press, Cambridge (a) ch. 2; (b) ch. 3; (c) ch. 7; (d) p. 314.
Pauling, L. (1935) Proc. Natl. Acad. Sci. USA, 21, 186.
Pauling, L. (1980-1981) Texas Rep. Biol. Med., 40, 1.
Preyer, W. (1867) Zbl. Med. Wiss., 259, 273.
Waymouth Reid, E. (1905) J. Physiol., 33, 12.
Zuntz, N. (1867) Zbl. Med. Wiss., 529.

Chapter 20

Pauling and the Magnetic Properties of Hemoglobins

Early in his career Linus Pauling was fascinated by hemoglobin and tried to find its detailed structure. His geometric interpretation of the Bohr effect, which he gave in 1935, has just been mentioned (Chapter 19). The following year, Pauling started publishing other findings, based upon the astute use of magnetic susceptibilities.

Before recounting the contributions of Linus Pauling to the elucidation of the electronic and geometric structures of oxy- and carboxyhemoglobin, a brief reminder on quantum chemistry is needed. Erwin Schrödinger had published his equation in 1926. Immediately, its appearance set off a flurry of very diverse applications, during the extremely active period 1926–1930. After Burrau (and others) had solved exactly the H_2^+ case, Heitler and London proposed an approximate treatment of the H_2 molecule (Heitler and London, 1927). At about the same time, Mulliken in the U.S. and Lennard-Jones in Britain (Lennard-Jones, 1929) developed an alternate and competing formalism, the method of molecular orbitals.

Consider the molecule of dioxygen O_2. Each of the two oxygen atoms has an electronic configuration $1s^2 2s^2 2p^4$. It will be recalled that atomic energy levels are ranked $1s < 2s < 2p$, that each energy level can accommodate a maximum of two electrons in which case the electronic spins are paired ($\uparrow\downarrow$) to obey the Pauli Principle. Filling of these energy levels occurs according to the *Aufbau* Principle, upwards in energy. And, when there are two levels with identical energies (so-called degenerate levels), Hund's rule specifies mono-electronic occupation of each level, with parallel spin vectors ($\uparrow\uparrow$) in order to minimize inter-electronic repulsion.

As two oxygen atoms come together, the atomic orbitals start to overlap. This mixture produces two molecular orbitals from two atomic orbitals: positive overlap leads to a bonding combination, with energy lowering; while negative overlap leads to an anti-bonding combination, with a raising of the energy by an equivalent amount.

Let us leave aside the $1s^2$ core electrons in each oxygen atom, and let us consider exclusively the valence electrons $2s$ and $2p$. The two spherical $2s$ atomic orbitals mix rather weakly with one another, and the resulting $(\sigma, \sigma\star)$ pair of molecular orbitals (MO's) are very much atom-like.

Labeling the intermolecular axis in O_2 the x axis, positive overlap of the two $2p_x$ AO's which point towards one another produces a strongly binding σ MO and a strongly anti-bonding $\sigma\star$ MO. The lateral overlap between the two $2p_y$ (or equivalently $2p_z$) AO's generates another two pairs of MO's: two degenerate bonding Π MO's and two degenerate anti-bonding $\Pi\star$ MO's.

Thus, the molecular orbital method leads to the following transform in the electronic configurations of the valence electrons:

$$(2s^2\,2p^4) + (2s^2\,2p^4) \rightarrow (\sigma_{2s})^2 + (\sigma_{2s}^{\star})^2 + (\sigma_{2p_x})^2 + (\pi_{2p_y})^2$$

$$+ (\pi_{2p_z})^2 + (\pi_{2\,p_y}^{\star})^1 + (\pi_{2p_z}^{\star})^1 + (\sigma_{2\,p_x}^{\star})^0$$

atomic orbitals $\qquad\qquad$ molecular orbitals

If one does not take into account the $2s$ 'lone-pair' electrons, the remaining eight-valence electrons go first into the available, σ, π_{2p_y} and π_{2p_z} MO's, which accommodate a total of six electrons. This leaves two electrons in each of the two degenerate anti-bonding MO's $\pi_{2\,p_y}^{\star}$ and $\pi_{2\,p_z}^{\star}$.

Thus, the ground state (lowest energy state) of the dioxygen molecule, according to Hund's rule, is a diradical (a triplet state). Furthermore, the bond order of O_2, as given by half the difference between the number of bonding (eight) and anti-bonding (four) electrons, is two. The Lewis structure of dioxygen is indeed written :O=O:.

This background is sufficient to explain the first contribution of Pauling: elucidation of the mode of binding of dioxygen to hemoglobin. A short note is nevertheless required here. I have used molecular orbital language in the above reminder, whereas Linus Pauling always favored the valence-bond approach. My justification is simplicity, and also the present widespread use of the molecular orbital approach. Simplicity: in this particular case – historically, one of the early triumphs of molecular orbital theory – the molecular orbital formalism gives dioxygen a triplet ground state, as observed, whereas the valence-bond formalism, at least in a simple-minded approach, is incapable of leading to this conclusion. Pauling notes this problem in his classic 1931 article on 'The nature of the chemical bond': he attempts to explain it away with some hand-waving, writing that:

"it seems probable that the additional degeneracy arising from the identity of the two atoms gives rise to a new type of bond, the *three-electron bond*, and that in normal O_2 there are one single bond and two three-electron bonds (. . .)" (Pauling, 1931).

Another note is in order before we proceed. The work done on hemoglobin by Pauling and Coryell in the late 1930s clearly finds its origin in the same 'The nature of the chemical bond' paper (Pauling, 1931). The whole second part of this article is devoted to showing that the magnetic moments of molecules and complex ions, available from magnetic susceptibility measurements with a Gouy balance, are determined by the number of unpaired electrons. As a 'finishing touch', Pauling himself writes:

"This makes it possible to determine from magnetic data which eigen functions are involved in bond formation, and so to decide between electron-pair bonds and ionic or ion-dipole bonds for various complexes" (Pauling, 1931).

This is the very approach which he would himself apply to hemoglobin five years afterwards.

After this preamble, we come to Linus Pauling's contributions to
the hemoglobin electronic structure proper. As Pauling himself
wrote in a recent historical essay:

"During much of my scientific career I have worked on the problem of the structure
and properties of the hemoglobin molecule, and on other problems that have a bearing on this one" (Pauling, 1980–1981).

One of the questions addressed by Pauling in the late 1930s was the
mode of attachment of the dioxygen O_2 molecule to hemoglobin,
when oxyhemoglobin forms: is it true chemical binding of dioxygen
to atoms in the protein, or a more physical interaction in which O_2
was adsorbed on the surface of the hemoglobin molecule or trapped
into a pocket? Pauling

" . . . thought that a decision could be made between the two by determining the
magnetic susceptibility of oxyhemoglobin. The dioxygen molecule has a large magnetic moment, corresponding to its two unpaired electrons and its $^3\Sigma$ normal 'ground'
state. Presumably these electrons would remain unpaired if dioxygen were held to the
hemoglobin molecule by physical forces, but would be paired if chemical bonds were
formed" (Pauling, 1980–1981).

Fortunately, one of Pauling's students, E.B. Wilson Jr., had earlier
built an apparatus for measuring magnetic susceptibilities. Pauling
assigned measurement of the hemoglobin susceptibility to a post-
doctoral fellow, Charles D. Coryell. The experiment was conclusive:
oxyhemoglobin is diamagnetic; whereas dioxygen is paramagnetic.
Hence, attachment of dioxygen to hemoglobin proceeds by a true
chemical bond.

This is a beautiful result, from a well-conceived experiment, and
Pauling reported it in an elegant manner. Linus Pauling is something of a polymath and Renaissance man. He is well-versed in a
number of fields, and is quite fond of history. Accordingly, his paper
with C.D. Coryell opens in this manner:

"Over ninety years ago, on November 8, 1845, Michael Faraday investigated the
magnetic properties of dried blood and made a note 'must try recent fluid blood'! If he
had determined the magnetic susceptibilities of arterial and venous blood, he would
have found them to differ by a large amount (as much as 20%, for completely oxygen-

ated and completely deoxygenated blood); this discovery, without doubt, would have excited much interest and would have influenced appreciably the course of research on blood and hemoglobin" (Coryell and Pauling, 1936b).

In their 1936 paper, Coryell and Pauling also discussed the nature of the bonding between the iron atom of the heme and either dioxygen or carbon monoxide. The experimental observation they built upon was that

"much to (their) surprise, (. . .) the hemoglobin molecule, without attached dioxygen or other ligand such as carbon monoxide, had a large magnetic moment, corresponding to the spins of four unpaired electrons per iron atom. It was known, of course, that compounds of iron(II) are of two types: the low-spin type, with zero magnetic moment, and the high-spin type, with the magnetic moment corresponding to four unpaired electron spins. (. . .) The high-spin complexes of ferrous iron, such as $[FeF_6]^{4-}$ and $[Fe(H_2O)_6]^{2+}$ involve bonds between the iron atom and the six ligated atoms that have a large amount of partial ionic character, whereas the low-spin complexes, such as the ferrocyanide ion $Fe(CN)_6{}^{4-}$, have bonds with high covalent character" (Pauling, 1980–1981).

The changes in magnetic properties of hemoglobin upon transformation into oxyhemoglobin or carbon monoxyhemoglobin allowed Pauling and Coryell to propose bent and linear geometries, respectively, for these two species, which they described with the following canonical forms (Pauling and Coryell, 1936):

$$(Hb.O_2): Fe-\overset{\ddot{O}:}{O} \longleftrightarrow Fe=\overset{\ddot{O}:}{O}$$
$$(Hb.CO): Fe-C\equiv O: \longleftrightarrow Fe=C=\overset{..}{O}:$$

We see in this work on hemoglobin the elements of Pauling's scientific style earlier mentioned (in Chapter 6, section 4): continuity between the chemical properties of large biomolecules and those of simpler substances such as $Fe(H_2O)_6^{2+}$ or $Fe(CN)_6^{4-}$ is assumed; structural explanations are favored; a simpler *analogon*, here the known low-spin and high-spin iron(II) complexes, is picked.

In this manner, Pauling and Coryell analyzed the magnetic properties of several iron(II)-containing species: ferroheme, several hemochromogens, hemoglobin, and carbon monoxyhemoglobin (Pauling and Coryell, 1936a, b). They concluded that "in two of these substances (ferroheme, hemoglobin) there are four unpaired electrons per heme, indicating that the bonds attaching the iron atoms to the rest of the molecule are essentially ionic in character, whereas the others contain no unpaired electrons, each iron atom being attached to six adjacent atoms by essentially covalent bonds" (Coryell et al., 1937).

Pauling and his co-workers went on to investigate in like manner the other members of the hemoglobin family (Coryell et al., 1937; 1939; Russell and Pauling, 1939).

REFERENCES

Coryell, C.D., Stitt, F. and Pauling L. (1937) J. Am. Chem. Soc., 59, 633.
Coryell, C.D., Pauling L. and Dodson, R.W. (1939) J. Phys. Chem., 43, 825.
Heitler, W. and London, F. (1927) Z. Phys., 44, 455.
Lennard-Jones, J.E. (1929) Trans. Faraday Soc., 25, 668.
Pauling, L. (1931) J. Am. Chem. Soc., 53, 1367.
Pauling, L. (1980–1981) Texas Rep. Biol. Med., 40, 1.
Pauling, L. and Coryell, C.D. (1936a) Proc. Natl. Acad. Sci. USA, 22, 159.
Pauling, L. and Coryell, C.D. (1936b) Proc. Natl. Acad. Sci. USA, 22, 210.
Russell, C.D. and Pauling, L. (1939) Proc. Natl. Acad. Sci. USA, 25, 517.

Chapter 21

Hemoglobin and the Beginnings of X-ray Work on Proteins

The first x-ray photographs from a protein crystal were taken in 1934 by Desmond Bernal, who had the genius to maintain the pepsin crystal in contact with its mother liquor; the full history is recounted elsewhere in this volume (Chapter 38). Among the observations made by Bernal and Dorothy Crowfoot and published in their letter to *Nature* (Bernal and Crowfoot, 1934) were the large M_r value of protein molecules, and the fact that the 'arrangement of atoms inside the protein molecules is also of a perfectly definite kind'.

This last finding, of proteins as highly organized molecules, must have been truly tantalizing. Accordingly, Bernal and Crowfoot concluded that

"now that a crystalline protein has been made to give X-ray photographs, it is clear that we have the means of checking them and, by examining the structure of all crystalline proteins, arriving at far more detailed conclusions about protein structure than previous physical or chemical methods have been able to give" (Bernal and Crowfoot, 1934).

This assertively ambitious conclusion appears from hindsight to be truly prophetic. It reflects the enthusiasm of Bernal and Crowfoot when they observed the beautiful photographs obtained of wet pepsin crystals.

Yet, hindsight also whispers to us that, had it not been for the advent of the computer, the enthusiasm, if well founded in principle, might have remained unfulfilled for very many years. As Max Perutz said almost 30 years later in his Nobel Lecture:

354 THE HEMOGLOBIN STORY

"In the X-ray diffraction pattern of protein crystals, the number of spots runs into tens of thousands. These all have to be measured in several isomorphous compounds, corrected by various geometric factors and finally used to build up an image by the superposition of tens of thousands of fringes. For instance, the calculation of a three-dimensional image of myoglobin at 2Å resolution involved the recording and measuring of about a quarter of a million spots, and in the final calculation about 5×10^9 figures had to be added or subtracted (Kendrew et al., 1960). *Clearly this would have been impossible before the advent of highspeed computers, and we have in fact been very fortunate, because the development of computers has always just kept in step with the expanding need of our X-ray analyses*" (Perutz, 1962; emphasis added).

Perutz nevertheless gives a lyrical testimony to the sheer delight from such work, reminiscent of the overwhelming and infective enthusiasm of Desmond Bernal upon viewing the first x-ray photographs from pepsin (on this point, see Hodgkin and Riley, 1968): Perutz expressed this inner glow from doing hard work on protein structure in a beautiful manner; he said, in the conclusion to his Nobel Lecture that the 'glaring sunlight of certain knowledge is dull and one feels most exhilarated by the twilight and expectancy of the dawn' (Perutz, 1962).

We have to go now, however briefly, into the principles of x-ray diffraction from protein crystals. Perhaps the simplest starting point is from a letter by P.P. Ewald of August 2, 1939 (quoted by Hodgkin and Riley, 1968):

"The idea was simply that a sphere of constant density gives a diffraction effect of a wavy kind and that by the general reciprocity between physical space v. Fourier space, a wavy density distribution in crystal space would give a sudden drop in intensity in Fourier space, i.e., with increasing order of diffraction. If the protein molecule might be conceived to have a structure of shells inside one another, like chinese carved ivory balls or certain radiolariae skeletons, such a wavy density distribution would be approximated".

The next level of sophistication is to consider the various atoms present in the protein molecule and whose electrons diffract the x-rays. As explained by Sir Lawrence Bragg in his Nobel Lecture:

"In some directions the atoms conspire to give a strong scattered beam, in others their effects almost annul each other by interference. The exact arrangement of the atoms is to be deduced by comparing the strength of the reflexions from different faces and in different orders" (Bragg, 1921-1922, quoted by Perutz, 1962).

As Perutz goes on explaining in his 1962 Nobel Lecture:

"Thus there should be a way of reversing the process of diffraction, and of getting back from the diffraction pattern to an image of the atomic arrangement. In order to produce such an image, each pair of symmetrically related spots in the x-ray pattern can be made to generate a set of fringes, each fringe having an amplitude proportional to the root of the intensity of the spot" (Perutz, 1962).

In general, an intensity I is the square of the absolute magnitude of the amplitude F: $I = |F^2|$. Hence, 'by taking the square root of I, one obtains $|F|$, but the phase angle cannot be determined experimentally by any direct method so far devised' (Hughes, 1968). As Perutz explains with respect to the superposition of fringes produced from the x-ray diffraction pattern required for reconstruction of the atomic arrangement:

"However, at this stage a complication arises. To obtain the *right* image, each set of fringes must be placed correctly relative to some arbitrarily chosen, common, origin. At this origin the amplitude of the fringe may show a crest or a trough, or some intermediate value which is referred to as the phase angle. (. . .) By itself, the x-ray pattern tells only about the amplitude, but not about the phase angles of the fringes to be generated by each pair of spots, so that half the information needed for production of the image is missing" (Perutz, 1962).

We will not go right away to the various solutions devised by protein x-ray crystallographers to this phase problem. And, before picking up the diachronic thread of x-ray studies of respiratory pigments, I wish to interject the finding of hemoglobin heterogeneity made by Haurowitz in the mid 1930s, i.e. at the time when x-ray work on proteins started.

Haurowitz tells the story so nicely that I shall quote him verbatim:

"Differences between the hemoglobins of different animal species had been claimed more than 100 years ago by E. Körber (Körber, 1866) in Dorpat, the capital of Estonia, which at that time was still part of the Russian empire. Körber's work was continued by F. von Krüger. After the First World War Estonia became an independent republic and dismissed professors who were not of Estonian origin. Professor von Krüger returned to the University of Rostok in Germany. Von Krüger had discovered that the hemoglobins of different animal species can be distinguished from each other by their resistance to 0.05 N NaOH. The time to convert the bright red solution of oxygenated hemoglobin (HbO_2) into the dark brown product of denaturation was designated by von Krüger as decomposition time. The decomposition times found by means of this very simple method were 1 minute for human adult hemoglobin, 3 minutes for canine Hb, 21 minutes for rabbit Hb, 80 minutes for horse Hb, and more than a day for the hemoglobins of sheep or cattle (Krüger, 1925). When I investigated the rate of denaturation spectrophotometrically I found that it proceeded in six different species as a first-order reaction suggesting the presence of a homogeneous hemoglobin in each of the different species (Haurowitz, 1929).

In 1910 Wakulenko, one of the co-workers of von Krüger in Dorpat, who had returned to Tomsk in Siberia, described a surprising reaction. He found that the hemoglobin of new-born children was about 100 times more resistant to denaturation by NaOH than the blood of the mothers (Wakulenko, 1910). When I applied the kinetic spectrophotometric method to the umbilical blood of new-born children I found a break in the curve indicating the presence of two different hemoglobins in the new-born child (Haurowitz, 1930). I succeeded in separating the two hemoglobins from each other and in crystallizing the fetal hemoglobin (Haurowitz, 1935). The discovery of the fetal hemoglobin was the first definite proof for the presence of more than one Hb in an individual.

My finding was at first not accepted because Barcroft, who at that time was investigating the hemoglobin of fetal sheep and their mothers, never found any difference between fetal and adult hemoglobin. I met Barcroft for the first time in 1935 at the International Congress for Physiology in Leningrad. He and I were invited to act as co-chairmen of a symposium on hemoglobin. When I introduced myself to the famous British physiologist he told me: 'I never believed in your finding of different fetal hemoglobins, but we repeated your experiments and by God you were right!' " (Haurowitz, 1979).

Perutz was later able to explain the resistance of fetal human hemoglobin to denaturation by alkali in molecular terms (Perutz, 1974). It is important to note Haurowitz's use in 1929 of a simple physicochemical (kinetic) technique as a criterion of homogeneity. His crystallization of the fetal hemoglobin in 1935 was more or less contemporary with Northrop's purification and crystallization of pepsin. He belonged then to the faculty, in the Department of Physiological Chemistry at the German University in Prague. Notice also in this account the stress put by Haurowitz, who was himself to experience at first-hand emigration because of political upheavals (he fled Prague to go to Istanbul in 1939), on similar moves which his predecessors in the field were forced to make.

While there have been quite a few studies on the impact of Jewish

emigration on American science in the 1930s and 1940s, or conversely on the effects on German science arising from this emigration, I know of no general study of the uprooting phenomenon: I
should not be surprised were one to find it beneficial to science *overall*; from the individual's positive psychological reaction at being
branded as an outcast; and from a scientific career offering somewhat easier sociological chances, less hindered by prejudice, to a
clever and enterprising young man.

I shall now return to my narrative of the x-ray elucidation of the
structure of the hemoglobins. The first x-ray photographs of hemoglobin had been taken by A.L. Patterson in the 1920s, and consisted
only of 'vague blurs' since dry crystals had been used rather than
hemoglobin crystals in contact with their mother liquor (Hodgkin,
1979). Shortly before World War II, Bernal, Fankuchen, and Perutz
took x-ray photographs of single crystals of horse methemoglobin
(Bernal et al., 1938).

Use of the Fourier series to obtain the phase constants was pioneered by Sir William Bragg in 1915. He showed

"the relationship between diffraction amplitudes and the amplitudes in the Fourier
representation of the electron density (. . .) but it was only late in the twenties that
the method began to be used on multiparameter structures" (Hughes, 1968).

The structure of copper sulfate was solved in 1934 by the 'direct'
Fourier approach: according to W.L. Bragg, this was one of the first
crystal structures thus solved (Hodgkin, 1979). This is where A.L.
Patterson re-enters the story, at least the part concerned with a general methodology for solving a crystal structure from the x-ray dif-
fractogram. As Dorothy Crowfoot-Hodgkin writes:

"A.L. Patterson had earlier (1930) considered working on protein crystals but
decided the methods available for the interpretation of x-ray diffraction phenomena
were inadequate. He left the Johnson Foundation where he was working in 1931 and
took a year off on his own resources – which extended to two – to investigate at M.I.T.
the properties and use of Fourier series. Out of this came the understanding of Fourier
series calculated with the squared amplitudes of the x-ray spectra as coefficients.
These series, which we all call by Patterson's name define distributions in which the
density corresponds with interatomic vectors in the crystal (Patterson, 1935). *No
protein crystal structures could have been solved without their use*" (Hodgkin, 1979;
emphasis added).

REFERENCES

Bernal, J.D. and Crowfoot, D. (1934) Nature, 133, 794.

Bernal, J.D., Fankuchen, I. and Perutz, M. (1938) Nature, 141, 523.

Bragg, W.L. (1915) Nobel Lecture, in Nobel Lectures – Chemistry. 1901–1921, Elsevier, Amsterdam, pp. 361–384.

Degel, P. (1932) Z. Vergleich. Physiol., 17, 337.

Haurowitz, F. (1929) Hoppe-Seyler's Z. Physiol. Chem., 183, 78.

Haurowitz, F. (1930) Hoppe-Seyler's Z. Physiol. Chem., 186, 141.

Haurowitz, F. (1935) Hoppe-Seyler's Z. Physiol. Chem., 232, 125.

Haurowitz, F. (1979) Ann. N.Y. Acad. Sci., 325, 37.

Hodgkin, D.C. (1979) Ann. N.Y. Acad. Sci., 325, 121.

Hodgkin, D.C. and Riley, D.P. (1968) in Structural Chemistry and Molecular Biology (Rich, A. and Davidson, N., Eds.), Freeman, San Francisco, p. 15.

Hughes, E.W. (1968) in Structural Chemistry and Molecular Biology (Rich, A. and Davidson, N., Eds.), Freeman, San Francisco, p. 617.

Kendrew, J.C., Dickerson, R.E., Strandberg, B.E., Hart, R.G., Davies, D.R., Phillips, D.C. and Shore, V.C. (1960) Nature, 185, 422.

Körber, E. (1866) Uber Differenzen der Blutfarbstoffes, Inaugural dissertation at the University of Dorpat; quoted by von Krüger (1932).

Krüger, F. von (1925) Z. Vergleich. Physiol., 2, 254.

Krüger, F. von (1932) Z. Vergleich. Physiol., 17, 337.

Patterson, A.L. (1935) Z. Krist., 90, 517, 543.

Perutz, M.F. (1962) X-ray analysis of haemoglobin, Nobel Lecture, December 11, 1962, in Nobel Lectures – Chemistry. 1942–1962, Elsevier, Amsterdam, p. 653.

Perutz, M.F. (1974) Nature, 247, 341.

Wakulenko, J.L. (1910) Communications from the Med.-Chem. Laboratory of the University of Tomsk, Siberia, quoted by Degel (1932).

Chapter 22

Determination of Phases in Single Crystals

1. Isomorphous replacement: principles

Any further progress in protein structure determination from that point encountered a major problem, i.e. the phase problem. At first, a promising approach, drawing upon the initial experience of Desmond Bernal with the wet pepsin crystals, seemd to be comparison of protein crystals in different states of hydration. The assumption was that this would leave unchanged the internal structure of the protein molecules within the crystal, only modify their relative distance and orientation. Exploratory experiments on wet and dry insulin crystals and for hemoglobin crystals in different shrinkage stages were encouraging:

"The peak pattern relative to the origin, (. . .), remained unchanged on drying or shrinking the crystals and encouraged the attempt to trace the molecular contribution as such to the x-ray scattering" (Crowfoot-Hodgkin, 1979).

Enters Max Perutz (1914-). He came from Vienna to Cambridge, England, in his early twenties because of dissatisfaction with the University of Vienna where he, in his own words, wasted five semesters in an exacting course of inorganic analysis (Perutz, 1962). With support from his affluent father – Perutz came from a family of textile manufacturers – he started his Ph.D. work under J. Desmond Bernal at the Cavendish Laboratory in September 1936:

"The scientific work of Perutz on the structure of haemoglobin started as a result of a conversation with F. Haurowitz in Prague, in September 1937. G.S. Adair made him the first crystals of horse haemoglobin, and Bernal and I. Fankuchen showed him how to take X-ray pictures and how to interpret them. Early in 1938, Bernal, Fankuchen,

and Perutz, 1938, published a joint paper on X-ray diffraction from crystals of haemoglobin and chymotrypsin. The chymotrypsin crystals were twinned and therefore
difficult to work with, and so Perutz continued with haemoglobin. D. Keilin, then
Professor of Biology and Parasitology at Cambridge, soon became interested in the
work and provided Perutz and his colleagues with the biochemical laboratory facilities which they lacked at the Cavendish. Thus from 1938 until the early fifties the
protein chemistry was done at Keilin's Molteno Institute and the X-ray work at the
Cavendish, with Perutz busily bridging the gap between biology and physics on his
bicycle" (Perutz, 1962).

He started applying himself to the dry/wet crystals scheme. As
Dorothy Crowfoot-Hodgkin writes:

"It was Max Perutz's major work for the next ten years of his life to test these ideas
within the framework of the crystal structure of horse methemoglobin, and particularly in relation to the centrosymmetrical projection based on the $h0l$ reflections
where the phase constants reduced to signs (Bragg and Perutz, 1952). He was interrupted by internment at the beginning of the war, and later by work on Habbakuk in
Canada, but by 1942 he had obtained a series of shrinkage states of the crystals that
made it possible to narrow down the solution of the sign contributions for the 001
reflections of hemoglobin from 64 to 8. Further studies on salt-free crystals ma(d)e it
possible to limit the alternatives to two, one of which gave a very improbable solution.
So he calculated a first one dimensional projection of all the electron density in the
hemoglobin molecule. I found a letter from W.L. Bragg in my files, dated 3rd August
1942, trying to stimulate me into further action on insulin in which he says, 'I have
been very interested in Perutz's latest work, and light seems to be beginning to break
in the case of hemoglobin.' In the following years Perutz was able to trace the relative
sign relations along parallel lines of $h0l$ reflections but the crystal shrinkage was in
one direction only (...) which made it extremely difficult to interrelate the phase
relations of reflections in different row lines (Kendrew, 1954). In retrospect, (...)
the major importance historically (of these researches) is that they required for their
achievement the measurement of the intensities of the x-ray intensities with the
highest possible accuracy, preferably on an absolute scale" (Crowfoot-Hodgkin,
1979).

A few words of explanation about several points in this paragraph
are now in order. 'Only the $h0l$ reflexions are "real", that is, can be
regarded as having relative phase angles limited to 0 or π, or positive
or negative signs, rather than general phases' (Kendrew et al., 1958).
The interruption of Max Perutz's work on hemoglobin was due to
British policy, at the outset of World War II, to intern in camps all
aliens because a few of them might be security risks, as spies for

Germany. The Habbakuk project had been conceived by Geoffrey Pike, and it was enthusiastically supported by Desmond Bernal: construction of a fantastic ship, an aircraft carrier, carved from an iceberg off the Canadian coast.

The whole period from 1940 till 1953 was thus relatively fruitless and frustrating to Max Perutz. Here is Perutz's narrative of the breakthrough which occurred in the early fifties:

"I had been doubtful if any heavy atom would change the intensities of hemoglobin successfully until I asked Bill Cochran if he would let me use his counter spectrometer – consisting of a Unicam oscillation camera with a geiger counter on one arm – to measure the absolute intensity of the hemoglobin reflections. I was surprised how small the absolute Fs were and did some simple calculations which showed that a heavy atom would produce changes that should be easily measured. The original purpose of this experiment had been quite a different one: to put the molecular transform derived from Bragg and my salt-water Fourier on an absolute scale. This happened in 1951 or 1952.

At that stage, I had no idea how I might attach a heavy atom to hemoglobin. As a side line I had done some work on the crystal structure of sickle cell hemoglobin. One day I received a set of reprints from the *Journal of General Physiology*, a journal that I would not normally have looked at, from an unknown man at Harvard called Austin Riggs. He had wondered whether hemoglobin A [normal] and hemoglobin S [sickle cell] differed in the number of reactive SH groups and had titrated them with paramercuribenzoate [PMB]. He also examined the effect of PMB on the oxygen equilibrium curve and found that heme–heme interaction was largely preserved.

I got very excited by this observation, because it suggested that you can attach molecules of PMB to hemoglobin without changing its structure significantly. I discussed the crystallization of PMB-hemoglobin with Vernon Ingram who had used this reagent before and kindly made the compound for me.

When I developed the first precession picture of PMB-hemoglobin and compared it with that of native hemoglobin, I saw that the two crystals were isomorphous and that the intensity changes were just of the magnitude that my measurements of the absolute intensities had led me to expect. Madly excited, I rushed up to Bragg's room and fetched him down to the basement dark room. Looking at the two pictures in the viewing screen, we were confident that the phase problem was solved" (Perutz, quoted by Crowfoot-Hodgkin, 1979).

Notice from this text how discovery came about in an unforeseeable manner, yet on prepared groundwork. Stage 1 was the back-of-the-envelope calculations which convinced Perutz that the heavy-atom method should work in principle. Stage 2 was finding out, almost by accident, how he could conceivably attach the required heavy atoms (at least two were needed) onto hemoglobin: using sulfhydryl groups. And stage 3 was simply carrying out this research program, once the problem had been solved in principle (Green et al., 1954; Bragg and Perutz, 1954).

Attachment of mercuric ions to hemoglobin drew upon existing chemical know-how:

"It was found in our laboratory that at low temperatures ($0°$ and $27°$) only about one half of the binding sites are available to mercuric ions. However, all of the binding sites (sulfhydryl groups) are titrable with mercuric ions at $38°$. This information has made it possible for Perutz to complete his isomorphous replacement studies of the horse hemoglobin molecule" (Murayama, 1964).

Just a few words to complement the indications given already. Electrons in atoms are responsible for x-ray scattering. The heavier the atom, the more intense the scattering. 'For example, the scattering power of platinum and carbon atoms, as used in phthalocyanin, are in the ratio of 78 to 6' (Ingram, 1961).

For the introduction of a heavy atom to help solve the phase problem, the stringent requirement is that the crystal of the substituted protein must be isomorphous with that of the parent protein. Then, a heavy atom H_1 will change the phase and amplitude of the diffraction wave in a given manner. Another heavy atom H_2, belonging also to a modified protein molecule whose crystal remains isomorphous with the parent crystal will modify the phase and amplitude in yet a different way.

The *magnitude* of the phase angle can be obtained, using each heavy atom as a common origin. The ambiguity in sign is then removed, using jointly the information from H_1 and H_2 substitution, and determining a vector function devised by M.G. Rossmann (Perutz, 1956; Rossmann, 1960).

2. Isomorphous replacement in practice: the example of myoglobin

In this manner, Perutz and his co-workers were able to determine unambiguously the signs of nearly all reflections in the *010* zone for the monoclinic crystals of horse hemoglobin (Green et al., 1954; Bragg and Perutz, 1954). Since myoglobin was a much simpler case – it is a macromolecule a quarter the size of hemoglobin – its x-ray diffraction pattern was fully deciphered earlier than that of its bulkier relative. For this reason, we shift now to this part of the myoglobin structure story having to do with the determination of phases for the x-ray reflections.

This was the work of John Cowdery Kendrew (b. 1917), a close associate of Max Perutz. Kendrew, at the start of World War II, had started research work on his Ph.D. at Cambridge, with E.A. Moelwyn-Hughes as his supervisor. He was then recruited into British Scientific Intelligence, working on radar, which was being developed at that time. For this purpose, he was assigned at first to the Air Ministry Research Establishment, in Bawdsey Manor, which Sir Robert Watson-Watt, scientific advisor to the Air Ministry, had set up in the years immediately before the war. He was later engaged in operational research at Royal Air Force headquarters: Coastal Command, Middle East, and South East Asia, in succession. Operational Research had also been initiated by Watson-Watt to study the use of radar by Fighter Command. Thus, it was natural for Kendrew to go from radar research to operational research. In Jones's book, describing the achievements of British Scientific Intelligence during the war, John Kendrew is singled out for the admirable clarity of the reports he wrote (Jones, 1978), a quality obvious also from his research papers. Kendrew's work with the military earned him the honorary rank of Wing Commander R.A.F.

"'During the war years his interest became more biological, and largely owing to the influence of J.D. Bernal and L. Pauling he decided to work on the structure of proteins. He returned to Cambridge in 1946 and, in the Cavendish Laboratory, began a collaboration with Max Perutz, under the direction of Sir Lawrence Bragg. He took his Ph.D. degree in 1949 and his D.Sc. in 1962. He and Perutz were the first two

364   THE HEMOGLOBIN STORY

members of the Medical Research Council Unit at the Cavendish Laboratory" (Kendrew, 1962). "In October 1947 (Perutz) was made head of this newly constituted unit for Molecular Biology, with J.C. Kendrew representing its entire staff" (Perutz, 1962).

However,

"in myoglobin it is not possible to apply the chemically straightforward method of attaching a "heavy atom" to the protein which was used by Perutz, namely to cause a mercury compound such as sodium p-chloromercuribenzoate (PCMB) to react with free sulfhydryl groups, because no myoglobin hitherto examined has been found to contain such groups (V.M. Ingram, unpublished data), all the sulphur in the molecule being present in the form of methionine residues. (...)

A more general approach has proved to be most fruitful; it consisted in making a wide survey of heavy ions which might be expected on chemical grounds to form complexes with various amino acid sidechains, in the hope of finding some which would become attached exclusively (or at least preferentially) at a single site on the molecule. (...)

Success in the search for suitable ligands has been established by X-ray methods. Simple comparison of the diffraction patterns of crystals made from treated and from untreated myoglobins is a conclusive indication of failure to combine, if the patterns are identical or nearly so. If, on the other hand, the diffraction patterns differ even though the unit cell dimensions have remained the same, it follows that some kind of combination has taken place, and it remains to discover whether the ligand is attached to one, two or more sites on the molecule. (...)

A search of the literature suggested that there were three classes of reaction undergone by the haem group in myoglobin which might provide ways of introducing heavy atoms at specific sites: namely, with imidazoles, with isocyanides, and with nitroso compounds. (...) [these did not give very useful results and the Kendrew group turned to the use of heavy metal ions.]

Potassium mercuri-iodide K_2HgI_4 was first investigated as possible heavy atom substituent. (...) mercuri-iodide does indeed combine, (...) it does so at two sites on the molecule. This behaviour is compatible with the idea that mercuri-iodide co-ordinates with methionine (...) (the) results suggest, though they do not prove that HgI_3^- is the group actually attached to the protein. (...)

There was no a priori reason to expect that potassium p-chloro-mercuribenzene sulphonate (PCMBS), which is primarily a reagent for sulfhydryl groups, would combine with myoglobin, which contains no such groups: in fact, however, it does so, though the nature of the groups on the protein which are involved is obscure (...) the diffraction data indicate that the PCMBS resides almost exclusively at a single site on the protein molecule (...) and that the sulphonic acid group plays an essential role (in the attachment). (...)

Myoglobin crystals grown from mother liquor containing one mole of silver nitrate per mole of protein gave diffraction patterns distinctly different from normal (...)

attachment takes place predominantly at a single site (. . .) histidine nitrogen, carboxyl groups, and amino groups might be involved. (. . .) (with) mercuri-ammine complexes $Hg(NH_3)^{2+}_2$ (. . .) combination of a mercury atom has taken place almost entirely at a single specific site on the protein surface. (. . .) (The signs of the $h0l$ reflections were determined from the potassium mercuri-iodide complex): the $|F|$ values (vary) as a function of the amount of mercuri-iodide added (in an approximately linear manner), as would be expected if progressively more of the heavy group were being added at a single site on the molecule. (. . .)

The value of $(\delta F)^2$ (taken between myoglobin with addition of 1.5 moles K_2HgI_4 and myoglobin itself) was calculated, and used as the terms of a difference-Patterson projection (. . .) this projection included all terms of spacing greater than 4 Å, with an appropriate temperature factor. It is (. . .) seen to contain a single prominent peak whose height is 39% of that of the origin. (. . .)

We are now in a position (i) to compute a difference-Fourier projection using the observed values of $|F|$ together with the signs of the corresponding structure factors, and (ii), to determine the signs of the F's themseves, in the usual manner – if F and δF, the calculated structure factor, have the same sign, the reflection will have become stronger, while if they have opposite signs the introduction of the heavy atom will have made it weaker. (. . .) (the authors then make use of similar information derived from the PCMBS complex): of the 226 reflections (in the region of reciprocal space investigated, i.e. out to 4 Å) the signs of 111 are definitely indicated by both complexes; 75 are definitely indicated by one, and ambiguously by the other; 30 are so small that their contributions to a Fourier synthesis would be negligible ($F < 100$). This leaves only 10 reflexions unaccounted for: in 4 (all fairly small) the indications from both complexes are ambiguous, while in the remaining 6 there is an apparent contradiction. All of the latter lie near the outer edge of the region of reciprocal state being studied.

The allocation of signs has since been confirmed for all significant reflections by independent studies of the mercury diammine and gold complexes. (. . .)" (Bluhm et al., 1958).

REFERENCES

Bluhm, M.M., Bodo, G., Dintzis, H.M. and Kendrew, J.C. (1958) Proc. Roy. Soc. London, A246, 369.
Bragg, W.L. and Perutz, M.F. (1952) Proc. Roy. Soc. London, A213, 425.
Bragg, W.L. and Perutz, M.F. (1954) Proc. Roy. Soc. London, A225, 315.
Crowfoot-Hodgkin, D. (1979) Ann. N.Y. Acad. Sci., 325, 121.
Green, D.W., Ingram, V.M. and Perutz, M.F. (1954) Proc. Roy. Soc. London, A225, 287.
Ingram, V.M. (1961) Hemoglobin and its Abnormalities, Thomas, Springfield, IL, p. 40.
Jones, R.V. (1978) Most Secret War. British Scientific Intelligence 1939-1945, Hamish Hamilton, London, p. 329.
Kendrew, J.C. (1954) Prog. Biophys. Biophys. Chem., 4, 244.
Kendrew, J.C. (1962) Nobel Lecture, December 11, in Nobel Lectures - Chemistry. 1942-1962, Elsevier, Amsterdam, p. 676 (and biography following his lecture).
Kendrew, J.C., Bodo, G., Dintzis, H.M., Parrish, R.G., Wyckoff, H. and Phillips, D.C. (1958) Nature, 181, 662.
Murayama, M. (1964) in Hemoglobin Its Precursors and Metabolites (Sunderman, F.W. and Sunderman Jr., F.W., Eds.), Lippincott, Philadelphia, p. 1.
Perutz, M.F. (1956) Acta Cryst., 9, 867.
Perutz, M.F. (1962) Nobel Lecture, December 11, in Nobel Lectures - Chemistry. 1942-1962, Elsevier, Amsterdam, p. 653 (and biography following his lecture).
Rossmann, M.G. (1960) Acta Cryst., 13, 221.

Chapter 23

The X-ray Structure of Sperm-Whale Myoglobin

John Kendrew chose to work on myoglobin:

"The choice of material was not so simple. One looked for a protein of low molecular weight, easily prepared in quantity, readily crystallized, and not already being studied by X-ray methods elsewhere. Myoglobin seemed to satisfy these criteria, and had the additional advantages of being closely related to hemoglobin, already the object of Perutz's attention for many years, and of sharing with hemoglobin a most important and interesting biological function, that of reversible combination with oxygen. As emerged more clearly later, myoglobin consists of a single polypeptide chain of about 150 amino acid residues, associated with a single haem group; its one-to-four relationship with haemoglobin already suggested in early days by a comparison of molecular weights, turned out to be not coincidental but a fundamental structural relationship, as has now been shown by comparing the molecular models of the two proteins" (Kendrew, 1962).

Where could Kendrew find an abundant source of easily crystallized myoglobin?

"First of all it was necessary to find some species whose myoglobin formed crystals suitable, both morphogically and structurally, to the purpose in hand; the search for this took us far and wide, through the world and through the animal kingdom, and eventually led us to the choice of the sperm whale, *Physeter catodon*, our material coming from Peru or from the Antarctic, with some close runner-ups including the myoglobin of the common seal (. . .)" (Kendrew, 1962).

Kendrew and his group had also to search far and wide for heavy atoms to be inserted into myoglobin; they tried several hundred potential ligands before finding suitable groups as outlined in the preceding chapter.

"First the heavy atom was located by carrying out a so-called difference Patterson synthesis: if all the heavy atoms are located at the same site on every molecule in the crystal, this synthesis will contain only one peak, from the position of which the x- and z-co-ordinates of the heavy atom can be deduced, and the signs of the $h0l$ reflexions determined. These signs were cross-checked by repeating the analysis for each separate isomorphous replacement in turn (. . .)" (Kendrew et al., 1958).

Now the three-dimensional Fourier synthesis can be initiated,

"knowledge of the signs of the $h0l$ reflexions to high resolution enabled us to determine the x- and z-co-ordinates of all the heavy atoms with some precision. This was the starting point for the three-dimensional analysis (. . .)" (Kendrew et al., 1958).

This is more easily said than done:

"The whole diffraction pattern of a myoglobin crystal consists of at least 25 000 reflexions. In 1955, when the three-dimensional work began, no computers existed which were fast enough to calculate Fourier syntheses containing so many terms; besides, the method was unproved and it seemed advisable to test it first on a smaller sample of data.

We may regard a typical X-ray photograph of a myoglobin crystal as a two-dimensional section through a three-dimensional array of reflexions; each reflexion corresponds to a single Fourier component, and the whole structure can be reconstructed by using all the components as terms of a Fourier synthesis. (. . .) the components of higher frequency (higher harmonics), which are responsible for filling in the fine details of the structure, lie to the outside of the pattern. Thus one can obtain a rendering of the molecule at low resolution by using simply those reflexions within a spherical surface at the centre of the pattern. By doubling the radius of the sphere (which now enclosed eight times as many reflexions) we double the resolution of the density distribution. We actually decided to undertake the solution of the structure in three stages; the first completed in 1957, involved 400 reflexions and gave a resolution of 6 Å; the second (1959) included nearly 10 000 reflexions and gave a resolution of 2 Å; the third (not yet complete) includes all the observable reflexions – about 25 000 – and gives a resolution of 1.4 Å (. . .) the three stages chosen would be expected to separate polypeptide chains, groups of atoms, and individual atoms, respectively" (Kendrew, 1962).

The computational task was awesome:

"At each stage of the myoglobin analysis the computers employed were among the most rapid available at the time (. . .) at the 2 Å stage the solution of the phase problem for 9600 reflexions involved the densitometry of some quarter of a million spots in all (. . .)" (Kendrew, 1962).

When the first stage of the analysis had been achieved, myoglobin was found

"to consist of a large number of rod-like segments, joined at the ends, and irregularly wandering through the structure; a single dense flattened disk in each molecule; and sundry connected regions of uniform density. These could be identified respectively with polypeptide chains, with the iron atom and its associated porphyrin ring, and with the liquid filling the interstices between neighbouring molecules. (. . .) The most striking features of the molecule were its irregularity and its total lack of symmetry (. . .)" (Kendrew, 1962).

A lot more computational effort was required to achieve 2 Å resolution:

"It was necessary to determine the phases of nearly 10 000 reflexions (. . .) this task represented about the extreme limit of what is practicable by photographic techniques, and the Fourier synthesis itself (excluding preparatory computations of considerable bulk and complexity) required about 12 hours of continuous computation on a very fast machine (EDSAC II)" (Kendrew, 1962).

Four quite important results sprang from the 2 Å resolution structure.

(*i*) It gave the sequence of the amino acids in the polypeptide chain, or nearly so, well enough that comparison with the sequence obtained chemically by the methodology evolved initially by Frederick Sanger for insulin (Chapter 16) could be fruitful. At John Kendrew's suggestion, Drs. W.H. Stein and Stanford Moore of the Rockefeller Institute in New York had put one of their graduate students, Allen Edmundson, on this project. Kendrew

"found that, by laying his peptides [obtained from tryptic digestion of myoglobin] along the partial and tentative sequence derived from the X-ray analysis, we were able in many cases to observe correspondences which confirmed both our identifications and his analysis, and to clear up ambiguities and confusions in each" (Kendrew, 1962).

(*ii*) It came out in

"striking confirmation of suggestions which had been made as much as thirty years earlier to the effect that histidine was the haem-linked group in haemoglobin and myoglobin" (Kendrew, 1962).

(*iii*) Another finding was totally unexpected:

"To our surprise we found that the iron atom lay more than 1/4 Å out of the plane of the group" (Kendrew, 1962).

(*iv*) A most impressive confirmation was vindication of the α-helical structures which had been postulated by Pauling and Corey a few years before on the basis of molecular modeling (see Chapter 15). For myoglobin,

"careful analysis of the density distribution, carried out on the computer, shows that the helical segments are nearly all precisely straight, and that their co-ordinates correspond to those given by Pauling and Corey within the limits of error of the analysis" (Kendrew, 1962).

And these were indeed right-handed helices.

Besides confirming the presence (the overbearing presence one might say since 118 out of the total of 151 amino acids were found by Kendrew and his group to make up eight segments of right-handed α-helices) of the α-helix, the Kendrew work revealed for the first time the detailed molecular architecture of a globular protein: the helical segments

"are joined by 2 sharp corners (containing no non-helical residues) and 5 non-helical segments (of 1 to 8 residues); there is also a non-helical tail of 5 residues at the carboxyl end of the chain. The whole is folded in a complex and unsymmetrical manner to form a flattened, roughly triangular prism with dimensions about 45×35×25 Å. The whole structure is extremely compact; there is no water inside the molecule, (. . .); there are no channels through it, and the volume of internal empty space is small. The haem group is disposed almost normally to the surface of the molecule, one of its edges (that containing the polar propionic acid groups) being at the surface and the rest buried deeply within.

Turning now to the side chains, it is found that almost all those containing polar groups are on the surface. Thus with very few exceptions all the lysine, arginine, glutamic, aspartic, histidine, serine, threonine, tyrosine, and tryptophan residues have their polar groups on the outside (. . .) The interior of the molecule, on the other hand, is almost entirely made up of non-polar residues, generally close-packed and in Van der Waals' contact with their neighbours. (. . .)

On the other (distal) side of the iron atom as the histidine, occupying its sixth co-ordination position, is a water molecule, as would be expected in ferrimyoglobin, the form of myoglobin used for X-ray analysis; beyond the water molecule, in a position suitable for hydrogen-bond formation, is a second histidine residue. It is noteworthy that the same arrangement of two histidines also exists in haemoglobin. For the rest the environment of the haem group is almost entirely non-polar; it is held in place by a large number of Van der Waals' interactions" (Kendrew, 1962).

The biggest surprise from the myoglobin structure was discovering how irregular it was. Despite the numerological speculations of Bergmann and Niemann, despite the cyclol structures of Dorothy Wrinch having been all but discarded by the 1950s, still many scientists expected proteins to turn out to be handsome Pythagorean bodies with an impressive regularity to them. That they should turn up with an odd, complex, convoluted and irregular structure came as a shock. When Kendrew saw the picture at 6 Å resolution, 'the most striking features of the molecules were its irregularity and its total lack of symmetry' (Kendrew, 1962). The original 1958 *Nature* article expresses a similar bewilderment more vividly yet:

"Perhaps the most remarkable features of the molecule are its complexity and its lack of symmetry. The arrangement seems to be almost totally lacking in the kind of regularities which one instinctively anticipates, and it is more complicated than has been predicated by any theory of protein structure" (Kendrew et al., 1958).

Indeed, one cannot help noticing in the above-quoted description by Kendrew of the globular myoglobin molecule that, immediately after referring to its folding in 'complex and irregular manner', Kendrew writes that it is shaped like a triangular prism: as if the crystallographer (and amateur art historian) John Kendrew could not resist re-injecting a little order and harmony into this chaotic jumble of atoms!

REFERENCES

Kendrew, J.C., Bodo, G., Dintzis, H.M., Parrish, R.G., Wyckoff, H. and Phillips, D.C. (1958) Nature, 181, 662.
Kendrew, J.C. (1962) Nobel Lecture, December 11, in Nobel Lectures - Chemistry. 1942-1962, Elsevier, Amsterdam, p. 676.

Chapter 24

From the Myoglobin to the Hemoglobin Structure

The myoglobin structure proved a great help in speeding to a conclusion the hemoglobin structure on which Perutz had been working since the eve of the war, during what finally added up to a total of 22 years. In 1960, Perutz and his co-workers were able to report in *Nature* their results on hemoglobin at 5.5 Å resolution. Despite the poor resolution, dictated by the sheer complexity of the problem to be solved for a molecule of that bulk (M_r 67 000), a blatant feature of the structure was the presence of four similar sub-units, 'each sub-unit closely resembling Kendrew's model of sperm-whale myoglobin (Kendrew et al., 1958)' (Perutz et al., 1960).

As Perutz wrote:

"Thanks to the similarity with myoglobin, the interpretation can be carried further than would have been possible on the basis of our results alone" (Perutz et al., 1960).

The four polypeptide chains present in hemoglobin had meanwhile been shown by Schroeder and his co-workers to consist of two pairs, two α and two β chains as they are now called (Schroeder et al., 1957; Schroeder and Matsuda, 1958). The x-ray diffraction patterns showed Perutz and his group a symmetrical arrangement, each half of the hemoglobin molecule consisting of an α and a β chain being related to the other half by a dyad axis of symmetry. Then, to reproduce the account Perutz gave in his Nobel lecture,

"the haemoglobin molecule is assembled by first matching each chain with its symmetrically related partner, then inverting the pair of white chains and placing it over

[373]

the top of the pair of black ones. The resulting arrangement is tetrahedral, the four subunits forming a compact spheroidal molecule with the haem groups arranged in separate pockets on the surface of the molecule" (Perutz, 1962).

This tetrahedral arrangement is at variance with Pauling's preference, in the mid-1930s, for a square planar disposition (Pauling, 1935). Perutz further found a paucity, not to say a non-existence, of contacts between like chains (α,α and β,β), 'suggesting rather tenuous linkages' (Perutz et al., 1960). Conversely, there were very many contacts between each α chain and its β neighbor:

"The surface contours of the white chains exactly fit those of the black, so that there is a large area of contact between them. This structural complementarity is one of the most striking features of the molecule" (Perutz et al., 1960).

For an extended discussion of this notion of complementarity, see Chapter 6 in this volume.

Perutz was quick to recognize another important consequence of the resemblance between the α and β chains of hemoglobin, on one hand, and the polypeptide chain of myoglobin on the other:

"The polypeptide chain-fold which Kendrew and his collaborators first discovered in sperm whale myoglobin has since been found also in seal myoglobin (Ingram et al., 1956). Its appearance in horse haemoglobin suggests that all haemoglobins and myoglobins of vertebrates follow the same pattern. How does this arise? It is scarcely conceivable that a three-dimensional template forces the chain to take up this fold. More probably the chain, once it is synthesized and provided with a haem group around which it can coil, takes up this configuration spontaneously, as the only one which satisfies the stereochemical requirements of its amino-acid sequence. This suggests the occurrence of similar sequences throughout this group of proteins, despite their marked differences in amino-acid content. This seems all the more likely, since their structural similarity suggests that they have developed from a common genetic precursor" (Perutz et al., 1960).

These are two notions of the first importance. The latter would give rise to a whole new field, chemical evolution based upon comparison of the amino acid sequence in homologous protein molecules. The former, the intuition of the three-dimensional conformation being dictated by the sequence of residues along the polypeptide chain, was brought home to Perutz by the 'recurrence of the same amino-acids in characteristic positions of the myoglobin and haemoglobin chains' (Antonini, 1964).

And Perutz confirmed the important principle of molecular architecture for globular proteins, already clearly seen by Kendrew: such molecules have a hydrophobic core. This is how Perutz described it a few years later, when he gave one of the Harvey Lectures in New York, on April 25th, 1968:

"The most striking feature common to all globin chains is the almost complete exclusion of polar residues from interior sites (Perutz et al., 1965), suggesting that the pattern of invariantly nonpolar side chains, together with the nonpolar parts of the porphyrin ring, may be decisive in determining the conformation of the globin chain. (. . .)
      The reasons for the interior location of the nonpolar side chains are similar to those underlying the insolubility of hydrocarbons in water or the high surface energies at interfaces between nonpolar substances and water. These were first explained by Frank and Evans (1945). The presence of a hydrocarbon immobilizes the water molecules in its immediate vicinity, with a resulting reduction in the entropy of the system; (. . .) Conversely, the packing of the nonpolar side chains away from contact with water results in a gain of entropy of the native, folded structure of the protein and its surrounding water relative to its unfolded, denatured form" (Perutz, 1969).

Perutz's was an instant Nobel Prize: he received it in 1962, in midstream as it were, since work on hemoglobin was still incomplete. At that time, as he stated in his award lecture, 'as it stands, the structure of oxyhaemoglobin leaves its physiological properties unexplained' (Perutz, 1962). This would henceforth be the set of questions to which Perutz and his collaborators addressed themselves with great vigor and, from hindsight, with important results.

REFERENCES

Antonini, E. (1964) in Oxygen in the Animal Organisms (Dickens, F. and Neil, E., Eds.), Pergamon, Oxford, p. 121.

Frank, H.W. and Evans, M.W. (1945) J. Chem. Phys., 13, 507.

Ingram, D.J.E., Gibson, J.F. and Perutz, M.F. (1956) Nature, 178, 905.

Kendrew, J.C., Bodo, G., Dintzis, H.M., Parrish, R.G., Wyckoff, H. and Phillips, D.C. (1958) Nature, 181, 662.

Perutz, M.F. (1962) Nobel Lecture, December 11, in Nobel Lectures - Chemistry. 1942-1962, Elsevier, Amsterdam, p. 653 (and biography following his lecture).

Perutz, M.F. (1969) in The Harvey Lectures, Series 63 (Harvey Society of New York), Academic Press, New York, p. 213.

Perutz, M.F., Rossmann, M.G., Cullis, A.F., Muirhead, H., Will, G. and North, A.C.T. (1960) Nature, 185, 416.

Perutz, M.F., Kendrew, J.C. and Watson, H.C. (1965) J. Mol. Biol., 13, 669.

Schroeder, W.A., Rhinesmith, H.S. and Pauling, L. (1957) J. Am. Chem. Soc., 79, 4682.

Schroeder, W.A. and Matsuda, G. (1958) J. Am. Chem. Soc., 80, 1521.

Chapter 25

The Hemoglobin Structure and Its Molecular Correlates for Biological Functions

The *Nature* article of February 13, 1960 on the structure of hemoglobin has this sentence, probably penned by Max Perutz himself:

"The most prominent feature of the Fourier synthesis consists of more or less cylindrical clouds of high density, like the vapour trails of an aeroplane; they are curved to form intricate three-dimensional figures" (Perutz et al., 1960).

The 'vapour trail' metaphor not only expands upon the usual and rather trite image of a 'cloud' of electron density. It begs for further analysis. Recourse to this metaphor contrasts with the more normal rhetoric to be expected from an x-ray crystallographer, who builds a *hard* balls-and-sticks model of the structures he has solved. It is as if, here, Perutz resorts to the analogy with an object so distant, so tenuous also, that it cannot be grasped: it can only be looked at, be wondered at for its delicacy of form and beauty. In this respect, the vapor trail analogy conveys well the awed sense of wonder which Perutz had upon discovering the structures of myoglobin – through the work of his close research associate John Kendrew – and, a little later, of hemoglobin itself.

Vapor trails stay in the sky for a little while and then gradually disappear. I do not wish to push the analogy too far. Let me only remark that for Perutz structure remains ancillary to function. Biological function is for him the important question, and the elucidation of the structure of hemoglobin, even though he devoted no less than 22 years of his life and career to it, is important to him only for the light it sheds on the physiological use of hemoglobin as oxygen carrier.

1. Cooperativity of dioxygen binding

Myoglobin and hemoglobin contrast in their O_2 affinity curves, when the amount of bound dioxygen is plotted against the partial pressure $p(O_2)$.

"The reaction of myoglobin with oxygen follows bimolecular kinetics, so that its oxygen equilibrium curve is a simple hyperbola (. . .)" (Perutz, 1969a).

The hemoglobin oxygenation curve differs markedly, it has a sigmoidal shape, as already mentioned. Visual comparison of the two curves shows a more abrupt initial rise for Mb; while the Hb oxygenation curve is characterized by a much steeper gradient over the physiologically significant range, when $p(O_2)$ goes from about 20 torrs (capillaries in a muscle at work) to about 100 torrs (alveolar pressure within the lungs).

The difference between the Mb and Hb oxygenation curves is related to the difference in their biological uses. Myoglobin serves to store oxygen in the tissues, in muscles especially. Hemoglobin transports oxygen from the lungs, where it binds it, to the tissues, where it releases it at the lower partial pressure present there:

"The affinity of haemoglobin for oxygen is much less than that of myoglobin at low O_2 pressures, so causing haemoglobin to release its oxygen much more thouroughly in the muscle tissues" (North and Lydon, 1984).

That Hb delivers dioxygen more effectively had been already described by the phenomenological Hill treatment, as related previously (Chapter 19). This means that oxygen binds to hemoglobin in a cooperative manner: binding of one oxygen molecule makes easier binding of the second oxygen molecule, and so on, till the fourth O_2 molecule is picked up; this last is also the easiest step. Denoting as Y the fractional saturation and as n the Hill coefficient, this is the quantitative analysis given in a recent biochemistry textbook:

"Assume that the alveolar $p(O_2)$ is 100 torrs, and that the $p(O_2)$ in the capillary of an active muscle is 20 torrs. Let $P_{50} = 30$ torrs, and take $n = 2.8$. Then Y in the alveolar capillaries will be 0.97, and Y in the muscle capillaries will be 0.25. The oxygen delivered will be proportional to the difference in Y, which is 0.72. Let us now make the same calculation for a hypothetical oxygen carrier for which P_{50} is also 30 torrs, but in which the binding of oxygen is not cooperative ($n = 1$). Then $Y_{\text{alveoli}} = 0.77$, and $Y_{\text{muscle}} = 0.41$, and so $Y = 0.36$. Thus, *the co-operative binding of oxygen by hemoglobin enables it to deliver twice as much oxygen as it would if the sites were independent*" (Stryer, 1981a).

What is the nature of the coupling between the four nearly equivalent hemoglobin sub-units?

The answer is morphological, there is a change in the shape of the hemoglobin molecule when it goes from the oxygen-free to the oxygenated state. This is an

"effect so substantial that crystals of deoxy-haemoglobin prepared in an oxygen-free environment will shatter when they are exposed to oxygen" (North and Lydon, 1984).

This difference between the crystals of the oxy- and deoxy-forms of hemoglobin had been reported as early as 1909 (Reichert and Brown, 1909) and 1938 (Haurowitz, 1938).

Before describing in molecular terms the transition from the deoxy- to the oxy-form, let us remind the reader of a conclusion Max Perutz reached in 1953:

"Carbonmonoxyhemoglobin and all the ferric derivatives of hemoglobin have the same structure as oxyhemoglobin; this implies that the valency and spin state of the iron atom have no influence on the conformation of the hemoglobin molecule" (Perutz, 1953; 1969a).

Accordingly, one can limit description of conformational changes attendant upon ligand binding to that or those undergone by the hemoglobin molecule from the deoxy- to the oxy-form.

"The most striking differences between the structures of deoxy- and liganded haemoglobin that were first seen to be important were the difference in position of each iron atom relative to the mean plane of its haem (· · ·)" (Baldwin, 1980).

In the deoxy-form, the iron atom is 0.55 Å from the mean plane of the porphyrin nitrogens and carbons. Upon oxygenation, the iron atom moves into the heme plane, in the center of the porphyrin cavity. This structural change, in tertiary structure, goes with an electronic change, as manifested in the magnetic susceptibility (Chapter 20): deoxy Hb is paramagnetic with four unpaired spins per iron atom, while oxy Hb is diamagnetic. As Perutz and TenEyck indicate:

"At that time the significance of this observation was unclear, but it now provides the key for the understanding of heme–heme interaction. R.J.P. Williams (1961) pointed out that the transition from high to low spin would be accompanied by a marked reduction in the radius of the iron atom and suggested that the change in the structure of hemoglobin on oxygenation 'is related to the change in the imidazole-iron distance or possibly to the concomitant change in the steric arrangement of the imidazole (of the heme-linked histidine) to the porphyrin plane' (Williams, 1961).

In crystallographic studies of ferric porphyrin complexes J.L. Hoard and his collaborators found that the transition from high to low spin is accompanied by a marked decrease in the radius of the iron atom; he predicted that (. . .) the substantial movement of the iron atom relative to the porphyrin ring accompanying its change of spin state on oxygenation could lead to cooperative movements in the protein framework of the hemoglobin molecule (Hoard, 1966).

In particular, he pointed out that the hole in the center of the porphyrin is just large enough to accommodate the low spin form of iron, allowing the heme to be planar" (Perutz and TenEyck, 1972).

There are heme motions as well, which were picked up more recently: on going from the deoxy- to the oxy- (or liganded) form of hemoglobin, the α and β hemes move further into the heme pockets by 0.5 and 1.5 Å, respectively (Baldwin, 1980).

Low-resolution pictures of hemoglobin at 3.5 and 2.8 Å - comparing the oxy-form of horse hemoglobin with the deoxy-form of human hemoglobin, at first - showed most interesting changes in the quaternary structure upon ligation (Muirhead and Perutz, 1963; Bolton and Perutz, 1970; Muirhead and Greer, 1970; Perutz et al., 1968):

"The contact $\alpha_1\beta_1$ undergoes slight changes which seem to involve a loosening, rather than a sliding, at certain areas of contact. There seem to be about 32 residues or 98 atoms in contact in deoxy, as compared with about 34 residues or 110 atoms in oxy. (. . .) The contact $\alpha_1\beta_2$ undergoes drastic changes as the two sub-units turn relative to each other by 13°. There are about 20 residues or 69 atoms in contact in deoxy, compared to 19 and 80 in oxy. The contact is dovetailed, so that the CD region of one chain fits into the FG region of the other (see Fig. 1 and Perutz 1969a, b). During the change of quaternary structure the dovetailing of CDβ with FGα remains much the same while that of CDα with FGβ changes, as shown diagrammatically in the figure. The hydrogen bond linking Asp G1(94)α to Asn G4(102)β in oxy is replaced by another between Tyr C7(42)α and Asp G1(99)β in deoxy" (Perutz and TenEyck, 1972).

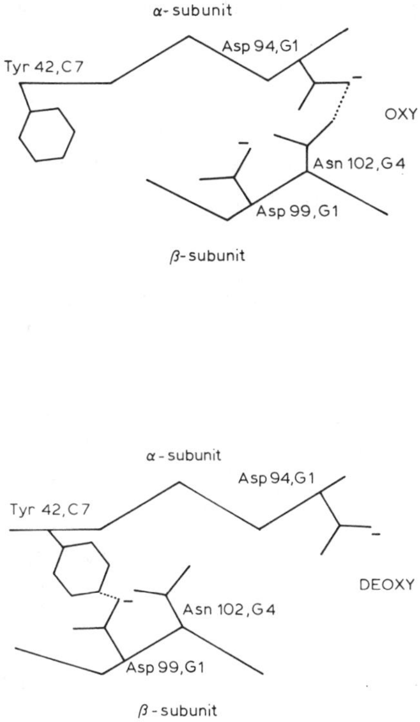

Fig. 7. The rotation at the contact $\alpha_1\beta_2$ causes a jump in the dovetailing of the CD region of α, relative to the FG region of β_2, and a switch of hydrogen bonds as shown in the figure (Perutz and TenEyck, 1972).

A word on the nomenclature. The four polypeptide chains of hemoglobin are labeled α_1, α_2, β_1 and β_2, since they exist in two pairs related by symmetry. By symmetry, the contacts $\alpha_1\beta_1$ and $\alpha_2\beta_2$ are the same. Likewise, the contacts $\alpha_1\beta_2$ and $\alpha_2\beta_1$ are the same. Each peptide chain is labeled according to the eight helical segments present in the myoglobin structure A through H. The intervening five non-helical regions are labeled AB through GH. The five residues at the carboxyl end are labeled HC1 through HC5. The changes in quaternary structure between the deoxy- and the liganded hemoglobin affect the $\alpha_1\beta_2$ (and $\alpha_2\beta_1$) contacts, which have been termed

" 'switch' regions, allowing only two stable packings between the dimers" (Baldwin, 1980).

The changes in tertiary and quaternary structure discussed so far are interrelated. The proximal histidine follows the motion of the iron atom as it steps down 0.55 Å into the porphyrin hole. This takes this histidine residue from an asymmetric to a more symmetric position:

"the change in the distance of the heme-linked histidine from the plane of the porphyrin is probably the main trigger responsible for the change in the tertiary structure of the α sub-units" (Perutz, 1970a; Perutz and TenEyck, 1972).

These conformational changes have been more accurately located, at the time of writing, in the F helix, the FG corner region of each sub-unit and the E helix of β (Baldwin, 1980).

A last important feature distinguishes the deoxy- and liganded forms of hemoglobin, the presence of salt bridges in the former but not in the latter:

"The constraints which clearly distinguish the deoxy from the oxy structure are the salt bridges formed by the C-terminal residues" (Perutz and TenEyck, 1972).

To be more explicit,

"In deoxyhaemoglobin, inter-dimer salt-bridges are formed involving the α-chain carboxyl-terminal residues Arg-α 141 (HC3). These cannot be formed in liganded haemoglobin because the space between dimers $\alpha_1\beta_1$ and $\alpha_2\beta_2$ is too small to accommodate the arginine side chains after the quaternary change. The β chain carboxy-terminal residues Hisβ146 (HC3) form inter-dimer salt-bridges with residues Lysα40 (C5) in deoxyhaemoglobin, but cannot do so in liganded haemoglobin because the residues are separated by 7 Å shifts which occur on the quaternary structure change" (Baldwin, 1980).

These structural changes constitute the factual basis upon which Perutz would base his explanation of the hemoglobin oxygen-binding cooperativity. The theoretical basis was provided for him by the appearance in 1965 of the article by Monod, Wyman and Changeux on allosteric interactions (Monod et al., 1965). Allosteric theory proposed an equilibrium between two states differing in their affinity for ligand. One form is referred to as tense (T), the other as relaxed (R). The low-affinity is the T form, the high-affinity is the R form. Perutz seized upon this theoretical treatment to identify the T form with deoxyhemoglobin, while the R form is oxy- (or liganded) hemoglobin. What is the source of the tension inherent in deoxyhemoglobin? Perutz is very affirmative:

"The constraints which clearly distinguish the deoxy from the oxy structure are the salt bridges formed by the C-terminal residues (. . .) selective removal or blockage of each salt bridge leads to a specific increase of oxygen affinity and reduction of Hill's constant even though the rupture does not significantly perturb the structure of the hemoglobin subunits. This means that the oxygen affinity and Hill's constant are directly dependent on the strength and number of salt bridges constraining the tetramer in the deoxy form. Removal of a sufficient number of salt bridges prevents formation of the quaternary deoxy structure, even in the total absence of heme ligand" (Perutz and TenEyck, 1972).

The experiments of selective removal or blockage of the salt bridges, performed by Perutz and his collaborators are crucial to test this role of the salt bridges in constraining the tense deoxy-form. Indeed, these experiments can now be related to the observed macroscopic difference between the crystals of both forms: this serves Perutz to clinch his case; and, accordingly, this is how he closes his 1972 article:

"When crystals of reduced hemoglobin are oxygenated, or crystals of oxyhemoglobin are reduced, they crack and their diffraction pattern disappears, which shows that the free energy changes of heme-heme interaction responsible for the change in quaternary structure ($\sim$ 13 kcal/mol) are greater than the lattice energy. However, if the C-terminal salt bridges of either the histidines or the arginines are removed, the diffraction pattern of oxy- or methemoglobin remains intact on reduction because the free-energy of heme-heme interaction then becomes comparable with, or smaller than, the lattice energy. Therefore salt affects the strength of the bonds between the subunits but not the structures of the oxy or deoxy tetramer" (Perutz and TenEyck, 1972).

With this identification of the difference between the deoxy- and the oxy-forms as stemming from an energy difference, which can be assigned in turn to a given part of the hemoglobin geometry, cooperativity of dioxygen binding is given a structural explanation (see Fig. 7).

"At each level the structure is essentially in one of two states, and the only energetically favourable states for the tetramer are symmetrical states in which all of the subunits have similar shapes. (. . .) starting from a fully deoxygenated tetramer, we see that binding the first oxygen is unfavourable because it tends to disrupt the pattern of interactions stabilizing the tetramer; the second oxygen causes a further, but lesser disruption; the final oxygen binds very favourably because it permits the restoration of a symmetrical tetramer, now in the fully oxygenated form. A corresponding pathway is followed in the reverse direction. The switch between one molecular state that has low affinity for oxygen and another with high affinity leads to the sigmoidal shape of the oxygenation curve" (North and Lydon, 1984).

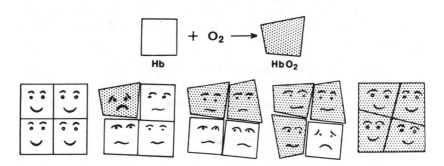

Fig. 8. The upper section of this schematic diagram shows that the binding of O_2 promotes a change of shape of the hemoglobin subunit to which it is attached. The lower section shows that the four subunits in a tetramer only fit together well if they all have the same shape. Mixed states are therefore unfavorable (from North and Lydon, 1984).

2. The Bohr effect

The shift of the hemoglobin oxygenation curve with pH is the Bohr effect. Already before the war, Wyman had interpreted it by the assumption of two proton-binding sites per heme (Wyman, 1939).

After the war, the two parameters for the Hill equation were determined accurately as a function of pH in the whole range from pH 4
till pH 11 (Antonini et al., 1962). While the Hill coefficient n
remains rather constant at a value slightly below 3, the oxygen affinity is at a maximum at pH 6.5 and decreases markedly on both sides
of this value. The reduction of oxygen affinity in acid medium is
physiologically significant: 'this promotes the release of oxygen in
working muscle tissues, where it is particularly needed' (North and
Lydon, 1984).

I shall borrow, because of its clarity, the description by Lubert
Stryer of the mechanism of the Bohr effect:

"(. . .) the pKs of some groups must be *raised* in the transition from oxy to deoxyhemoglobin; an increase in pK means stronger binding of H^+. About 0.5 H^+ is taken up
by hemoglobin for each molecule of O_2 that is released. (. . .)

Which groups have their pKs raised? (. . .) the most plausible candidates for the
groups participating in the Bohr effect are histidine, (. . .) and the terminal amino
group, which normally have pK values near 7.

The identification of the specific groups responsible for the Bohr effect was
achieved by combining information derived from chemical and x-ray studies. The
carboxyl-terminal histidines of the β chains (histidine 146β) were implicated by x-ray
studies. This suggestion was tested by preparing a hemoglobin devoid of histidine 146
on its β chains. Carboxypeptidase B, a specific proteolytic enzyme that splits the
peptide bond involving carboxyl-terminal basic amino acids in some polypeptide
chains, was used for this purpose. The modified hemoglobin produced in this way had
only half of the Bohr effect of normal hemoglobin. Hence, *histidine 146 on each β
chain appears to make a major contribution to the Bohr effect.*

The role of the terminal amino groups of hemoglobin in the Bohr effect was ascertained by specifically modifying them with cyanate to form a *carbamoylated* derivative, which can no longer bind H^+. The Bohr effect was unchanged when the terminal
amino groups of only the β chain were carbamoylated. In contrast, the Bohr effect was
significantly reduced when the terminal amino groups of the α chains were carbamoylated.

$$R-NH_2 \; + \; {}^-NCO + H^+ \; \rightarrow \; R-\overset{\displaystyle H}{\underset{\displaystyle \underset{O}{\parallel}}{N}}-\overset{}{C}-NH_2$$

terminal   cyanate   carbamoylated
amino   derivative
group

The x-ray results provide a concrete picture of how these groups might function in the Bohr effect. In oxyhemoglobin, histidine 146β rotates freely, whereas, in deoxyhemoglobin, this terminal residue participates in a number of interactions. Of particular significance is the interaction of its imidazole ring with the negatively charged aspartate on the same β chain. The close proximity of this negatively charged group enhances the likelihood that the imidazole group will bind a proton. In other words, the proximity of aspartate 94 raises the pK of histidine 146. Thus, *in the transition from oxy to deoxyhemoglobin, histidine 146 acquires a greater affinity for H^+ because its local charge environment is altered.*

The environment of the terminal amino groups of the α chains changes in a similar way on deoxygenation. In oxyhemoglobin, these groups are free. In deoxyhemoglobin, the terminal amino group of one α chain interacts with the carboxyl-terminal group of the other α chain. *The proximity of a negatively charged carboxylate residue in deoxyhemoglobin increases the affinity of this terminal amino group for H^+.* X-ray results have implicated a third group in the Bohr effect, namely histidine 122 of the α chains" (Stryer, 1981b).

To sum up, in the words of the scientists who first elucidated this mechanism,

"the main contribution to the Bohr effect arises from the opening and closing of the salt bridges made by the C-terminal residues (Perutz, 1970b)" (Perutz and TenEyck, 1972).

The above description should be complemented in the following manner:

"A. Arnone showed that the contribution of the terminal amino group to the Bohr effect is due, *not* to its binding to the terminal carboxyl of the opposite α-chain, but to a chloride ion. Its contribution disappears in the absence of chloride (Arnone et al., 1979). My idea that this 122 α contributes was wrong. On the other hand Lysine 82 β makes an important contribution, again only in the presence of chloride (Perutz et al., 1980).

Not only was the mechanism of the Bohr effect thus explained. Other ligands have interesting physiological effects when they attach themselves to hemoglobin. Under physiological conditions (pH, ionic strength), 2,3-diphosphoglycerate (DPG) binds to deoxyhemoglobin, one-to-one. Ruth and Reinhold Benesch showed also that this substance reduces the oxygen affinity of hemoglobin by as much as a factor 26 (Benesch and Benesch, 1969); this phenomenon has a role in high altitude adaptation, in hypoxic disorders, etc. Perutz also able to explain, from the structure of the DPG-hemoglobin complex (DPG sits in a central cavity where it interacts predominantly with the β chains), the mechanism of DPG action (Perutz and TenEyck, 1972).

3. Concluding remarks

Max Perutz is not only a fine scientist, he writes well. He writes a crisp prose. He does so from within the conventions of scientific English, but manages nevertheless to be stylish, without idiosyncrasy or show-off. His mark is conciseness and simplicity. Let me give an example, which will also serve me as a kernel around which to wrap a few conclusions to this chapter.

My quote is again from the 1972 article with L.F. TenEyck; the authors are discussing the role of salt bridges in cooperative effects:

"(. . .) the subunit contacts in the quaternary deoxy structure are stressed even in the absence of oxygen, and (. . .) to remain in that conformation the tetramer must be clamped by the salt bridges. *As soon as enough of the clamps are released the structure clicks to the quaternary oxy form*" (Perutz and TenEyck, 1972; emphasis added).

I find this last sentence delightful. It rings simple and true. The alliteration between the two key words, *clamps* and *clicks*, endows it with harmony. Moreover, this short paragraph buoyed by its felicities of style, has unmistakable meaning taken as a whole: the hemoglobin story has reached the stage of mechanism. We can almost hear, with our reader's ears, the metallic click of the machinery as it clamps itself in its pre-set stressed condition.

With the work of Max Perutz, the hemoglobin functions become the smooth result of interconnected parts, working complementarily to one another and in unison, as a machine. Using the x-ray diffraction patterns, Perutz and his co-workers have established the topography of the active parts, in this gigantic piece of machinery, and they have mapped out in detail their interactions.

Their tactic for doing so is of utmost simplicity, in principle: comparison between molecular assemblies. For instance, the comparison between myoglobin and hemoglobin, coming from various animal species, has enabled them to single out amino acid residues occupying key positions in the polypeptide chain. Others, Pauling and Zuckerkandl to start with, would rush along the way thus pointed, and develop a molecular evolution based upon comparison of respiratory pigments from various species.

Another example is that of the comparisons between deoxy- and oxyhemoglobin, detailed in sections 1 and 2 above: these comparisons have served Perutz to relate atomic motions, displacements and rotations of entire parts of the macromolecule, with well-defined physiological functions.

It is noteworthy that the mechanisms thus established, entirely reliably, are static. It could not have been otherwise: because x-rays were the tool of investigation, because also of the sheer complexity of the problem, combined with the concepts and methods available then. Only now, some 20 years later, does a combination of molecular dynamics calculations and magnetic resonance evidence start to shed some light on the internal dynamics of protein molecules and their role in biological functions. In a way, not only Perutz, but also Monod, Wyman, Changeux, Manfred Eigen, and all those who have contributed to the analysis of hemoglobin cooperativity were very lucky that this system was so well-behaved, so highly simple that a static view was sufficient; and that simple comparison of initial and final states, with no need to consider any intermediate state, was sufficient to unravel the essence of mechanism for the important physiological effects.

Consider now the chronology of the hemoglobin studies, outlined below in schematic and hence highly disputable form:

chemical physiology	ca. 1870	F. Hoppe-Seyler
phenomenology	ca. 1910	A.V. Hill
theorizing	1965	J. Monod
		J. Wyman
		J.P. Changeux
mechanism		
static	ca. 1970	M.F. Perutz
dynamic	(in progress)	M. Karplus and others

About a century was needed, from the first firm conceptualization of hemoglobin as a molecule to the convincing detailed explanation, at the atomic level, of how this molecule works as oxygen carrier in the various parts of the organism.

Another concluding remark is the cross-fertilization from various fields of science necessary to bring these hemoglobin studies to fruition. Hematology and genetics contributed with the problem posed by the abnormal hemoglobins, such as Hb(S). The crystallographers could draw upon results from very varied sources: measurements of magnetic susceptibilities; molecular architecture and model building, which provided them with α helices as an important building block; inorganic chemistry, and the insight it gave them into ionic radii for low- and high-spin atoms; synthetic organic chemistry, with methods for heavy atom labeling; physical chemistry, with the analysis of coupled equilibrium and kinetic equations; applied mathematics, with development of the powerful high-speed computers demanded by analysis of the diffraction patterns with their myriads of spots to be translated back into molecular information, etc.

What I find remarkable in Max Perutz is, not only that he has had the prowess to master all these methods and make them converge in his hands into a solution; but also, beyond these already remarkable achievements, that he was not content to gather the molecular data and leave it to others to interpret. What I find truly remarkable is that Max Perutz, if we refer back to the chronological sequence just alluded to, has seen fit, or more exactly has felt that it was his special responsibility, to bring the story to a close; by bridging chemical physiology and his structural determinations into a set of well-established mechanisms. This dual aspect, structural and mechanistic, makes his contribution memorable.

Plate 9. Max Perutz at work during the 1960s. (Photograph by the *Times*, courtesy Prof. M.F. Perutz.)

REFERENCES

Arnone, A., O'Donnell, S., Mandaro, R. and Schuster, T.M. (1979) J. Biol. Chem., 254, 12204.

Baldwin, J. (1980) TIBS, 224.

Benesch, R. and Benesch, R.E. (1969) Nature, 221, 618.

Bolton, W. and Perutz, M.F. (1970) Nature, 228, 551.

Haurowitz, F. (1938) Z. Pysiol. Chem., 254, 266.

Hoard, J.L. (1966) in Hemes and Hemoproteins (Chance, B., Estabrook, R.W. and Yonetani, T., Eds.), Academic Press, New York, p. 9.

Monod, J., Wyman, J. and Changeux, J.P. (1965) J. Mol. Biol., 12, 88.

Muirhead, H. and Geer, J. (1970) Nature, 228, 516.

Muirhead, H. and Perutz, M.F. (1963) Nature, 199, 633.

North, A.C.T. and Lydon, J.E. (1984) Contemp. Phys., 25, 381.

Perutz, M.F. (1953) Proc. Roy. Soc. London, B141, 69.

Perutz, M.F. (1969a) in The Harvey Lectures, 1967-1968, Series 63 (Harvey Society of New York) Academic Press, New York, p. 213.

Perutz, M.F. (1969b) Proc. Roy. Soc. London, B173, 113.

Perutz, M.F. (1970a) Nature, 228, 726.

Perutz, M.F. (1970b) Nature, 228, 734.

Perutz, M.F. (1985) Letter of March 1st to Pierre Laszlo.

Perutz, M.F. and TenEyck, L.F. (1972) Cold Spring Harbor Symp. Quant. Biol., 36, 295.

Perutz, M.F., Rossmann, M.G., Cullis, A.F., Muirhead, H., Will, G. and North, A.C.T. (1960) Nature, 185, 416.

Perutz, M.F., Muirhead, H., Cox, J.M. and Goaman, L.C.G. (1968) Nature, 219, 131.

Perutz, M.F., Kilmartin, J.V., Nishikura, K., Fogg, J.H., Butler, P.J.G. and Rollena, H.S. (1980) J. Mol. Biol., 138, 649.

Reichert, A.T. and Brown, A.P. (1909) The Crystallography of Hemoglobins, Carnegie Institution, Washington, DC.

Stryer, L. (1981) Biochemistry, 2nd ed., Freeman, San Francisco, (a) p. 69; (b) pp. 80-82.

Williams, R.J.P. (1961) Fed. Proc., 20, No. 3 (suppl. 10), 5.

Chapter 26

Molecular Disease: Sickle-Cell Hemoglobin

Sickle-cell anemia is a very serious disease, among Blacks (0.25% of American Blacks, for which reliable statistics exist), Greeks, Indians and several populations in Africa. It is a debilitating, sometimes deadly condition. Only 20% of those suffering from it live to have children, and seldom do they go past the age of 40. Yet, and rather paradoxically, victims of sickle-cell anemia are protected from malaria. The malaria vector does not find enough nourishment in the blood of patients affected by sickle-cell anemia. Among symptoms of the disease are swollen hands and feet, and painful joints. Sudden deaths can occur, as the result of effort. For instance, newly enlisted Black soldiers in the U.S. have been known to collapse upon performing strenuous exercise. The observation of erythrocytes with the peculiar, sickle-like shape, was first made by Herrick in 1910 (Herrick, 1910). Following which,

"the sickling phenomenon was recognized as an inherited condition by Taliaferro and Huck in 1923, who considered it to be a dominant Mendelian character (Taliaferro and Huck, 1923). (. . .) In 1949, Neel and Beet, independently (Neel, 1949; Beet, 1949) put forward the now accepted view that individuals with the severe sickle cell anemia were homozygous for an abnormal gene and that sickle-cell trait carriers were heterozygous, having one normal and one abnormal gene. (. . .)
    The reason for the sickling of deoxygenated red cells containing sickle-cell hemoglobin (S) was obscure until Harris (1950) showed the presence of birefringent bodies in deoxygenated solutions of this hemoglobin" (Ingram, 1963a).

In 1949, just one year earlier, Pauling and Itano made the determining step of identifying sickle-cell anemia as a molecular disease (Pauling et al., 1949). The story of how Linus Pauling became involved is the following:

"The discovery of the abnormal human hemoglobins was the result of my having been appointed to a committee. (. . .) After a strenuous day listening to representatives of medical schools and medical research institutes in New York City, the members of the Medical Advisory Committee were having dinner togetber at a club in the city (Pauling, 1955). Dr. William B. Castle, Professor of Medicine at Harvard University, began talking about the disease sickle-cell anemia, with which he had some experience. (. . .) when Dr. Castle said that the red cells in the blood of a patient with this disease are sickled in the venous blood but not in the arterial blood, the idea occurred to me that sickle-cell anemia might be a disease of the hemoglobin molecule. (. . .) I thought at once of this possibility: that the hemoglobin molecules of these patients might have, as a result of a gene mutation, a structure formed so that one portion of the surface of the molecule would be sufficiently complementary to another portion to permit the molecules to aggregate into long chains. Further, these long chains would then line up side by side to form a needle-shaped crystal, which, as its length grew greater than the diameter of the red cell, would twist the red cell out of shape. (. . .) The attachment of the dioxygen molecules to the iron atoms of the heme groups might interfere with the approach to one another of complementary regions of adjacent molecules close enough to permit a strong interaction to take place, so that oxygenation of the hemoglobin would reverse the sickling process" (Pauling, 1980–1981).

A rare example of creative committee work!

Notice here the recurrence of Pauling's notion of complementarity. I have had occasion to detail it earlier in this volume (Chapter 6). Suffice it to say here that Pauling's flight of imagination, during this post-committee meeting dinner, reversed its usual route: during the same period of the Forties, Linus Pauling kept referring in his concept of biological complementarity to the *analogon* of a crystal from which a fragment had been detached, in a *gedanken* experiment. Bringing together the two moieties, whose three-dimensional surfaces would match one another perfectly, like die and cast, served as an image for complementarity of biomolecules (Chapter 6). Here, it is the opposite: the biomolecules revert to the inanimate nature of a crystal because they are endowed with complementary surfaces. Mutual fit of these brings about a self-association akin to crystal growth. It was very natural for Linus Pauling, given his obsession with hemoglobin and given the mental image he kept of complementarity of biological objects at the microscopic level, that he should think immediately of such an explanation.

Pauling, Itano, and their co-workers Singer and Wells proceeded

to compare the physicochemical characteristics of normal hemoglobin (A) and of sickle-cell hemoglobin (S). They were able to find a
key distinction: electrophoretic mobility. This reflected a difference
in the distribution of electric charges between Hb(A) and Hb(S)
(Pauling et al., 1949). As Ingram writes:

"It was likely from Pauling and Itano's work that an amino acid with a charged side
chain was involved in the alteration and that glutamic acid, aspartic acid, lysine,
histidine, or arginine might be involved. These all occur sufficiently frequently,
except for arginine, to make analyses of the total hydrolysate an unreliable tool for the
detection of the appearance or disappearance of one of these amino acids" (Ingram,
1961a).

A problem set is a problem solved. And the solution to this problem
of sickle-cell hemoglobin was to be given during the next few years,
even though the difference which had to be pin-pointed was
extremely minute.

"Next, Perutz and Mitchinson (1951), Perutz and Mitchinson (1950), Perutz et al.
(1951) demonstrated the very low solubility of reduced hemoglobin S in salt solutions.
We can now see the reason for the sickling phenomenon: the hemoglobin S in the red
cell remains in solution as the normally soluble oxyhemoglobin S as long as the cells
are in an oxygen rich environment. When the partial pressure of oxygen is reduced,
either artificially or in the tissues, the hemoglobin protein inside the cell precipitates
to form paracrystalline aggregates of tactoids which distort the cell to the sickle-cell
shape" (Ingram, 1963a).

Needless to say, Perutz tried to find out from x-ray diffraction the
structural difference between the two hemoglobins, A and S. There
had to be one, given the difference in the electrophoretic mobilities
(Pauling et al., 1949). Perutz could not see any difference in the
x-ray diffraction patterns of the two hemoglobins, the normal and
the pathological (Perutz et al., 1951). This showed that sickle-cell
anemia was not associated with a gross structural change in the
hemoglobin molecule. As Ingram writes (1963b):

"Perutz's result was an important one, because following the demonstration of an
electrophoretic difference between hemoglobins A and S the possibility existed that
this was due, not to a change in amino acid sequence, but solely to a rearrangement of
the same peptide chain in a different folding so as to uncover or to mask charged
groups in the protein molecule".

Indeed this was a very real possibility, not only the 'straw man' it looks like from our post-hoc perspective. Till the unraveling of the first protein structure (myoglobin) by Kendrew, and its comparison with the hemoglobin structure (Perutz), in the late 1950s (see Chapter 24), the one-to-one relationship of amino acid sequence to conformation was, at best, speculative.

Scheinberg and his collaborators addressed themselves, in the late 1950s, to the elucidation of the electrophoretic difference that existed between hemoglobins A and S. At pH 8.6 a decrease of two units of negative charge per molecule is observed: the most likely origins are loss of a negative group or gain of a positively charged group by two of the polypeptide chains. To distinguish between these two possibilities, Scheinberg performed electrophoresis at two widely differing pH values, pH 3 – where carboxyl groups are uncharged, because they are protonated – and pH 12 – where the lysine-positive groups are uncharged. He found that hemoglobins A and S did not differ in electrophoretic mobility at pH 3. The inference was that they differed only by the loss of a carboxyl group (Scheinberg, 1958).

The structural problem posed by hemoglobin S was now very close to its solution, which was provided by Ingram's painstaking 'two-dimensional fingerprints' technique, in the first step of which

"the mixture of tryptic peptides from one of the hemoglobins is applied to thick filter paper moistened with a volatile buffer of pyridine and acetic acid at pH 6.4 and submitted to electrophoresis in a sandwich of plate glass. As a result of this simple procedure, the peptides are now spread out along the line of the electric field. After removal of the volatile buffer in preparation for the *second step*, the paper is allowed to hang and dip with one edge into a chromatographic solvent so that the line of peptide spots form a horizontal starting line near the bottom. After the solvent, *n*-butanol:acetic acid:water, has reached the top of the paper, the paper can be dried and the individual peptides can be localized with the ninhydrin reagent, or with a number of other reagents specific for certain amino acids – tryptophan, tyrosine, histidine or arginine" (Ingram, 1961b).

Ingram found in this manner a difference between the two octapeptides:

Hb(A) Val–His–Leu–Thr–Pro–Glu–Glu–Lys
Hb(S) Val–His–Leu–Thr–Pro–Val–Glu–Lys

"The difference between these peptides turns out to be a replacement of the glutamic acid in the sixth position of the β peptide chain of hemoglobin A by valine in the sickle cell hemoglobin" (Ingram, 1958; 1959; 1963c).

REFERENCES

Beet, E.A. (1949) Ann. Eugenics, 14, 279.

Harris, J.W. (1950) Proc. Soc. Exp. Biol. Med., 75, 197.

Herrick, J.B. (1910) Arch. Int. Med., 6, 517.

Ingram, V.M. (1958) Biochim. Biophys. Acta, 28, 539.

Ingram, V.M. (1959) Biochim. Biophys. Acta, 36, 402.

Ingram, V.M. (1961) Hemoglobin and its Abnormalities, Thomas, Springfield, IL, (a) p. 76; (b) p. 85.

Ingram, V.M. (1963) The Hemoglobins in Genetics and Evolution, Columbia University Press, New York, (a) p. 38; (b) p. 41; (c) p. 46.

Neel, J.V. (1949) Science, 110, 64.

Pauling, L. (1955) in The Harvey Lectures (1953-1954), Series 49, Academic Press, New York, p. 216.

Pauling, L. (1980-1981) Texas Rep. Biol. Med., 40, 1.

Pauling, L., Itano, H.A., Singer, S.J. and Wells, L.C. (1949) Science, 110, 543.

Perutz, M.F. and Mitchison, J.M. (1950) Nature, 166, 677.

Perutz, M.F., Liquori, A.M. and Eirich, F. (1951) Nature, 167, 929.

Scheinberg, I.H. (1958) Conference on Hemoglobin, National Academy of Sciences, Washington, DC, p. 227; quoted by Ingram, 1961.

Taliaferro, W.H. and Huck, J.G. (1923) Genetics, 8, 594.

PART IV

Enzymes as Proteins

(Chapters 27–37)

Chapter 27

Terminology: Ferment, Diastase, Enzyme

The history of the word 'enzyme' is apparently uncomplicated: of rather recent introduction (Kühne, 1876), it gained widespread use in the early 1900s and gradually replaced its two earlier rivals, 'ferment' and 'diastase', the latter having been used by mostly French biochemists. Nevertheless, closer scrutiny at this shifting vocabulary will show the underlying concepts during the last two centuries.

Enzymes have been associated since Antiquity with fermentation processes, especially alcoholic fermentation, putrefaction, and the raising of baker's dough. The word 'ferment' is very graphic. It originates in the Latin *fermentatio*, with an identical meaning to ours, which in turn comes from the Latin verb *feruēo*, to boil (the present English words *fervid, fervor, effervescent*, etc. are all derived from it). Many languages, such as Celtic, Vedic, Old Germanic, had cognate words with the same meaning of 'to boil, to seethe, to froth up, to bubble'. Hence, fermentations were associated in many cultures with effervescence, with bubbling, with letting off gases.

The modern study of fermentations started in the 18th century. The *Encyclopédie* of Diderot and d'Alembert thus defines '*Ferment* or *Leaven* (in Chemistry)':

"One names thus a body presently fermenting which when it is mixed exactly and in small quantity into a considerable mass of fermentable matter determines in this matter the action of fermentation" (Diderot et al., 1751-1780).

In this definition is the concept of the ferment (later to be called the enzyme) as a fermentation operator, of which a small quantity only has to be added for fermentation to occur.

To shift now to a later period, Anselme Payen and Jean-François Persoz, two French chemists interested in practical applications, isolated a substance from germinating seeds of barley, in 1833. They found that this compound was capable, just like strong mineral acid, to hydrolyse starch into sugar. They argued that this compound, which therefore they named 'diastase', acted by piercing the sheath of starch globules prior to chopping starch into sucrose (Payen and Persoz, 1833).

As they wrote (Payen and Persoz, 1833):

"Cette singulière propriété de séparation nous a déterminés à donner à la substance qui la possède le nom de *diastase* qui exprime précisément ce fait".
(This singular property of separation determined us to give to the substance having it, the name of *diastase* which expresses precisely this fact.)

The term was quickly adopted: not only by the French but also by British scientists, especially in the applied fields. For instance, Baxter (1846) translates the findings of Payen and Persoz as follows:

"During the germination, some of the elements (. . .) in the grain form a fresh compound, which acts as a ferment. This compound is called (. . .) diastase, the effect of which (. . .) is to turn all the starch (. . .) first into gum and then into sugar".

Another author concerns himself also with agriculture:

"Neither potatoes nor cereals contain diastase before germination" (Watts, 1863–1872).

While at the end of the century the *lysis* properties of the diastase assume increased importance:

"An extremely active poison, delicate, resembling the diastases or venoms" (The Lancet, 1894).

The word 'enzyme', which was to gain widespread use as biochemistry became established in the 1900s, was introduced by Wilhelm Friedrich Kühne, in 1876. Kühne, who held the chair of Physiology at Heidelberg, had mastered the new technique of pancreatic fistula introduced by Claude Bernard in Paris. He could thus isolate both 'organized and non-organized' ferments which could digest proteins. On February 4, 1876, Kühne presented a paper 'On the behavior of several organized and as well unformed ferments' to the Heidelberger Naturhistorischen und Medizinischen Verein. This article deals with what we now know as proteolytic enzymes. It opens thus:

"Mr. W. Kühne reports upon the behavior of several organized and as well unformed ferments. To dispel misunderstanding and to avoid obnoxious periphrases, we propose to bring forward, for the unformed and non organized ferments whose working one can study without and outside the presence of organisms, the designation of *Enzymes*.
More precisely, to take the example of the egg white-digesting pancreatic enzyme, which also produces splitting of the albuminous bodies, it is given the name of *Trypsin*" (Kühne, 1876).

The important idea here is that extracellular agents such as trypsin or pepsin can be active outside the presence of an organism and thus differ from the 'organized' substances present in the protoplasm of the cell. As writes Fruton (1982) 'it is ironic that a few years ago (Gutfreund, 1976) Kühne was hailed for the introduction of a word that assumed great importance after 1930 to denote intracellular agents he explicitly excludes from his definition'.

Historians of biochemistry have also diverged about the significance of the gradual replacement of the name 'ferment' by 'enzyme'. At the conference on 'The Historical Development of Bioenergetics', sponsored by the American Academy of Arts and Sciences in Boston in 1973, Robert E. Kohler argued that

"originally the two terms meant something quite distinct, their meaning shifted a great deal and certainly during the period from 1900 to 1915 were not denoting the same thing. This view was opposed by such distinguished biochemists as Nobel Prize winners H.A. Krebs and F. Lipmann who maintained that these terms were essentially synonymous after 1900. Whereas the English workers employed the term 'enzyme', the German biochemists continued to use the name 'ferment' " (Teich, 1981).

The evidence pro and con has been studied carefully and it now appears that Kohler had a point:

" 'Ferment' " writes Teich (1981), "was a narrow and specific term because it historically developed out of and was applied to the 'specific' process of alcoholic fermentation or processes allied to it. (. . .) 'Enzyme' is a broad and non-specific term in the
sense that it concerns not only fermentation but all catalytic chemical changes taking
place in biological systems".

Kohler himself writes – and I tend to agree with him:

"The elevation of the enzyme concept from the relatively minor role it played in the
older physiology to a central theoretical position in the new biochemistry was both a
profound conceptual innovation and a crucial factor in the emergence of a new discipline with its own sense of historical identity" (Kohler, 1975).

I find an indisputable ancestor to the word 'enzyme', perhaps the
earliest recorded so far, in the *Encyclopédie* of Diderot and d'Alembert. Already in 1753, in the entry for chemistry written by Gabriel-
François Venel (1723–1775), this chemist refers to *zimotechnie* as
'the art of conducing certain vegetable substances to fermentation,
including the wine-making art'.

After its coinage by Kühne, the word 'enzyme' was quickly
adopted outside Germany:

"I would suggest the desirability of adopting this term (German *enzym*) into English,
with a slight change of orthography, as 'enzymes', and also of coining from this root
the cognate words which are requisite for clear and concise description. The action of
an enzyme may be designated *enzymosis* and the nature of the action may be spoken
of as *enzymic*" (Roberts, 1881).

And, to give testimony to the generalized use by 1930 or so, here is a
short quotation from a paper by two British biologists and biochemists of the first rank:

"Each digestive enzyme is a definite substance with the property of bringing about, or
enormously speeding up, a particular chemical reaction" (Haldane and Huxley,
1927).

After chronicling the appearance of the term 'enzyme', we shall now
summarize the history of the earlier notion, that of ferments conceived as dissolution operators.

REFERENCES

Baxter, J. (1846) Libr. Pract. Agric. (ed. 4), I. 19.

Diderot, D. et al. (1751–1780) Encyclopédie ou Dictionnaire Raisonné des Sciences, des Arts et des Métiers (35 vols., Neuchâtel, etc.; my translation): entry CHIMIE, vol. 3, p. 420; entry FERMENT, vol. 14, p. 18; entry ESTOMAC, vol. 13, p. 103; entry ALKAHEST, vol. 2, p. 120; entry FERMENTATION, vol. 14, p. 32.

Fruton, J.S. (1982) J. Clin. Chem. Clin. Biochem., 20, 243.

Gutfreund, H. (1976) FEBS Lett., 62 (suppl.), E.1.

Haldane, J.B.S. and Huxley, J.S. (1927) Anim. Biol., 4, 106.

Kohler, R.E. (1975) J. Hist. Biol., 8, 275.

Kühne, W.F. (1876) Heidelberger Naturhist. Med. Ver., 2 (translation Sophie and Pierre Laszlo).

The Lancet (1894) 3 Nov., 1045.

Payen, A. and Persoz, J.F. (1833) Ann. Chim., 53, 73.

Roberts, W. (1881) Proc. Roy. Soc., 32, 146.

Teich, M. (1981) Hist. Phil. Life Sci., 3, 193.

Watts, H. (1863–1872) Dict. Chem., II. 319.

Chapter 28

Dissolution Operators

In earlier times, many physicians had investigated the anatomy of the stomach in relation to its digestive function. A key experiment was performed by the French physical scientist René Antoine Ferchault de Réaumur (1683-1757) in 1752. He made his pet falcon swallow small perforated boxes containing pieces of meat. The regurgitated meat was retrieved half-digested. From hindsight, the experiment is clever, nicely designed and quite demonstrative. However, contemporary scientists diverged as to its interpretation. For instance, the *Encyclopédie*, in its entry on 'digestion', sees this experiment merely as ruling out the mechanical grinding and crushing of food by the stomach as a pre-requisite to digestion. Elsewhere, it also blinds itself as to the full significance of this key experiment. The *Encyclopédie* entry on 'the stomach' (signed by Baron de Haller) claims that gastric juices are produced in the arteries leading to that organ. It goes to some trouble also to deny an acidic character to gastric juices, which are assimilated to a slightly alkaline mucus, and writes:

"Experiments performed on the falcon and on various animals by other authors more or less concur with similar results" (Diderot et al., 1751-1780),

whereas Réaumur had found that digestion is done by a sour and salty gastric juice, of acidic character.

This hesitant, vague and contradictory stand of the *Encyclopédistes* reflects the absence of a contemporary consensus about fermentations and digestions. The context, to summarize it briefly, was the more or less discredited doctrine of Van Helmont, and the efforts of contemporary chemists such as H.M. Rouelle (1716-1778)

Gabriel-François Venel (1723-1775) to conceptualize solubility, fermentation and digestion as a generic class of phenomena.

I cannot resist a short quotation from the 'fermentation' entry of the *Encyclopédie*, not only because of its polemical nature but because it inveighs against an intuition which we now know from hindsight to have been singularly percipient:

"In this manner, the chemical sect, after having made to depend on *fermentation*, or on some analogous physical power, the main changes undergone inside the primitive humors, went further: they fancied they could transfer in all the organs in which are prepared these [humors] which derive from these [primitive humors], the ferments from the laboratory and make them operate all secretions in their variety; they imagined that in each duct there are distinct leavens which change the fluids which abound (in these sites) by commingling with them, and by the attendant effects, i.e. always by a *fermentation* or an effervescence: but nothing proves this notion (...)" (Diderot et al., 1751-1780).

The 'proof' - that is to say the experiment which historians of science nowadays hail as the first firm indication that the chemical processes of digestion could be made to occur outside of the stomach - was provided 30 years after the Réaumur experiment. From hindsight, this appears as the next logical step. And this advance was performed by the Italian Lazzaro Spallanzani (1729-1799), who showed that 'meat would digest outside the animal if it had been soaked in gastric juice' (Northrop, 1961). His experiment is a follow-up on Réaumur's: he collected gastric juice from birds. He mixed it together with meat or grain inside sealed metallic tubes. He incubated these 10 days at body temperature (by the simple trick of wearing the little containers under his armpits). After this lapse of time, the avian gastric juices had made the food liquefy. This was the first in vitro digestion.

Johann Nepomuk Eberle (1798-1834) performed an extraction of gastric mucosa with dilute hydrochloric acid. The resulting solution also had digestive properties in vitro (Eberle, 1834). As writes Teich (1981),

"this provided a more convenient method to study the chemical changes incident to digestion than the one employed by William Beaumont (1785-1853). This American army surgeon made use of a gastric fistula which a trapper, Alexis St. Martin, acquired after being accidentally wounded by a gunshot. (. . .) Beaumont's collection of gastric juice from a living man and its utilization in experimental studies of digestion removed, as it were, the last vestiges of ephemerality from gastric juice. (. . .) Beaumont made the suggestion, among others, that gastric juice contained in addition to free muriatic acid (hydrochloric acid) 'some other active *chemical* principles' " (Beaumont, 1838; quoted by Teich, 1981).

The early (18th and 19th century) view of ferments did not limit itself to their property as dissolution operators. The question of their origin was also deemed important.

REFERENCES

Beaumont, W. (1838) Experiments and Observations on the Gastric Juice and the Physiology of Digestion, reprinted from the Plattsburgh edition, Maclachlan and Stewart, Edinburgh; Simphin, Marshall, London, p. 299.

Diderot, D. et al. (1751-1780) Encyclopédie ou Dictionnaire Raisonné des Sciences, des Arts et des Métiers (35 vols., Neuchâtel, etc.; my translation): entry CHIMIE, vol. 3, p. 420; entry FERMENT, vol. 14, p. 18; entry ESTOMAC, vol. 13, p. 103; entry ALKAHEST, vol. 2, p. 120; entry FERMENTATION, vol. 14, p. 32.

Eberle, J.N. (1834) Physiologie der Verdauung nach Versuchen auf natürlichem und künstlichem Wege, Etlinger, Würzburg, p. 78.

Northrop, J.H. (1961) Annu. Rev. Biochem., 30, 1.

Teich, M. (1981) Hist. Phil. Life Sci., 3, 193.

Chapter 29

Animal or Vegetable Origin of Ferments

During much of the late 18th and early 19th century, scientists interested in fermentation wanted to know if the ferment belonged to the animal or vegetable kingdom. Their concern expressed the dominant *episteme* (in the sense of Michel Foucault): classification. They divided fermentations into three classes: vinous, acetic, and putrefaction. In the same categorizing spirit, they wanted to identify the fermentation operator as to its class of origin. Lavoisier's work provided them with a set of criteria.

As noted by Henry Guerlac, Antoine-Laurent Lavoisier (1743–1794) did not speculate on the nature of the ferment,

"the underlying cause of a fermentation, but confined himself to the chemical changes. Vinous or alcoholic fermentation, marked by a violent intestine motion, he saw to be a chemical reaction involving rearrangement of the three essential 'organic' principles: oxygen, hydrogen, and carbon. He assumed an equality or 'equation' between the amount of these elements in the original sugar and the amount in the end products: alcohol, carbon dioxide, and acetic acid. His experiments seemed to bear out his assumption. His data, however, were wholly unreliable, although the end result was correct" (Guerlac, 1975).

Just as a short digression, there are two fascinating footnotes to Lavoisier's work on fermentation: the first use of the word 'equation' in the sense of a chemical equation (Partington, 1948); this is where he wrote his 'nothing is created' paragraph on the conservation of mass.

Lavoisier's work on fermentations was highly influential. It gave the field an initial impetus and a methodology to follow. It also proved to be stultifying with respect to isolation and characterization of ferments.

[411]

Observations and experiments by three French scientists in sequence, in Lavoisier's wake, Antoine François de Fourcroy (1755–1809), Joseph Louis Gay-Lussac (1778–1850), and Nicolas Appert (1750–1841) established that vinous and acetic fermentations and putrefaction require oxygen (Fourcroy, 1788; Gay-Lussac, 1810; Appert, 1810). Hence, the invention of preserves by Appert: food could be stored without fermentation or putrefaction if bottled in the absence of air.

Conversely, Lavoisier had clearly shown incorporation of oxygen during acetic fermentation: it was 'nothing more than the acidification or oxygenation of the alcoholic liquor produced during fermentation by means of the absorption of oxygen' (Teich, 1981).

As for putrefaction, Lavoisier had proposed elemental composition of a substance as an index to its origin, from animals or from vegetables. There were two criteria. Animal substances contained nitrogen (*azote*) while vegetable substances did not, or only to a much lesser degree. Furthermore (second criterion), compounds of animal origin contained also, besides nitrogen, phosphorus and sulfur. Hence, the outcome of putrefaction was, for a vegetable substance, evolution of hydrogen gas and carbonic acid gas, whereas the decay of animal matter led to ammonia, phosphines or hydrogen sulfide (to use the modern nomenclature).

These criteria of Lavoisier helped the following investigators to establish whether fermentation operators belonged to the animal or to the vegetable sphere. For instance, Louis Jacques Thénard (1777–1857) studied alcoholic fermentation by yeast, and became convinced that the ferment was of animal origin (Thénard, 1803): 'its carbon became to a certain extent a component part of carbonic acid and its nitrogen was supposed to have been absorbed by the alcohol' (Teich, 1981). 15 years later, a German pharmacist and brewer settled in Bohemia, Christian Polykarp Friedrich Erxleben (1765–1831), expressed the opposite view: the ferment was of vegetable origin:

"the fermentation (. . .) appears in no way to be a mere chemical *operation* but much rather in part *a plant process* and must be considered as the great chain in nature which brings about union of the activities which we call chemical processes with those of vegetation" (Erxleben, 1818).

In the same Lavoisier tradition, both with respect to formulation of the scientific problems raised by fermentations and to the solutions offered, one finds also a concept of the ferments as transcending the distinction between animal and vegetable life.

"Thus in 1799 the Italian G.IV.M. Fabbroni (1752-1822), likening alcoholic fermentation to the decomposition of carbonates by acids with evolution of carbonic acids, thought it was induced by a particular kind of matter in plant 'utricles' (plant cells?) called 'vegeto-animal substance' (Fourcroy, 1799) (. . .) this term had earlier been applied by H.M. Rouelle to gluten, the substance obtained by washing out the starch from wheat flour (Rouelle, 1773)" (Teich, 1981).

The latter discovery by Payen and Persoz in 1833 of the 'diastase', capable when applied to starch to transform it into sugar, is in line with this identification by Fabbroni of fermentation with the action of a gluten on sugar.

REFERENCES

Appert, N. (1810) L'Art de conserver, pendant plusieurs années, toutes les substances animales et végétales, Paris.

Erxleben, C.P.F. (1818) Ueber Guete und Staerke des Biers, und die Mittel diese Eigenschaften richtig zu wuerdingen, Haase, Prague, p. 69, quoted and translated by Teich (1981).

Fourcroy, A.F. de (1788) Elements of Natural History and of Chemistry, 2nd ed., Robinson and Robinson, London, p. 137 (transl. W. Nicholson).

Fourcroy, A.F. de (1799) Ann. Chim., 21, 299.

Gay-Lussac, J.L. (1810) Ann. Chim., 76, 245.

Guerlac, H. (1975) Antoine-Laurent Lavoisier: Chemist and Revolutionary, Scribner, New York, pp. 118–119.

Partington, J.R. (1948) A Short History of Chemistry, MacMillan, London, 2nd ed.

Rouelle, H.M. (1773) J. Med. Chir. Pharm., 39, 250; 40, 59.

Teich, M. (1981) Hist. Phil. Life Sci., 3, 193.

Thénard, L.J. (1813) Ann. Chim., 85, 61.

Chapter 30

A Disquisition on an Unfavorable Zeitgeist

From the above, it would seem that at the turn of the 19th century the first isolation of a ferment was imminent. The studies of Lavoisier had pointed to the importance of elemental composition for establishing the animal or vegetable source of a ferment. Such a research program should have led quickly to isolation and characterization of numerous ferments. Yet, only in 1833 did Payen and Persoz discover diastase, and the collective discovery of pepsin (see below) was made in 1837. Such a delay occurred, despite numerous scientists of the first rank – such as Lavoisier, Fourcroy, Thénard, Gay-Lussac in France, Spallanzani in Italy, Prout in England, Tiedemann and Gmelin in Germany – getting interested in the chemistry of fermentation and digestion. However, the emphasis was placed on the changes in elemental composition taking place during these transformations and not on *how* these changes came to be effected. The reason was Lavoisier's durable and overwhelming influence.

For instance, the *Conversations on Chemistry* (a minor classic in the popularization of science, whose originality lies also in that it had been written by a woman) devotes indeed considerable attention to fermentations. The relevant section of the fourth edition, that of 1813, starts by explaining how the decomposition of vegetables is a step-wise process, governed by the affinities so that in the long run chemicals are brought 'to their most simple order of combinations', namely water and carbonic acid. After this summary of the Lavoisier work on putrefaction the author gives a definition of fermentations very close to the etymology of the word, with which we started (Ch. 27):

"The decomposition of vegetables is always attended by a violent internal motion,

[415]

produced by the disunion of one order of particles, and the combination of another. This is called FERMENTATION" (*Conversations*, 1813).

Heat and water are, we are told, pre-requisites for fermentation. The author then turns to alcoholic fermentation, going into sufficient detail such as to distinguish between its various modes, beer generated from malt, wine from grape; distilled spirits (brandy, rum) are mentioned. A key exchange follows:

"*Caroline*: I am astonished to hear that so powerful a liquid as spirit of wine should be obtained from so mild a substance as sugar.

Mrs. B.: Can you tell me in what the principal difference consists between alcohol and sugar?

Caroline: Let me reflect ... Sugar consists of carbon, hydrogen, and oxygen. If carbonic acid be subtracted from it, during the formation of alcohol, the latter will contain less carbon and oxygen than sugar does; therefore hydrogen must be the prevailing principle of alcohol.

Mrs. B.: It is exactly so. And this very large proportion of hydrogen accounts for the lightness and combustible property of alcohol, and of spirits in general, all of which consist of alcohol variously modified.

Emily: And can sugar be recomposed from the combination of alcohol and carbonic acid?

Mrs. B.: Chemists have never been able to succeed in effecting this; but from analogy, I should suppose such a recomposition possible" (*Conversations*, 1813).

The notions in the background of this conversation are quickly identified: identification by Lavoisier of carbon dioxide and ethanol as products of sugar fermentation, and his analogy between fermentation and a slow combustion; the demonstration by Gay-Lussac of mass conservation, with the weight of the sugar almost exactly the same as that of its two decomposition products, ethanol and carbon dioxide; and perhaps also the concept of a reversible chemical equilibrium. But the only operators invoked are heat and humidity. Ferments, per se, are conspicuously absent from fermentations in this didactic work clearly influenced by Lavoisier.

Ironically, after this long parenthesis opened by Lavoisier, ferments when they re-enter the scene are ushered in by the philosophy of Romanticism.

REFERENCE

Anonymous [Jane Marcet, 1769 – 1858] (1813) Conversations on Chemistry in which the Elements of that Science are Familiarly Explained and Illustrated by Experiments, 2 vols., 4th ed., Longman, Hurst, Rees, Orme, Brown, London; vol. 2., Ch. 21, On the Decomposition of Vegetables.

Chapter 31

Relation of Microorganisms to Fermentation and Putrefaction

Louis Pasteur (1822-1895) is generally credited with establishing the microbiological origin of fermentations. Yet, he was antedated in this respect by another Frenchman, Cagniard-Latour, and by two Germans, Schwann and Kützing. Furthermore, these three scientists reported their independent discoveries more or less simultaneously. This chapter will consider the background (philosophical, ideological, technological) to this multiple discovery, and the response by leading contemporary scientists, Berzelius, Liebig, and Wöhler.

Cagniard-Latour, Schwann, and Kützing worked in the long shadow of the Dutch natural scientist Van Leeuwenhoek who, as early as 1680, had used the microscope to examine yeast and establish the globular shape of its particles.

Charles Cagniard-Latour (1777-1859) had been interested in the alcoholic fermentation for over 25 years when he made his discovery. During all these years, he had sought to study the yeast particles under the microscope. Suddenly, in the 1830s, improved microscopy made it possible for him to see the yeast ferment at 300-400 magnification. He could see individual yeast cells (as we now know them) as globules with shape rather than, as earlier, as mere particles. He saw the budding of yeast cells, which told him this was living matter. 'From the absence of locomotion he deduced that the yeast organism was a living plant rather than an animal, capable of breaking up sugar into alcohol and carbonic acid' (Teich, 1981). As early as November 23, 1836, he described multiplication of yeast in the newspaper *L'Institut* (Florkin, 1968). But the main report of his observations, presented to the French Academy of Sciences on June 12, 1837 was published the following year (Cagniard-Latour, 1838).

Theodor Schwann (1810-1881) was much younger than Cagniard-Latour. He was at the start of his career, and was working in Johannes Müller's laboratory in Berlin. Like Cagniard-Latour's, Schwann's work was also based on the progress of microscopy, which renewed interest in observation of infusoria and in the related question of spontaneous generation. I have earlier mentioned how the experiments of Appert and those of Gay-Lussac had identified the presence of oxygen as necessary for fermentation and putrefaction.

"Schwann reported first on this experiment concerning spontaneous generation to the (Annual) Meeting of German Naturalists and Physicians, held at Jena in September 1836. No infusoria appeared in a vessel containing infusion of organic material if it was placed in boiling water" (Teich, 1981).

Schwann then turned to the role of air in the vessel: if it was heated, fermentation and putrefaction did not occur. This experiment showed Schwann that

"air normally contained the germs of mould and infusoria which on heating had been destroyed. (. . .) what was called putrefaction was brought about by these germs feeding on the organic substance and in the process decomposing it" (Teich, 1981).

Schwann moved on to alcoholic fermentation:

"to his great surprise, he observes that heating of the air which he makes pass through a boiled suspension of yeast in a sugar solution inhibits fermentation. As early as January 1836, he notes in his laboratory notebook his conclusion that alcoholic fermentation is performed by a living organism" (Florkin, 1968).

Schwann later demonstrated that yeast was a fungus-like organism, and that carbonic acid of the alcoholic fermentation was liberated around the yeast cells, feeding on sugar and nitrogen-containing sources (Schwann, 1837).

Friedrich Traugott Kützing (1807–1893) was an algologist, of the same generation as Theodor Schwann. He 'did not doubt that alcoholic and acetic fermentation was a chemical activity of low plant life' (Teich, 1981). Kützing's contribution was to extend Cagniard-Latour's and Schwann's observations and conclusions, which he duplicated on yeast, to 'mother of vinegar', responsible for acetic fermentation (Kützing, 1837).

These three contributions (from Cagniard-Latour, Schwann, and Kützing) occurred in the span of just a few months and qualify as a multiple discovery. Historians of science diverge as to the generality and significance of such multiple discoveries. William F. Ogburn and Dorothy S. Thomas, who published a list of 148 multiple, simultaneous discoveries in various fields, ascribed these to cultural maturation reaching a given stage in a particular society (Ogburn and Thomas, 1922). In essence, this amounts to positive restatement of the Hegelian *Zeitgeist*. Merton went on to the provocative notion of *all* discoveries being multiples, which he explained by the reward system endemic to a scientific community. Questions of priority are all-important: scientists vie for priority in discovery in order to gain prestige and to draw various advantages from the attendant institutional recognition (Merton, 1973). At the other extreme, there are scientific historians who maintain that discoveries are unique and singular, often the work of a lone genius (Mendel, for instance). They discount a substantial number of multiple discoveries as artifacts, generated by failure of communication among the scientists involved and also by our a posteriori compression of longish periods of several months, sometimes a year or two, into a semblance of simultaneity (Brannigan, 1981).

In the present context - the history of the gradual emergence of enzymes as molecular objects - it is neither desirable nor easy to adjudicate between these rival claims. An explanation by the advent of a new, powerful technology (a significantly improved microscopy) can be put forward: if this was a necessary condition for the discovery under discussion, was it sufficient to make it happen?

I do not think so. Indeed, even in the context of a 'social maturation' hypothesis, technical progress cannot be considered in isolation. I believe the philosophical and ideological background to have been extremely important to the issue at hand.

However, it is extremely difficult to gauge its exact significance or tenor, for each of the three co-discoverers. Take the example of Theodor Schwann, probably the best documented. Some historians of science, such as Georges Canguilhem, stress the influence on his thinking of *Naturphilosophie*. Lorenz Oken belonged to the Romantic school of *Naturphilosopie* founded by Schelling. Both

Schleiden and Johannes Müller were steeped, during their formative years, in this natural philosophy. And Theodor Schwann, in turn, was markedly influenced at the start of his scientific career both by Schleiden and by Johannes Müller, the latter was his mentor and supervisor (Canguilhem, 1965). Besides Canguilhem, Debru, another French scientific historian, writes thus:

"This chemical degradation (anaerobic fermentation) appears as tightly coupled to the energetics of the microscopic organism, subordinated to its needs. It does not place itself any longer on the side of an organic chemistry of living matter in its dying and decomposing mode. It belongs, to the contrary, to a living chemistry, in the full sense of the word. This inversion is crucial. This new perspective in the study of fermentation has been glimpsed before Pasteur by Cagniard-Latour, by Schwann and Kützing (. . .)" (Debru, 1983, p. 38).

Canguilhem's and Debru's are descriptions of Schwann as deeply influenced by *Naturphilosophie* at the time he discovered the role of yeast in alcoholic fermentation.

Everett Mendelsohn makes the opposite argument: in reaction to Johannes Müller's vitalism, Theodor Schwann sought quite consciously to banish vital forces from biology, and to explore, and experiment with, physicochemical explanations for biological phenomena. Mendelsohn agrees with Canguilhem on Müller being a vitalist in the early 1830s when Schwann, who had already been his student in Bonn, joined him in Berlin, first to complete his medical degree and then as an assistant. He refers to a

"letter which Schwann wrote to his brother Peter, a student of philosophy and theology, in 1835, which shows Schwann still searching and examining the status of vital forces" (Mendelsohn, 1965).

Mendelsohn thus softens the stand, later taken by Schwann and credited him by many of his contemporaries, of his having been the main proponent of reductionism in biology from when he joined Müller's laboratory. The balanced conclusion of Mendelsohn is that

"fairly early in his scientific career (just how early is not yet certain) Schwann began asking, and finding answers to a series of questions which were designed to explore –

to expose – the concept of a vital force. The commitment he developed was aimed at banishing vital forces from biology" (Mendelsohn, 1965).

It is somewhat ironical, in view of Schwann's post hoc stance as an apostle of reductionism, that this single discovery of his, in 1836–1837, reinjected biological organisms at the core of the phenomenon of fermentations. It is doubly ironical, since he was led into it by a research program aimed at ruling out spontaneous generation among organic matter undergoing putrefaction.

To return now to the question of the determinants of the multiple discovery, I believe that a good case can be made for implicating, besides the progress of microscopy (science as observation), the ideological issue of spontaneous generation: just as it later made Pasteur establish the role of microorganisms, it led Theodor Schwann to discover the microbiological origin of fermentation. This sort of stuttering of history indicates to me that spontaneous generation and, more generally, mechanism versus vitalism were such burning issues in the early 1830s – witness the discussions in Balzac's *Magic Skin*, to choose a remote and thus convincing example – that in 1836–1837 the intellectual climate was ripe for the multiple discovery to be made.

And indeed, as soon as it had been made, it drew heated criticism. It re-installed organisms at the center of a phenomenon (fermentation) which had been held for a generation or two to be reducible to chemical changes. Pure and applied chemists were appalled and very upset. One of them was none other than Jöns Jacob Berzelius (1779–1848):

"He admitted that Kützing's work could have contributed to the understanding of low plants but completely rejected his attempt to relate the inanimate and animate. He saw in it the detrimental expression of philosophical influence on science thought to have been overcome a long time ago (*Naturphilosophie*)" (Teich, 1981).

Berzelius jeered at the new concept of yeast, which he saw as a scientific-poetic fiction:

"If the coalescence of the globules of yeast can be ascribed to the presence of vegetable life, the same reason might well be assumed for the coalescence of globules of clay or calcium phosphate" (Berzelius, quoted by Buchner, 1907).

Wöhler and Liebig were incensed. They went to the extent of publishing, anonymously, a spoof upon Schwann's discovery in which they claimed to have observed with a microscope infusoria swallowing sugar from the solution and evacuating alcohol at the other end of the digestive tube as excrement, while carbonic anhydride was coming out of a Champagne bottle-shaped bladder:

"In a word, these infusoria gobble sugar, and discharge ethyl alcohol from the intestine and carbon dioxide, from the urinary organs".

Liebig published in 1839 a more circumspect and carefully argued retort (Liebig, 1839). His theory rested on a series of experiments with nutrition of animals; his forerunners had been Tiedemann and Gmelin, in 1826. These research workers had used chemical analysis to establish the balance sheet of nutrients for various animals (Holmes, 1979). Their work had belonged squarely to the tradition of the Lavoisier study of fermentations. Justus Liebig also built upon the prior research by G.J. Mulder who 'had established by combustion analysis that albumin, fibrin, and casein have the same elementary composition (. . .) $C_{40}H_{62}N_{10}O_{12}$, which he named the "protein" radical' (Holmes, 1979).

For Liebig, animal heat was generated by a slow combustion. Respiratory oxygen served to burn in the blood the non-nitrogenous components, carbohydrates and fats, of the diet. Because of the Mulder analyses, Liebig had become convinced that the food proteins were converted to blood and tissue proteins with little chemical change. He ascribed a central role to degradation of the tissue proteins, especially those of muscle. Liebig called this process a *Stoffwechsel*, i.e. a tissue change or metamorphosis (Holmes, 1979). The term is revealing: Liebig is working in the conceptual framework of an animal *economy*, in which intakes are matched with excreta, in consonance with the paradigm started by Lavoisier.

In Liebig's view, these vital muscle proteins were attacked by respiratory oxygen and converted into the degradation products uric acid, urea and ammonia, which were then excreted in the urine. Liebig also adopted the distinction, dating back to Lavoisier's time, between putrefaction (fetid smell) and fermentation of vegetable substances (smell-less).

As to the nature of the ferment, of the substance responsible for fermentation, Liebig thought it formed by concentrating 'all nitrogen from the nitrogenous (albuminous) plant substances' (Teich, 1981). The mode of action of this ferment was a transfer of excitation: the ferment was in an inherently unstable, agitated state, and, by contact with sugar molecules would cause them to agitate wildly and split off ethanol and carbon dioxide. We find here again an intrusion by language into the conceptual sphere: Liebig must have been influenced unconsciously by the etymology of the word 'ferment' (see Ch. 27) in devising this model of fermentation.

"It should be added that neither Friedrich Wöhler, Liebig's closest scientific companion, nor Berzelius received favorably what had come from Giessen on the subject of fermentation. In view of the fact that he rejected Berzelius's concept of catalysis, Liebig was criticized for his failure to explain the way the substance undergoing putrefaction actually acts as a ferment" (Teich, 1981).

As for the *coup de grâce*,

"the decisive blow to Liebig's idea that fermentation is caused by the oxidation of albuminoid substances was delivered by Pasteur when he showed that yeast grows in the absence of such substances and that it ferments best in the absence of oxygen" (Fruton, 1976).

Any modern textbook defines enzymes as biological catalysts: how and when was this idea born?

426 REFERENCES

REFERENCES

Brannigan, A. (1981) The Social Basis of Scientific Discoveries, Cambridge University Press, Cambridge.

Buchner, E. (1907) Nobel Lecture, December 11; Nobel Lectures - Chemistry, 1901-1921, Elsevier, Amsterdam, pp. 97-122.

Cagniard-Latour, C. (1838) Ann. Chem., 68, 206.

Canguilhem, G. (1965) La Connaissance de la Vie, Vrin, Paris, pp. 58-63.

Debru, Cl. (1983) L'esprit des Protéines. Histoire et Philosophie Biochimiques, Hermann, Paris.

Florkin, M. (1968) Théodore Schwann, in Florilège des Sciences en Belgique, Académie Royale de Belgique, Classe des Sciences, Brussels, p. 929.

Fruton, J.S. (1976) Science, 192, 327.

Holmes, F.L. (1979) Ann. N.Y. Acad. Sci., 325, 171.

Kützing, F. (1837) J. Prakt. Chem. (Erdmann), 2, 385.

Liebig, J. von (1839) Ann. Pharm., 30, 250.

Mendelsohn, E. (1965) Br. J. Hist. Sci., 2, pp. 208-214.

Merton, R.K. (1973) The Sociology of Science, Theoretical and Empirical Investigations, University of Chicago Press, Chicago.

Ogburn, W.F. and Thomas, D.S. (1922) Pol. Sci. Quart., 37, 83.

Schwann, Th. (1837) Ann. Phys. Chem. (Poggendorff), 11, 184.

Teich, M. (1981) Hist. Phil. Life Sci., 3, 193.

Chapter 32

Ferments as Catalysts

By a quirk of fate, Berzelius devised the concept of catalysis, invented the name, and included ferments among catalysts almost exactly at the same time when Cagniard-Latour, Schwann and Kützing showed ferments to be produced by micro-organisms. The years 1835–1837 saw both the designation of enzymes as catalysts and proof for the biological origin of enzymes.

A convenient and logical starting point of the chain of observations and experiments which were to culminate in Berzelius's definition of catalysis in 1835 and 1836 is the work of Kirchoff in the 1810s. Kirchoff had decomposed starch with sulfuric acid into dextrin and saccharose in 1811. This was

"one of the early examples of a chemical operation apparently in contradiction to the known laws of chemical combination. The significant feature was that for the change to succeed the acid had to be present though it was not used up in the reaction" (Teich, 1981).

Later, in 1814, Kirchoff was able to obtain the same hydrolysis products from starch in the presence of a malt extract.

Thénard, in 1818, performed the decomposition of hydrogen peroxide in the presence of various substances, such as platinum metal or manganese dioxide. Thénard had earlier (1813) discovered that ammonia decomposed upon contact with heated iron, copper, silver, gold or platinum, in order of decreasing reactivity (Thénard, 1813). Humphry Davy at the end of his work on flame in 1817 made a similar observation:

"A preheated length of fine platinum wire – silver and gold are devoid of this property – could be made to glow in mixtures of air with coal gas, carbon monoxide, alcohol

[427]

vapor or similar substances, and (. . .) the glowing continued until all the inflammable matter was exhausted" (Collins, 1976).

Humphry Davy believed this to be a special case of combustion and did not pursue the matter further (Collins, 1975). But his cousin Edmund Davy, who had worked for him as an assistant at the Royal Institution before his appointment as professor of chemistry at the Cork Institution, picked up this research. Edmund Davy

"found that, on boiling with alcohol or ether, a strong aqueous solution of platinum sulfate yielded a slightly contaminated sample of finely divided platinum nitrite. When this powder was exposed to alcohol vapor *at room temperature*, sufficient heat was generated 'to reduce and ignite the metal, and to continue it in a state of ignition, until the alcohol is consumed' (Davy, 1820)" (Collins, 1976).

In the early 1820s, the German chemist Johann Wolfgang Döbereiner (1780-1849), who had been interested for several years in platinum chemistry, read the 1820 paper by Edmund Davy, repeated his experiments, ascertained that Davy's powder was platinum suboxide and made the following crucial observation:

"The platinum suboxide, moreover, does not undergo any change during this transformation of the alcohol (into acetic acid and water, if enough oxygen is present) and can immediately be used again to acidify fresh, perhaps limitless, quantities of alcohol (. . .)" (Döbereiner, 1822).

Döbereiner 'spent the remaining months of 1821 repeating his experiments, concentrating particularly on the oxidation products of alcohol and *trying to draw a rather dubious parallel with the fermentation of sugar*' (Collins, 1976). During the Christmas holiday of 1822, he showed his friend Goethe, who was then a Privy Councillor and Minister of State to Duke Carl-August in Weimar, these experiments. Döbereiner continued his work, evolved in 1823 a procedure for making finely divided platinum metal. On July 27, 1823, Döbereiner

"prepared some of this powder on filter paper and exposed it to hydrogen gas. (. . .) So strong was the power of this reaction that 'by mere contact with this platinum powder, the inflammability of the hydrogen is raised to such an extent that it extracts all the oxygen from a mixture of 99 volumes of nitrogen with one volume of oxygen in a few minutes' (Döbereiner, 1823)" (Collins, 1976).

Two days afterwards, Döbereiner wrote to Goethe, bursting with excitement about this new experiment. Thus, what we now know as heterogeneous catalysis had been discovered. The next strand in the story came a decade later when Mitscherlich, a student of Berzelius, found in 1834 that ethanol dehydrates into diethyl ether in the presence of 50% aqueous sulfuric acid, which is left intact at the end of the reaction.

The genius of Berzelius was, when he compiled his annual report on scientific advances for the Swedish Academy of Sciences in 1835, to recognize a common concept in all these phenomena – for which he devised the term catalysis (Berzelius, 1836a). As Berzelius wrote in another context:

"Many bodies (. . .) have the property of exerting on other bodies an action which is very different from chemical affinity. By means of this action they produce decomposition in bodies, and form new bodies into the composition of which they do not enter. This new power, hitherto unknown, is common both in organic and inorganic nature. I shall (. . .) call it catalytic power. I shall also call catalysis the decomposition of bodies by this force" (Berzelius, 1836b).

As writes Teich (1981),

"it says much for Berzelius's imagination that while reflecting on the catalytic participation of diastase in the conversion of starch in living plants he visualized catalytic reactions playing an essential part in the chemistry of living systems".

This sentence echoes the earlier assessment by Eduard Buchner:

"Berzelius assumed that yeasts caused the decomposition of sugar catalytically, simply by its presence as a contact substance or catalyst. There seemed to be analogies with many processes – for instance, with the action of very finely divided platinum on hydrogen peroxide which, in the presence of that contact substance, rapidly decomposes into water and oxygen, whilst the platinum apparently remains unchanged" (Buchner, 1907).

And, since I have been quoting from Buchner's Nobel Lecture, here is the tribute to Berzelius from another Nobel laureate on the same occasion:

"Berzelius, a century ago, pointed out that these enzymes were similar to the catalysts of the chemist and suggested that they be considered as special catalysts formed by the cells. This hypothesis was far ahead of its time, and met with great opposition (. . .)" (Northrop, 1946).

Nevertheless, Berzelius had not created in a vacuum. The prehistory of the concept of catalysis includes the Chinese concept of *Tsoo Mei* (a marriage broker). Catalysis draws also upon the Arabic notion of the elixir: Al Alfani, in a 14th century Arabian manuscript, writes of the 'Xerion, aliksir, noble stone, magisterium, that heals the sick and turns base metals into gold, without in itself undergoing the least change' (Moore, 1962). And there is the *alkahest*, the universal dissolvent of the alchemists: not only could it dissolve anything, but it could be recovered intact afterwards. Van Helmont provided the most comprehensive description of its purported properties. Here is how the *Encyclopédie* refers to one of its properties, in the entry on *alkahest*:

"The *alkahest* suffers no change neither loss of power when dissolving the bodies upon which it acts, which is why it does not undergo any reaction on their part, being the only solvent (*menstrue*) unalterable in nature" (Diderot et al., 1751–1780).

And one might make an excellent case for Goethe (who, as alluded to in the paragraphs on the discovery of heterogeneous catalysis by Döbereiner, was both a witness and a participant in this discovery) being the inventor of catalysis: in his novel *Elective Affinities* (1809), the arrival on the scene of a character by the name of Mittler ('the intermediate') triggers all sorts of catastrophes . . .

Due to the prestige of Berzelius, the concept of catalysis became quickly adopted. In order to set the stage for the Buchner experiment to be described in the next chapter, I shall use the entry Payen (of Payen and Persoz fame; see Ch. 27) wrote on 'chemical fermentation' for the *Encyclopédie du XIXe siècle*:

"Fermentation (chemistry). One should understand by this word a spontaneous reaction, a chemical alteration excited within a mass of organic matter by the sole presence of another substance, without the latter borrowing or lending anything to the body it decomposes. This active substance, the *ferment*, acts thus somewhat like the Volta battery; it separates, by splitting, complex material into simpler material. In ordinary chemical reactions, we see a body combine with another to form a new compound; or a body displaces another from a combination, where he replaces it by virtue of a greater affinity. We predict, and we explain, these facts by the intervention of this molecular force which attends every chemical reaction, by the affinity which binds together the molecules of the various substances. Fermentation, on the contrary, cannot be explained by any of these forces. The ferment acts by drawing benefit from absorption of the force by which were united the particles of the body which undergoes its influence (. . .)" (Payen, 1876).

This rather concise text on 'fermentation' presents in fact what can be considered as the late 19th century paradigm of enzymes as catalysts.

I shall close this section with another 'multiple': several scientists isolated pepsin, the active principle of gastric juice, in the mid-1830s. These were Theodor Schwann in 1835, and Cagniard-Latour also in 1835. Theodor Schwann refined the technique introduced by Eberle the preceding year, extracting the mucosa of the stomach with dilute HCl. Schwann named the protein he had isolated, and which was capable of dissolving unsoluble proteins, *pepsin* and considered it to be a catalyst, akin to the Persoz and Payen diastase.

That the soluble ferment pepsin was the active principle of gastric juice was not universally accepted. Claude Bernard, for instance, wrote his medical dissertation in 1843 on gastric juices and their role in nutrition. He concluded to a mixture of several acids (predominant in lactic acid rather than in hydrochloric!) and came out strongly against the hypothesis of a gastric ferment such as pepsin (Debru, 1983, p. 20; Sernka, 1979).

We move now by half a century to relate what probably was the key experiment in the emergence of enzymes as molecules.

REFERENCES

Berzelius, J.J. (1836a) Jahres-Ber., 15, 237.

Berzelius, J.J. (1836b) Edinb. New. Phil. J., 21, 223.

Büchner, E. (1907) Nobel Lecture, December 11; Nobel Lectures - Chemistry, 1901-1921, Elsevier, Amsterdam, pp. 97-122.

Collins, P. (1975) Ambix, 22, 205.

Collins, P. (1976) Ambix, 23, 96.

Davy, E. (1820) Phil. Trans., 110, 108.

Debru, Cl. (1983) L'esprit des Protéines. Histoire et Philosophie Biochimiques, Hermann, Paris.

Diderot, D. et al. (1751-1780) Encyclopédie ou Dictionnaire Raisonné des Sciences, des Arts et des Métiers (35 vols., Neuchâtel, etc.; my translation): entry CHIMIE, vol. 3, p. 420; entry FERMENT, vol. 14, p. 18; entry ESTOMAC, vol. 13, p. 103; entry ALKAHEST, vol. 2, p. 120; entry FERMENTATION, vol. 14, p. 32.

Döbereiner, J.W. (1822) Ann Phys., 72, 194 (in Collins's 1976 translation).

Döbereiner, J.W. (1823) Ann. Phys., 74, 272 (in Collins's 1976 translation).

Moore, W.J. (1962) Physical Chemistry, 4th ed., Prentice Hall, Englewood Cliffs, NJ, p. 300.

Northrop, J.H. (1946) Nobel Lecture, December 12; Nobel Lectures - Chemistry, 1942-1962, Elsevier, Amsterdam, p. 124.

Payen, A. (1876) Encyclopédie du XIX Siècle, 4th ed., Paris, vol. 20 (my translation).

Sernka, T.J. (1979) Persp. Biol. Med., 523.

Teich, M. (1981) Hist. Phil. Life Sci., 3, 193.

Thénard, L.J. (1813) Ann. Chim., 85, 61.

Chapter 33

The Buchner Experiment

The seemingly simple 1897 discovery of cell-free fermentation by Eduard Buchner can assume many meanings. To a Whig historian of science in the rationalist tradition, it is a major step in the freeing of biological chemistry from vitalist dogma. To an institutional historian of science such as Robert Kohler, it signals the possibility of erecting a biochemical science (Kohler 1971; 1972; 1975). To a mainstream historian of science such as Joseph Fruton, this lucky find by a none-too-gifted scientist only slightly shifted the balance of power between biologists and chemists (Fruton, 1972; 1976; 1982). To an epistemologist, this episode might support a transdisciplinary view of scientific progress, in which the important breakthroughs are made by outsiders to the field. To a chemist, Buchner's discovery helped to reveal the grand unity of catalysts, whether of inorganic or organic origin: this original guess of Berzelius had been masterly. To an applied scientist, the finding of Eduard Buchner paved the way to biotechnologies. That Buchner came up with cell-free fermentation may even have meaning to psychologists, who might explain it by rivalry between siblings and by self-assertion: Eduard was working in a subordinate capacity in the laboratory of his brother Hans, 10 years his senior, who was overshadowing him in every respect. And my list is open-ended . . .

I wish to concentrate in this section on the single topic of interest to this volume: the advancement of biochemistry as a molecular science. I shall overlook the other aspects, however fascinating.

Numerous epistemological blocks, or hurdles, stood in Eduard Buchner's way. First, the triple discovery by Cagniard-Latour, Schwann, and Kützing (1837) had shown yeast cells to be the agents of alcoholic fermentation. Eduard Buchner refutes apparently this statement: he proved in 1897 that yeast cells are not the (primary, or ultimate) agents of alcoholic fermentation. The malicious satire against Schwann's 1837 discovery written by Wöhler, with Liebig's help, was unwillingly accurate to a fault: yeast cells did feed on sugar and they did excrete ethanol, and carbon dioxide was indeed bubbling out of them!

The 1897 discovery of Eduard Buchner flew in the face of the protoplasmic theory of life: after his 1837 discovery of the cellular origin of fermentation, Theodor Schwann had gone on to study digestion and discover pepsin, and to numerous classic investigations culminating in his concept of the cell as embodied in his book of 1838-1839. For Schwann, alcohol fermentation as performed by yeast cells was a model for the chemical changes in cells of higher organisms. Alcoholic fermentation led him to the idea of metabolic changes in the cell (the word 'metabolism' was invented by Schwann, as well). Schwann's view of the cell as an 'organized unit of metabolic activity' to use a felicitous description (Fruton, 1976) became widely accepted. It led to the protoplasmic theory of life. To quote again Fruton (1976):

"The dynamic portion of the cell became protoplasm, whose most characteristic property was its resemblance to albumin. This albuminous character of protoplasm was elevated, in Huxley's famous phrase, to the physical basis of life".

Not only was the protoplasmic theory ambitious and far-reaching. In subsequent years, it became very explicit as well. It gave explanations of various life processes in the protoplasm. Here is an example of the speculations made about life's machinery during the very period of Buchner's formative years (1884-1893) or immediately before: the physiologist Eduard Pflüger suggested in 1875 that 'intracellular oxidation is effected by a labile energy-rich protoplasmic protein that contains cyano groups, which combine explosively with molecular oxygen to liberate CO_2' (Fruton, 1976). This idea was later modified: O. Loew proposed the aldehydic rather than the cyano function as the site of oxygen attack (1881-1896).

A second epistemological difficulty bore Pasteur's name. Pasteur had tried systematically to do what Buchner found by accident, and Pasteur had failed. Louis Pasteur indeed tried to answer the question of cell-free fermentation. He performed experiments to find out if a ferment made by yeast cells was responsible for alcoholic fermentation. Pasteur had not been more fortunate than Marcelin Berthelot, Adolf Mayer, Naegeli and Loew, who had also tried to isolate an enzyme from yeast cells. They had all failed. In his Nobel lecture (Buchner, 1907), Buchner quotes Pasteur as follows:

"In what does for me the chemical process of sugar decomposition consist, and what is its intrinsic cause? I confess that I am completely in the dark about it. Can we say that the yeast nourishes itself on the sugar, only to give it off again as an excrement in the form of alcohol and carbon dioxide? Or must we say that the yeast in its developments produces a substance of the nature of peptase which acts on the sugar and disappears as soon as it has exhausted itself, since we find no substance of this kind in the fermentation liquids? I have no answer to the substance of these hypotheses. I neither accept them nor do I reject them, and I shall always try not to go beyond the facts" (Pasteur, 1860).

(Peptase, in Pasteur's time, was synonymous with pepsin.)

Not only was Pasteur unsuccessful in his attempt at isolating a ferment active in alcoholic fermentation. His support for the biological rather than the chemical view of fermentation was quite literally forbidding and acted as a brake on research.

"Regarding alcoholic fermentation, Pasteur furnished renewed experimental evidence that it was an adjunct of yeast (cellular) life. The other side that led Pasteur to support the biological view of fermentation was his discovery that racemic compounds could be dissolved biologically" (Teich, 1981).

As expressed by Buchner (1907) in no uncertain, if ironical and morose, terms,

"it was only the systematic and striking experiments of Louis Pasteur, extending over a decade, which finally led to the recognition that in Nature without living organisms, without live yeast, no fermentation exists. This put an end to all disputes. Fermentation was seen to be a physiological act inseparably linked with the life processes of yeast".

Which brings us to a third and major difficulty, institutional as much as epistemological. In the 1880s and 1890s, study of the alcoholic fermentation - which was held to be a physiological act by essence, as I have just made clear with the quote from Buchner – was the proper province of the physiologists. Eduard Buchner was not a physiologist. He was an organic chemist, whose career had been supervised almost from the start by Adolf von Baeyer in Munich. True, as an organic chemist, Buchner would have made an acceptable recruit for an institute of physiology: physiological chemistry was on the rise, and German physiologists in the 1880s and 1890s no longer felt much threatened by it (Kohler, 1982). But Buchner did not opt for such a (necessarily subservient) collaboration with the professional physiologists. He tried to be autonomous, with his own research program of chemical fermentations. This was not easy.

During the period 1891–1893, he could start, with Baeyer's support, a small laboratory. He made there his first experiments on the rupture of yeast cells. However, the Board of the Laboratory

"was of the opinion that 'nothing will be achieved by this' – the grinding of the yeast cells had already been described during the past 40 years, which latter statement was confirmed by accurate study of the literature – (and) the studies on the contents of yeast cells were set aside for three years" (Buchner's biography at the end of Buchner, 1907).

Fortunately, Buchner was helped at this stage by Theodor Curtius (1857–1928) and Hans Frn. von Pechmann (1850–1902), who had been fellow assistants, who were friends, and who took him under their wing when they migrated to the Universities of Kiel and of Tübingen, respectively, which had given them 'calls'. One will have gathered that Eduard Buchner did not belong to the mainstream of German physiological chemistry in the final years of the 19th century.

Not being *persona grata* in physiological chemistry, he found a temporary shelter within immunological biology. At the end of 1896, Eduard Buchner returned from Tübingen to Munich, to the Institute of Hygiene, where his older brother Hans was on the Board of Directors. Eduard was thus able to get funding and laboratory space to resume his researches on the contents of yeast cells.

Hans Buchner was a physician and microbiologist, a former student of Carl von Naegeli. In 1888, Hans Buchner had shown the existence in blood serum of a bactericidal substance, inactivated by heat, which he thought was a 'living protein' of the plasma. Hans Buchner

"experimented with the pathogenic properties of bacterial proteins, including those from killed cells, in 1890 – the year Koch discovered tuberculine. Toxic, these proteins reveal themselves also, in 1891, as bearing immune functions with the discovery of the antitoxins for diphtheria and tetanos by Behring and Kitasato. But the antitoxines are extracellular substances" (Debru, 1983).

And

"Buchner's experiment grew accidentally out of work with his brother Hans on the immunological proteins of bacteria. The new immunology of protein toxins and antitoxins, which was flourishing in the 1890s, was an impressive model for the biochemists' new faith in the importance of enzymes. It looked as if all the important phenomena of life and disease could be understood in terms of these large protein molecules, with their specific vital activities" (Kohler, 1975).

So that, as writes excellently Debru (1983),

"biochemistry owes the occasion of its consolidation to surviving ambiguities between protoplasmic, intracellular and molecular, exocellular [concepts]".

Before coming to the description of the Buchner experiment itself, I wish to stress again his outsider status: not only was he an organic chemist outside the institutional fold of contemporary physiological chemistry; in his brother's laboratory, he was also an outsider, as a non-physician and a non-immunologist.

Moreover, his brother Hans was older by 10 years and (much) better established. Eduard had done most of his career under Hans's direct supervision: even at the start of his studies at the Polytechnic in Munich, at the age of 24, Eduard had been tutored, guided, and ordered about by his older brother. There is a fascinating tidbit, in this regard, in Buchner's Nobel Lecture. Close to the beginning this paragraph occurs:

"The part played by yeast remained for a long time obscure. It was believed that its appearance was of a secondary nature and it was regarded as an inferior kind of precipitation product. The old name for yeast, '*Faex cerevisiae*', and the expression which has taken root in our language, '*die Hefe des Volkes*' (the yeast of the people), meaning the outcasts of the nation, also point to this view" (Buchner's, 1907).

The key words in this paragraph, are *obscure, secondary, inferior, outcasts*: all of which could have been applied to Eduard Buchner self-appraisal at the time just prior to his historical experiment of 1897! We come now to Buchner's experiment proper. Here is a narrative, summarized from his own words:

"In 1878, Naegeli and Löw had declared: 'The difficulties of yeast analysis, when the compounds, not the elements, are involved, consist in the fact that the cells, owing to their smallness, cannot be pulverized, ruptured or burst, so as to separate mechanically the contents from the membrane' (Naegeli and Löw, 1878).

If one part of quartz sand and one-fifth part of diatomite [celite], by weight, are added to yeast, the initially dust-dry substance can be ground in a large mortar with a heavy long-shafted pestle within a few minutes. The mass becomes dark-grey and doughy.

(. . .) If the thick dough is now wrapped in strong canvas and placed in the hydraulic press, a liquid juice is squeezed out when a pressure increased gradually up to 90 kg per square centimeter is applied. Within a few hours 500 cc of liquid can be obtained from 1000 g of yeast, so that considerably over half of the total cell content is expressed. (. . .) 'Pressed yeast juice' (. . .) on heating, soon separates flakes of coagulated protein and on further heating this formation may be so extensive that when the container is inverted scarcely any liquid flows out. The presence of coagulable protein in the interior of micro-organisms has thus been established for the first time. If the expressed juice (*Pressaft*) is diluted with a quantity of water and hydrogen peroxide is added, a violent foaming begins as a result of oxygen formation. By this means we prove the presence of catalase, an enzyme, discovered by O. Löw, which is known to be present in almost all liquids of vegetable or animal origin, e.g. in blood. If sugar solution is added to freshly expressed yeast juice, a strong formation of gas sets in after a little while. (. . .) When sugar is dissolved in expressed juice at blood heat, the phenomena are visible even after about a quarter of an hour. Careful investigations have shown that the formation of carbon dioxide is accompanied by that of alcohol, and indeed in just the same proportions as in fermentation with live yeast. (. . .) Toluene actually does prevent the action of live yeast on sugar, though not the action of expressed juice. (. . .) one can add the live yeast cells to large quantities of alcohol and acetone and finally wash them with ether. The air-dried 'permanent yeast' thus obtained (. . .) is incapable of growth but, when sugar solution is added, can produce an extremely powerful fermentation. (. . .) The active agent in the expressed juice appears (. . .) to be a chemical substance, an enzyme which I have called 'zymase' " (Buchner, 1907).

This a posteriori account leaves out the serendipitous nature of the discovery. Eduard Buchner's goal, in crushing yeast, seems to have been exploratory experiments to prepare rupture of the cells of pathogenic bacteria, in order to test his brother's theories about bacterial toxins and antitoxins (Debru, 1983). The reason why Eduard Buchner, in his 1896 experiments, added concentrated glucose to the yeast juice (the *Pressaft*) was to prevent decomposition of the proteic solution he had obtained (Teich, 1981; Debru, 1983).

A decade after his discovery, Buchner could make the solemn statement in his Nobel Lecture that 'a separation of the fermentation effect from the live yeast cells can be carried out' (Buchner, 1907). A physical separation had led to a conceptual separation: this is the important tenor of this sentence.

The driving force of fermentation is not a vital force. Life does not impel into motion the bubbling of CO_2, the effervescence and the other attributes of alcoholic fermentation. The words 'alcoholic fermentation' had seemingly implied to the biologists and physiological chemists of the second half of the nineteenth century the mysterious agency of a living factory, of a conglomerate of active biological particles embedded in a fluid matrix, the protoplasm. The protoplasm was capable of oxygenation reactions or contained oxygenating devices. And yet inert chemicals - inanimate substances as Buchner might have called them - were capable by themselves of doing the job.

Buchner's discovery is as important, conceptually, as Friedrich Wöhler's urea synthesis of 1838, to which it is clearly related. Wöhler had shown that synthesis of organic compounds did not require a living organism and could be done in the test tube. Buchner showed that metabolism of organic compounds did not require a living organism and could be done in the test tube.

Wöhler's work led to the rapid growth of a new and powerful field, organic synthesis. Likewise, Buchner's work resulted in the accelerated development of biochemistry as a powerful discipline, drawing its object from biology and its methods from chemistry.

True, Buchner's discovery did not appear in a conceptual vacuum. It was consistent with the Moritz Traube enzyme theory of fermentation (Traube, 1858), which had been supported by Berthelot and by a number of other influential scientists, such as Claude Bernard, Schönbein and Schaer, Felix Hoppe-Seyler, G. Hüfner, and Liebig. Liebig, who had earlier challenged the Schwann biological interpretation of alcoholic fermentation and had proposed a rival theory of the oxidation of albuminoid substances brought about by transfer of mechanical agitation from the ferment, had rallied to the more incisive views of Traube, for whom the ferments were 'defined compounds of proteinic origin with the property of transmitting oxygen and participating in a sequence of oxydation and reduction' (Teich, 1981). But no one had been capable of isolating these ferments prior to Buchner, with a single exception, which he himself recalls:

"Something similar to what has just been discussed regarding the alcoholic fermentation of sugar was established as far back as 1890 in respect of what is called urea fermentation, i.e. the conversion of urea into ammonium carbonate. At that time P. Miquel in Paris showed that this process, which gradually sets in in discharged urine, is activated not directly by the vital activity of the bacteria which appear, but by the agency of a separable enzyme, urease" (Buchner, 1907).

And Buchner - whose Nobel Lecture thus purports to set the record straight - goes on to mention that in the 1890s Emil Fischer was doing somewhat *related work*. Eduard Buchner is indeed at pains to claim priority for his initiative of crushing cells in order to burst out their contents:

"It should be mentioned that in 1894 Emil Fischer found in low organisms various enzymes which had hitherto remained unknown, such as maltase in beer yeast, lactase in lactose yeast and (...) monilia invertase in *Monilia candida*, using also a process for rupturing the cells. (...) However, my work was by no means instigated by Emil Fischer's investigations, since it began as early as 1893 with the development of the crushing process, while Fischer's first publications appeared in 1894" (Buchner, 1907).

Scientific historians are divided upon the significance of Buchner's discovery of cell-free fermentation. Kohler makes it a central event in the emergence of biochemistry:

"The elevation of the enzyme concept from the relatively minor role it played in the older physiology to a central theoretical position in the new biochemistry was both a profound conceptual innovation and a crucial factor in the emergence of a new discipline with its own sense of historical identity" (Kohler, 1975).

Debru gives comparable status to the event:

"Buchner thus proposes an entirely chemical definition of the intracellular functioning. Life is no longer a defining entity. Cellular physiology becomes the sum total of
elementary chemical events involving chemically defined and theoretically isolable
substances, even if they have not all been actually isolated. Affirmation of the reduci-
bility of complex processes to a series of simple events, effected by isolable agents,
such as an enzymatically catalysed reaction, is a fundamental concept. This presump-
tion of simplicity is a birth act for biological chemistry. As a chemical object, its study
is entirely based on the principles of chemistry. As an object of biological origin, its
study constitutes biochemistry" (Debru, 1983, p. 88; my translation).

Fruton has taken an opposing stand, deflating the myth from hindsight that Buchner's discovery 'scotched "vitalism", established the
chemical view, and set modern biochemistry on its course' (Kohler,
1975). Fruton is clearly unimpressed by Buchner as a scientist,

"an organic chemist of modest talent who stumbled into his (discovery) while helping
his brother Hans, a bacteriologist, to prepare microbial extracts that Hans thought
might contain antitoxins" (Fruton, 1976).

Fruton also points out that Buchner's discovery did not clinch the
case, that 'the validity of his claim was questioned by many biologists' (Fruton, 1976), such as Neumeister, and that 'although many
chemists and biochemists welcomed Buchner's achievement, the
debate over the role of intracellular enzymes continued, especially in
the field of biological oxydations' (Fruton, 1982).

Buchner had punctured a hole in the vitalistic conception of
enzymes by the physiological chemists. But it was not conclusive.
Someone else had to reiterate the molecular nature of enzymes with
greater emphasis and from another angle.

REFERENCES

Buchner, E. (1907) Nobel Lecture, December 11, Nobel Lectures - Chemistry, 1901-1921, Elsevier, Amsterdam, pp. 97-122.

Debru, Cl. (1983) L'esprit des Protéines. Histoire et Philosophie Biochimiques, Hermann, Paris.

Fruton, J.S. (1972) Molecules and Life. Historical Essays on the Interplay of Chemistry and Biochemistry, Wiley-Interscience, New York.

Fruton, J.S. (1976) Science, 192, 327.

Fruton, J.S. (1982) J. Clin. Chem. Clin. Biochem., 20, 243.

Kohler, R.E. (1971) J. Hist. Biol., 4, 35.

Kohler, R.E. (1972) J. Hist. Biol., 5, 327.

Kohler, R.E. (1975) J. Hist. Biol., 8, 275.

Kohler, R.E. (1982) From Medical Chemistry to Biochemistry. The Making of a Biomedical Discipline, Cambridge University Press, Cambridge.

Naegeli, C. von and Loew, O. (1878) Sitzungsber. Bayer. Akad. Wiss. Math. Phys. Kl., 161.

Pasteur, L. (1860) Ann. Chim. Phys., 58, 3, 360.

Teich, M. (1981) Hist. Phil. Life Sci., 3, 193.

Traube, M. (1858) Ann. Phys., 103, 331.

Chapter 34

The Aftermath of the Buchner Experiment. Crystallization of Enzymes

Buchner's discovery met with the approval of his fellow organic chemists and with the disbelief of many biologists. Physiological chemists were spurred to study the chemical changes wrought by zymase, the ferment(s) isolated by Buchner. 'Thirty years later, their efforts had led to the recognition that zymase is not a unitary chemical principle but a mixture of 12 separate catalytic proteins' (Fruton, 1976). Harden also discovered, during the period 1905–1908, that cell-free alcoholic fermentation required phosphate and an organic co-enzyme (Kohler, 1974; Stephenson, 1939).

During the next decade, in 1917, the American chemist James B. Sumner (1887–1955) decided he would isolate an enzyme. He was 30 and had been since 1914 an assistant professor at Cornell University. In 1926, he succeeded in isolating and crystallizing urease. In 1929 (he was 42), Cornell made him a full professor. Sumner was a former student of Otto K. Folin at Harvard Medical School. At Cornell, 'Sumner was never able to advance beyond second fiddle in an ensemble dominated by chemists and physiologists' (Kohler, 1982). He had a severe disability: I take this account from his biography at the back of the printed version of his Nobel Lecture:

"(Sumner) was interested in fire-arms and often went hunting. While grouse hunting at the age of 17, he was accidentally shot in the left arm by a companion; as a consequence, his arm had to be amputated just below the elbow. Having been left-handed, he then had to learn to do things with his right hand. The loss of his arm made him exert every effort to excel in all sorts of athletic sports, such as tennis, skiing, skating, billiards, and clay-pigeon shooting" (Sumner, 1946).

James B. Sumner majored in chemistry at Harvard College in 1910.

He entered graduate school in 1912 and opted for biochemistry with Otto Folin at Harvard. In view of his disability, Folin advised him to become a lawyer, 'since he thought that a one-armed man could never make a success of chemistry' (Sumner, 1946). But Sumner was obdurate and persistent (a quality he would put to advantage in his urease isolation), and got his Ph.D. already in June 1914.

The account Sumner gives in 1946 of his work is a gem of science popularization, written in a direct and chatty style. He uses these words, which assume special value because of his accident, to explain his resolve: 'I decided to take a long shot' (Sumner, 1946). And he was clearly stimulated by opposition to his scheme:

"A number of persons advised me that my attempts to isolate an enzyme were foolish, but this advice made me feel all the more certain that if successful the quest would be worthwhile" (Sumner, 1946).

Sumner picked urease for his target because of previous familiarity and because he could use the jack bean, *Canavalia ensiformis*, as a rich source of this enzyme. This was a stroke of luck: 'Willstätter was unfortunate in his choice of saccharase as an enzyme to isolate. I was fortunate in choosing urease' (Sumner 1946).

Sumner initiated his work in the fall of 1917. He met with repeated failure and frustration at nearly every step. 'At times,' he would say in his Nobel Lecture, 'I grew discouraged and temporarily abandoned the quest, but always returned to it again' (Sumner, 1946). Sumner must have been extraordinarily persistent and strong-willed, as this episode shows:

"In 1921, when his research was still in its early stages, he had been granted an American–Belgian fellowship and decided to go to Brussels to work with Jean Effront, who had written several books on enzymes. The plan fell through, however, because Effront thought Sumner's idea of isolating urease was ridiculous" (Sumner, 1946).

Effront's lack of enthusiasm was quite representative of the skepticism of the biochemists. This skepticism was rooted in a double misconception about the nature of enzymatic catalysis, and about the nature of enzymes.

Enzymatic catalysis was thought then to be a physical rather than a chemical phenomenon, 'arising from physical adsorption on surfaces rather than from specific chemical combination' (Fruton, 1976). Enzymes were generally held not to be proteins, as a result of a large body of work by Richard Willstätter (1872–1942) and his school, which 'led to the conclusion that enzymes were a special class of unknown compounds, and certainly not proteins' (Northrop, 1946). Willstätter held the pernicious doctrine that enzymes were *not* proteins and ought to be purified from proteins.

"He believed that enzymes, like other bioactive substances known at that time, such as the hormones epinephrine and thyroxine, are small organic molecules, and that the presence of proteins in enzyme preparations is a consequence of the ready adsorption of such molecules by colloids" (Fruton, 1982).

Willstätter is undisputably the villain in this part of the story, and his winning the Nobel Prize in 1930 may have somewhat influenced the Nobel Committee in delaying the award of the Nobel Prize to Sumner, who got it in 1946 for a discovery made 20 years earlier, and to Northrop, also acknowledged in 1946 only for work completed in 1929. Richard Willstätter wrote in 1933 that while his laboratory 'aimed at gradual and, if possible, complete liberation of enzymes from protein, these American colleagues proceeded in exactly the opposite direction' (Willstätter, 1933). This was a gratuitous, polemical jab, but the chemist Willstätter was building on the prejudices of the biochemists themselves, the British biochemist Bayliss for instance who arrogantly stated in 1914 that enzymes were *not* proteins (Bayliss, 1914).

Not surprisingly, Sumner opened his Nobel Lecture, in didactic and dramatic manner, by asking 'Why was it difficult to isolate an enzyme?', to which he gives these answers: 'inertia of man's ideas', 'little time for research, not much apparatus, research money, or assistance'. Beyond these all-too-usual answers, Sumner also mentioned the presence in the cell of natural protectors, which made it difficult to maintain the integrity of an enzyme once it had been removed from their shielding action.

But Sumner singled out, as a chief culprit,

"the influence exerted by Willstätter and his school, who held that enzymes were neither lipids, carbohydrates nor proteins and who believed enzymes to exist in excessively low concentrations in plants and animals" (Sumner, 1946).

How did Sumner isolate urease? To summarize 9 years' work in as many sentences: he performed a selective extraction of urease from jack beans using ethanol; some mysterious intuition whispered to him to try acetone instead of ethanol ('It seemed to me of interest to employ dilute acetone instead of 30% alcohol', Sumner 1946):

"Accordingly I diluted 316 ml of pure acetone to 1000 ml and used this as the means of extracting the urease. (. . .) The acetone extract was chilled in our newly acquired ice chest overnight. (. . .) upon observing a drop of the liquid under the microscope it was seen to contain many tiny crystals. These were of a shape that I had never observed previously. I centrifuged off some of the crystals and observed that they dissolved readily in water. I then tested this water solution. It gave tests for protein and possessed a very high urease activity. I then telephoned my wife, 'I have crystallized the first enzyme' " (Sumner, 1946).

'I have crystallized the first enzyme': what an incredible statement in content, however simple the form. The old dichotomy between animate and inanimate matter had been bridged. Enzymes, which at a time were considered to be preciously nurtured in the womb of living cells, with their mysterious, seemingly unreachable, hyperdelicate, and terribly complex workings, suddenly could be handled like any inorganic crystals.

This was a scientific revolution, to borrow from Thomas S. Kuhn. In his classic book, Kuhn has likened a major scientific discovery to a 'Gestalt switch', changing from one paradigm – here, enzymatic catalysis as a holistic property arising in some way from the colloidal nature of the protoplasm – to another, our modern paradigm of enzymes as molecules (Kuhn, 1970).

By astonishing coincidence, Sumner says the same thing! He does it in his usual, plain and delightful manner:

"In obtaining the crystals I felt much the same as a person does who is trying vainly to place in position a piece of a machine. Suddenly the piece slides in as if covered with butter. One knows that it is now where it belongs" (Sumner, 1946).

A thematic analysis of these sentences will contrast 'trying vainly', which points to the by now familiar Sumnerian sense of frustration, with 'slides in', 'as if covered with butter', 'it is now where it belongs', three expressions of the satisfaction felt when one's action leads swiftly and smoothly to success. We have already (Chapter 6) commented upon this same notion of interlocking or assembly of parts as representative of the harmony of nature.

It is now too late to ascertain (he died in 1955) whether Sumner in 1946 was reminiscing about his exact feelings in 1926, and if he was indeed aware of the main tenets of *Gestalt* psychology; F. Köhler's book, *The Mentality of Apes*, had just been published the year before (1925). What we know is that, once Sumner had convinced himself he had really isolated urease in crystalline form, he read a paper on this matter at Clifton Springs, New York and published an article in the August issue of the *Journal of Biological Chemistry* for 1926 (Sumner, 1926a).

It was a *Gestalt* switch – by hindsight. Sumner's discovery was not accepted readily. Many biochemists, especially in Europe, remained unconvinced (Debru, 1983). As writes Fruton (1982), 'the issue remained in doubt, even after Sumner had reported in 1926 the isolation of urease as a crystalline globulin'. Elsewhere, Fruton explains that enzymes, in the 1920s, were

"widely thought to be small bioactive molecules adsorbed on nonspecific colloidal carriers (and that) (. . .) the predominance of such views during the 1920s suggests an explanation for the dismissal of Sumner's claim to have isolated the enzyme urease in the form of a crystalline protein" (Fruton, 1976).

A very interesting reaction to the breakthrough by Sumner was that of John Northrop: he was stimulated by it ('Sumner's results encouraged me to take up the pepsin problem again'; Northrop, 1946); but he was also rather critical ('Sumner at that time had no convincing proof of the purity of his preparation'; Northrop, 1961).

John Northrop (1891–   ) studied zoology and chemistry as an undergraduate at Columbia (1908–1912) under T.H. Morgan and J.M. Nelson (the latter of whom recruited him for biochemistry;

Plate 10. J.H. Northrop, 1946. (Courtesy Rockefeller University Archives.)

Kohler, 1982), and went on to acquire a Master of Arts (1913) and a Ph.D. in chemistry (1915). From there, he went to Jacques Loeb's laboratory at the Rockefeller Institute, where he stayed till his retirement. Morgan tried to pry John Northrop loose from the Rockefeller Institute to CalTech, in 1928, without success (Kohler, 1982). Only in 1949, did Northrop accept a professorship at the University of California, in Berkeley, but he did not renounce his appointment at Rockefeller. Loeb communicated to Northrop his interest in the chemistry of proteins, together with his physico-chemical approach – highly heterodox at the time, for this was the Age of Colloids: 'proteins at that time were considered to be "colloidal complexes" of no definite molecular weight, which combined with electrolyte by a process of adsorption' (Northrop, 1961).

Northrop started about 1920 his attempts to isolate and crystallize pepsin. These were unfruitful. Northrop was further discouraged (I quoted him earlier to this effect) by the disbelief by Willstätter and his school in the proteic nature of enzymes. After 1926, Northrop picked up this research problem again. His solution was to use large quantities of material. Northrop blames earlier failures, especially Pekelharing's in 1896, on the fact that

"nearly all of the work was carried out with dilute solutions. (. . .) Had Pekelharing filtered his pepsin preparation and then dissolved it in a very small amount of water, instead of centrifuging the precipitate and dissolving it in a large amount of water, I am quite sure he would have crystallized the enzyme nearly fifty years ago",

said the Rockefeller chemist in Stockholm on December 12, 1946 (Northrop, 1946). With this simple trick, pepsin isolation becomes relatively straightforward, Northrop implies.

The difficulty was elsewhere: establishing the purity of the protein. There, Northrop proved an excellent adept of Loeb's physico-chemical strategy. He made use of an arsenal of (at the time) very modern techniques of analysis and characterization. He used The Svedberg's ultracentrifuge as a criterion for the homogeneity of his pepsin solutions. He used Tiselius's electrophoresis, also as an homogeneity criterion. He related the enzymatic activity to the protein by diffusion measurements, denaturation and hydrolysis of the proteins, etc.

Northrop also made use of an original and clever method of proving purity, based upon solubility measurements.

"A pure protein has a definite constant solubility. Most protein preparations contain a group of similar proteins, and the solubility therefore increases with the quantity of solid. This is understandable since it is now known that animals produce slightly different proteins at different ages" (Northrop, 1961).

As he explained in his Stockholm Lecture:

"According to Willard Gibbs's phase rule, a system consisting of a single solid in equilibrium with its solution is of fixed composition. That is to say, if more of the same solid is added no change in concentration will occur. If any other substance is added, however, a change will occur. The test is exceedingly specific, even more so than the serological tests as Landsteiner and Heidelberger pointed out. It will, in fact, distinguish between optical isomers, even though the two isomers have exactly the same solubility. (. . .) From the experimental point of view it has the great advantage that the test may be carried out in concentrated salt solution, in which the proteins are most stable" (Northrop, 1946).

Using this whole battery of tests, Northrop was able to isolate, crystallize, and declare chemically pure the enzyme pepsin in 1929 (Northrop, 1930; 1935).

A comment is in order. I wish to draw attention to a common feature of the scientists whose work has just been presented. All of them had their initial training in chemistry. Eduard Buchner was an organic chemist – and we saw that, however useful this upbringing may have been to his isolation of 'zymase', if placed him outside the dominant strain of physiological chemists. The early part of his career was made difficult by this status, or lack thereof. Setting up a well-equipped lab to work on his chosen research goals was well-nigh impossible.

Arthur Harden, who continued the work on 'zymase', was also an organic chemist. Kohler's perceptive comment is worth repeating:

"Allen Macfadyen, Harden's biological mentor, was interested in the immunoproteins of bacteria, like his counterpart Hans Buchner. Harden picked up their [Eduard and Hans's] work on zymase, going on to a successful career in biochemistry, while Macfadyen, like Hans Buchner, pursued a fruitless immunochemical vision. One sees in both cases the fortuitous partnership of a chemist with analytical skills and a bacteriologist with a visionary sense of the opportunities of the active proteins. Both episodes tell us a good deal about the conceptual and social regrouping that took place around 1900" (Kohler, 1975).

James B. Sumner also had his initial training in chemistry. John H. Northrop, likewise, came to biochemistry from chemistry.

These four men (E. Buchner, A. Harden, J.B. Sumner, J.H. Northrop) are representative of the seminal contribution of chemical concepts, especially molecular concepts, and of chemical methods, such as purification, crystallization, criteria of purity, etc., to biochemistry in the bud.

Ever since the start of chemical physiology in the 19th century, there had been a permanent tension, enlivened by occasional and recurring clashes, between the chemists and the biologists. Fruton (1976) gives a sample of the acerbic verbal exchanges, starting with Justus Liebig's stating in 1844 that 'in chemical-physiological work, physiology is not threatened most by chemists, but by physiologists and physicians' (Liebig, 1874), going on to du Bois-Reymond commenting 4 years later on Liebig that 'I consider his physiological fantasies as worthless and pernicious'(du Bois-Reymond, 1927), and on again in the early 20th century to Frederick Gowland Hopkins who complained that 'it is a rare thing in this country [the U.K.] to meet a professed biologist . . . who has taken the trouble so to equip himself in organic chemistry as to understand fully an important fact of metabolism in terms of structural formulae' (Hopkins, 1913), with the parting shot being Marcel Florkin's, who commented that 'chemists cannot provide any proper methods to tackle the specific problems of biochemistry, the molecular structure of cells, and never showed any interest in the problem' (Florkin, 1972).

I see fit to recognize that there is much truth in the statement of my esteemed friend the late Marcel Florkin. To give it flesh, when Northrop in the late 1920s was intent upon showing that he had *pure* pepsin on hand, he could not draw upon the existing techniques of chemistry, they were all meant for the small molecules of organic chemistry. Nevertheless, *pace* Florkin, the chemical instincts of Northrop served him right. And he went to the appropriate newly introduced biophysical techniques, such as electrophoresis and the ultracentrifuge, to fulfill his needs. Moreover, he had the imagination and intellectual rigor to make use of Willard Gibbs's phase rule – a chemical method if there is one – in a brilliant manner.

The organic chemist I am cannot resist stating here my sympathies. To show where they are, I cannot do better than quoting at some length from the delightful small reminiscing essay which Northrop published in 1961 in the *Annual Review of Biochemistry*:

"This antagonism [of chemists and biologists] is a natural consequence of different languages and different ways of thought. The biologist uses the language of semantics and is satisfied with a descriptive and qualitative theory. The chemist leans toward the language of mathematics and requires quantitative results. There is still another difference: chemists (and physicists) have great respect for the Reverend Occam's razor and endeavor to limit their assumptions to the minimal number essential for an explanation, in accordance with the principle of conservation of hypotheses; whereas some biologists have no respect for the Reverend's weapon and fearlessly bolster an ailing (and unnecessary) assumption by another similar one. As a result, the chemist, who thinks he stands on firm ground, is frequently astonished to find himself facing a whole company of unnecessary assumptions which he is expected to disprove, rather than lop off with William of Occam's weapon.

This incompatibility of temperament renders collaboration between chemists and biologists a difficult matter.

It is not surprising, therefore, that the history of biochemistry is a chronicle of a series of controversies, in several of which I have been more or less engaged. These controversies exhibit a common pattern. There is a complicated hypothesis, which usually entails an element of mystery and several unnecessary assumptions. This is opposed by a more simple explanation, which contains no unnecessary assumptions. The complicated one is always the popular one at first, but the simpler one, as a rule, eventually is found to be correct. This process frequently requires 10 to 20 years. The reason for this long time lag was explained by Max Planck. He remarked that scientists never change their mind, but eventually they die" (Northrop, 1961).

This, by way of introducing the last topic in this section, the proteic nature of enzymes. This statement has become the central and founding dogma of enzymology for a clear historical reason: because Willstätter and his school maintained the opposite viewpoint in the 1920s. I shall quote again Northrop here:

"In the meantime, the chemical nature of the enzymes themselves remained in dispute. *Their general properties were those of proteins, and about 1900 a protein was isolated from gastric juice by Pekelharing and Ringer, who considered it to be the enzyme pepsin.* This conclusion was not accepted, and as a result of repeated failure to isolate any enzyme, they were considered (by fallacious reasoning, this time by chemists [Willstätter et al.] to be neither fats, carbohydrates, nor proteins, but an entirely new and unknown class of compounds).

It was at this point that I became involved in the controversy. I had spent several years in the Department of Zoology at Columbia University, where I had the good fortune to study under [T.H.] Morgan, [E.B.] Wilson and [G.] Calkins, and to be associated with Muller, Altenberg, Bridges, and [J.M.] Sturtevant. I had then transferred to the Department of Chemistry (where I found Michael Heidelberger and George Scatchard). I worked with Nelson and Kendall and became fascinated by the enzyme puzzle. I then moved to Jacques Loeb's laboratory at the Rockfeller Institute as W.B. Cutting Traveling Fellow of Columbia University.

Loeb believed that the properties of living things could be completely explained in terms of physics and chemistry. (. . .) At the time I entered the laboratory, he was about to attack the problem of the peculiar properties of proteins – in particular their osmotic pressure, viscosity, and swelling. Proteins at that time were considered to be 'colloidal complexes' of no definite molecular weight, which combined with electrolytes by a process of adsorption. This explanation again contained unnecessary assumptions which could not survive the Reverend's [Occam] weapon. Loeb was able, with the help of Donnan's theory of membrane equilibrium, to explain all the complicated phenomena of osmotic pressure, viscosity, and swelling (Loeb, 1924). (. . .)

Loeb's interest and success with the chemistry of proteins led me to consider the proteic nature of enzymes, and I repeated Pekelharing and Ringer's experiments without being able to carry them further. (. . .) Sumner's work encouraged me to repeat the pepsin experiments again, this time on a much larger scale. I finally succeeded in isolating and crystallizing *a protein that is the enzyme pepsin.* This protein is probably the same as that isolated by Ringer and Pekelharing 30 years before and buried under a series of unnecessary assumptions. The pepsin protein was also considered a 'carrier' for years, and it required a long series of quantitative experiments (. . .) (Northrop et al., 1948; Sumner and Somers, 1947) before the simple explanation that *the protein was the enzyme* gained general acceptance.

The controversy over the chemical nature of enzymes, which had raged for over a century, was thus ended" (Northrop, 1961; emphasis added).

Northrop is right to complain about the unnecessary assumptions which plagued this question. He also indicts Willstätter for delaying by many decades recognition of the proteic nature of enzymes. This had appeared as a likely hypothesis *even before* the work by Ringer and Pekelharing Northrop refers to:

"Of the true chemical nature of the enzymes we are ignorant. *They are probably proteins"* (Allbutt, 1896; emphasis added).

Recognition of the proteic nature of enzymes was protracted for the very reasons denounced by Northrop. As soon as the reports by Sumner and by Northrop appeared (Sumner, 1926a, b; Sumner and Hand, 1928; Northrop, 1929; 1930; Northrop and Kunitz, 1931) enzymes became generally accepted as proteins. For instance, as early as 1931, Hsien Wu makes such a statement at the start of his lucid article on protein denaturation (Wu, 1931).

We have reached the stage in this history of enzymes, where they are established as proteic molecules. The focus can shift now to where the action is, to the conceiving and description of their active sites.

REFERENCES

Allbutt (1896) Allbutt's Syst. Med., I, 520.

Bayliss, W.M. (1914) The Nature of Enzyme Action, Longmans, London, 3rd ed.

Debru, Cl. (1983) L'esprit des Protéines. Histoire et Philosophie Biochimiques, Hermann, Paris.

du Bois-Reymond, E. (1927) Zwei Grosse Naturforscher des 19. Jahrhundert, Barth, Leipzig, p. 19.

Florkin, M. (1972) A History of Biochemistry, Elsevier, Amsterdam, parts 1 and 2.

Fruton, J.S. (1976) Science, 192, 327.

Fruton, J.S. (1982) J. Clin. Chem. Clin. Biochem., 20, 243.

Hopkins, F.G. (1913) Nature (London), 92, 214.

Kohler, R.E. (1974) Bull. Hist. Med., 48, 22.

Kohler, R.E. (1975) J. Hist. Biol., 8, 275.

Kohler, R.E. (1982) From Medical Chemistry to Biochemistry. The Making of a Biomedical Discipline, Cambridge University Press, Cambridge.

Kuhn, T.S. (1970) The Structure of Scientific Revolutions, University of Chicago Press, Chicago, 2nd ed.

Liebig, J. von (1874) Reden und Abhandlungen, Winter, Leipzig, p. 81.

Loeb, J. (1924) Proteins and the Theory of Colloidal Behavior, McGraw-Hill, New York.

Northrop, J.H. (1929) Science, 69, 580.

Northrop, J.H. (1930) J. Gen. Physiol., 13, 739.

Northrop, J.H. (1935) Biol. Rev. (Biol. Proc.) Cambridge Philos. Soc., 10, 263.

Northrop, J.H. (1946) Nobel Lecture, December 12, Nobel Lecture - Chemistry, 1942-1962, Elsevier, Amsterdam, p. 124.

Northrop, J.H. (1961) Annu. Rev. Biochem., 30, 1.

Northrop, J.H. and Kunitz, M. (1931) Science, 73, 262.

Northrop, J.H., Kunitz, M. and Herriott, R.M. (1948) Crystalline Enzymes, Columbia University Press, New York.

Stephenson, M. (1939) Bacterial Metabolism, 2nd ed., Longmans, Green, London, pp. 7-9.

Sumner, J.B. (1926a) J. Biol. Chem., 69, 435.

Sumner, J.B. (1926b) J. Biol. Chem., 70, 97.

Sumner, J.B. (1946) Nobel Lecture, December 12, Nobel Lectures - Chemistry, 1942-1962, Elsevier, Amsterdam, p. 114.

Sumner, J.B. and Hand, D.B. (1928) J. Biol. Chem., 76, 149.

Sumner, J.B. and Somers, G.F. (1947) Chemistry and Methods of Enzymes, Academic Press, New York.

Willstätter, R. (1933) Chem. Rev., 13, 501.

Wu, H. (1931) Chinese J. Phys., 5, 321.

Chapter 35

Enzymes as Receptors Complementary to their Substrate

The immediate precursors to Emil Fischer's key-and-lock analogy for the enzyme-substrate interaction (see Chapter 6) were the Cambridge physiologist John Newport Langley (1852–1925) and the German physician and immunologist Paul Ehrlich (1854–1915). Langley's work was initiated by Luchsinger who showed in 1877 that atropine and pilocarpine had antagonistic actions on the sweat glands of the cat's paw. Luchsinger gave a graphic and chemical wording to his observations:

"There exists between pilocarpin and atropin a true mutual antagonism, their actions summing themselves algebraically like wave crests and hollows (*wie 'Wellenberg und Wellenthal'*), like plus and minus. The final result depends simply and solely upon the relative number of the molecules of the poisons present" (Luchsinger, 1877).

John Langley set upon investigating the 'mutual antagonism' of these two drugs on the sub-maxillary gland of the cat. He concluded that 'there is some substance or substances in the nerve endings or gland cells with which both atropin and pilocarpin are capable of forming compounds' (Langley, 1878). In John Parascandola's words, 'this statement contains the germs of the receptor theory, but Langley did not follow it up for another quarter of a century' (Parascandola, 1980a).

At the same time, Paul Ehrlich was completing his M.D. at the University of Leipzig with experimental work on histological staining. Ehrlich was very much interested in the specific staining of the various parts of a tissue by a dye. 'He noted that some dyes (. . .) become fixed to fabrics by combining with certain constituents of the material to form insoluble salt-like compounds termed "lakes" '

(Parascandola, 1980a). In Pflüger's tradition, Ehrlich used a series of redox indicator dyes to determine the intracellular location of the respiratory processes.

Dyes were powerful microscopic tools for Ehrlich's studies. Because of their chemical properties (acidity and basicity, oxydation potential), they provided Ehrlich with a flexible and powerful probe of intracellular chemistry. In such a manner, Ehrlich studied microbiology, hematology, neurology, and, as mentioned above, the physiology of intracellular oxydation. He presented part of this work as his 'Habilitation-schrift' at the University of Berlin in 1885.

"He adopted Edward Pflüger's view that protoplasm can be envisioned as a giant molecule consisting of a chemical nucleus of special structure which is responsible for the specific functions of a particular cell (e.g., a liver cell) with attached chemical side chains" (Parascandola, 1980a).

These side chains assumed special importance for Paul Ehrlich. He theorized that they were active in cellular nutrition and oxidation. In 1897, Ehrlich ascribed to these side chains neutralization of toxins from pathogenic bacteria. (The reader will recall that the 1890s were a period of great activity in the new field of immunology of toxins and antitoxins.) Whenever one of these side chains would combine with a toxin, this would render the side chain useless for its normal physiological function, such as nutrition or oxidation. The cell would respond to this crippling of a number of its side chains by producing more. These new side chains would break off into the blood, and these 'excess side chains in the blood are what we call antibodies or antitoxins' (Parascandola, 1980a).

Ehrlich imagined this mechanism under the influence of the general biological principle devised by his cousin and teacher Carl Weigert: a damaged cell or tissue will hyper-regenerate a missing part (Weigert, 1896; Rubin, 1980). And Ehrlich made use of the new stereochemical language introduced by Emil Fischer in 1894 to state (in 1897, the same year as Eduard Buchner's experiment) his belief that the toxin–antitoxin interaction was chemical in nature. Ehrlich writes that

"one must accept that this ability to bind the antibodies can be reduced to the existence of a specific atomic grouping which shows a maximal specific affinity for a particular atomic grouping in the antitoxin complex and inserts easily into it, just like the key inside the lock to resort to a well-known comparison due to Fischer" (Ehrlich, 1897).

In so doing, Ehrlich jettisons the notion of a destruction of the toxin by the specific antitoxin in favor of a temporary neutralization of the one by the other. Paul Ehrlich, however, had 'doubts about the probability of a "true chemical combination" taking place between the drug and the cell' (Parascandola, 1980a) and his doubts were shared by many of his contemporaries.

The story of molecular receptors now reverts to John Langley who, in 1905, became more precise in formulating his concept of the action of drugs:

"(. . .) neither the poisons nor the nervous impulse act directly on the contractile substance of the muscle but on some accessory substance. Since this accessory substance is the recipient of stimuli which it transfers to the contractile material, we may speak of it as the *receptive substance* of the muscle".

And Langley went on to formulate his generalization:

"Thus there is evidence that the majority of substances which are ordinarily supposed to act upon nerve-endings (as nicotine, curare, atropine, pilocarpine, strychnine) act upon the receptive substance of the cells. And as adrenalin, an internal secretion, acts upon receptive substances, it is probable that secretin, thyroidin, and the internal secretion formed by the generative organs, also act on receptive substances, although in these cases the cells may be unconnected with nerve fibres" (Langley, 1905).

Paul Ehrlich also went one step further and started to speak in the early 1900s of receptors. The Croonian Lecture he gave in 1900, at the invitation of the Royal Society in London, was illustrated with diagrams showing the complementary male–female attachment of antibodies (labeled as receptors) to the antigenic side chains on the cell surface (Ehrlich, 1900). By June of 1907 Ehrlich had generalized this notion (Parascandola, 1980a) to the statement that he had

"now formed the opinion that some of the chemically defined substances are attached to the cell by atom groupings that are analogues to toxin receptors; these atom groupings I will distinguish from the toxin-receptors by the name of 'chemoreceptors' ".

Ehrlich's antecedents were clearly organic chemical. During his student years in Breslau and in Strasbourg, he had 'experienced an intense interest in chemistry, especially organic chemistry' (Rubin, 1980). His conceptualization of immunology bore the mark of organic chemistry. Ehrlich himself made explicit this analogy: 'Complex organic compounds have repeatedly been shown to contain groups which confer definite properties upon the whole molecule' (Ehrlich, 1904). He also said (Ehrlich, 1956–1960): 'My real natural endowment lies in the domain of chemistry; and mine is, indeed, (. . .) a kind of visual, three-dimensional [*eine chemisch-plastische*] chemistry'. And Ehrlich's fascination with dyes, from very early on, was legendary: 'Do you remember Ehrlich as a student in Breslau? We laughed at him because he was always running around among us with blue, yellow, red and green fingers' (Temkin, 1950).

Yet another potent contributor to the receptor concept was none other than Svante Arrhenius. Paul Ehrlich and the Danish immunologist T. Madsen had turned to him, in order to elucidate the mechanism of the antigen–antibody interaction (Haurowitz, 1979). More exactly, a controversy had started in 1903 between Ehrlich and Madsen, on one hand, and Arrhenius on the other. Rubin (1980) has excellently identified the nature of the disagreement: differing schools of thought. Ehrlich was committed to his organic chemical point of view, while Arrhenius stuck to rigorous physicochemical concepts (mass law action, rate laws, reversibility, etc.). Arrhenius was moved to write this strong statement in his 1915 book *Quantitative Laws in Biological Chemistry*:

"I am convinced that biological chemistry cannot develop into a real science without the aid of the exact methods offered by physical chemistry. The aversion shown by biochemists, who have in most cases a medical education, to exact methods is very easily understood. They are not acquainted with such elementary notions as 'experimental errors', 'probable errors', and so forth, which are necessary for drawing valid conclusions from experiments. The physical chemists have found that the biochemical theories, which are still accepted in medical circles, are founded on an absolutely unreliable basis and must be replaced by other notions agreeing with the fundamental laws of general chemistry" (Arrhenius, 1915).

Arrhenius published his results, framed in molecular language, in a book appropriately entitled *Immunochemistry* (Arrhenius, 1907). Ehrlich's thinking continued its maturation and shortly thereafter, in 1909, he coined a principle which, for its slogan value, he worded in Latin (Ehrlich 1909; 1913):

'*Corpora non agunt nisi fixata*'

Thus, in the span of a generation (1878–1909), immunological studies produce the notion of

"a chemical complex which could not be dissociated under physiological conditions and, in so doing, they base themselves upon total reduction of the phenomena of specificity to chemical laws (...) the influence of organic chemistry, which associates a specific reaction to a given atomic grouping localized on a molecule, is here determining" (Debru, 1983).

By contrast with the careful, deliberate progression of Paul Ehrlich, Emil Fischer moved like a flash. He is remembered mainly, with respect to enzyme receptors, for his key-and-lock analogy. But his departure point as long as 90 years ago, was a correlation between molecular structure and biological activity (for earlier attempts to relate structure and activity, see Parascandola, 1980a).

Fischer was interested in alcoholic fermentation, and he wanted to ascertain if yeast would differentiate between the various hexoses. He found indeed (see Chapter 6) that yeast would ferment four sugars – fructose, glucose, mannose readily; galactose more sluggishly – as the D-enantiomers only. He pointed out the connection between this observation and a joint structural feature of fructose, glucose, mannose: these hexoses share identical configurations at C-3, C-4 and C-5. From there on, continuing the logical chain of inferences, Emil Fischer proposed that the chemical groups responsible for yeast activity had a configuration complementary to those of the sugar molecules they acted upon. This is where the key-and-lock analogy came in (Fischer, 1894a, b; 1895).

And Fischer was prompt in welcoming Eduard Buchner's cell-free fermentation as (another nail in the coffin of the biological theory of fermentation?) confirming his chemical view of alcoholic fermentation (Fischer, 1898–1899). Emil Fischer had also the foresight to write as follows, on the probable proteic nature of the yeast enzyme(s):

"Even if these substances are not yet known in the pure state, their analogy with proteic matter is so great and their formation from it is so likely that they [the enzymes] have undoubtedly to be considered as molecular objects optically active, and thus asymmetric" (Fischer, 1898-1899).

How prescient!

The next stage in the enzyme-receptor story was provided by J.H. Quastel, then in Cambridge, putting forward 'the concept of active centers to explain enzyme activity (1926; 1927)' and going on to discover (1928) that 'the activity of an enzyme could be inhibited by an analog of the substrate with competition between substrate and analog taking place, an early model being the competition between succinate and malonate for succinic dehydrogenase' (Quastel, 1980).

The concept of an active center was an idea awaiting crystallization and formulation from the mid-1920s on: the comparison between the size of a protein molecule and the size of a small organic substrate molecule made it inevitable, as soon as the proteic nature of enzymes had been established firmly, and as soon as the molecular size of proteins became available from hydrodynamic determinations of the Stokes radius, or from combining the molecular weight determined by The Svedberg with the ultracentrifuge and the mean density.

The next logical step was to define *how* the enzyme binds the substrate in this active center. 'Enzymes are very efficient catalysts, and thus, must stabilize the reactants in the transition state relative to the transition state of the uncatalysed reaction' (Byers, 1978). This idea was proposed as early as 1930 by J.B.S. Haldane (1892-1964), who expressed it in most delightful manner:

"Using Fischer's lock and key simile, the key does not fit the lock perfectly, but exercises a certain strain on it" (Haldane, 1930).

This notion of the transition state being stabilized and immobilized within the active site of the enzyme was given powerful support when Linus Pauling reenacted it (Pauling, 1946; 1948a) and expressed his belief that 'enzymes are molecules that are complementary in structure to the activated complexes of the reactions

they catalyze' (Pauling, 1948b). This brilliantly simple idea continues to be forgotten, and tends to spring up again with each new
scientific generation: 16 years separate its formulation by Haldane
from its reinforcement by Pauling, and then, 20 years or so after
Pauling's statements, it was again re-discovered by Wolfenden.
Richard Wolfenden has exploited this self-evident principle in an
astute and most productive manner, designing transition-state analogs which therefore can serve as potent enzyme inhibitors. Over 60
potential transition-state inhibitors have been reported, some of
which are bound to the enzyme by factors of 10^4–10^5 more strongly
than substrate (Wolfenden 1969; 1972; 1976).

We have almost reached the contemporary view of enzymatic
receptors as clefts which evolutionary pressure has optimized into a
shape congruent and complementary to that of the substrate, which
is steered into position, i.a. by electrostatic forces from the α-helices'
strong dipole moments. That the substrate must be attached in a
tense state akin to the transition state is obvious when one thinks of
it.

Another very obvious (to us) property of the enzyme is that, since
the active site is chiral, it should discriminate between enantiotopic
groups of atoms. Enantiotopic (Mislow and Raban, 1967) groups of
atoms, the reader will recall, are related to one another by symmetry
operations S_n, i.e. by rotation and reflection. Hence they reside in
enantiomeric environments. Examples of enantiotopic groups (in
italics) are the methylenic protons of ethanol H_3C-CH_2-OH and the
methylenecarboxylate groups of citrate:

$$CH_2COOH$$
$$|$$
$$HOOC\text{-}C\text{-}OH$$
$$|$$
$$CH_2COOH$$

In other words, if one considers a tetracoordinate carbon atom of the
type Ca_2bd, bearing two a groups which are the same and another
two groups b and d which are different (from one another, as well as
from the a groups), the plane Cbd is a plane of symmetry: reflection
symmetry interchanges the two a groups, which are thus enantiotopic. A simple criterion to apply is that of isotopic substitution.

The two methylenic hydrogens in methylene chloride CH_2Cl_2 are homotopic, i.e. identical: they cannot be distinguished, whether the environment is achiral or chiral. Replacement of one of these hydrogens by (say) deuterium generates a single stereoisomer $CHDCl_2$. By contrast, replacement of one protium by one deuterium in the methylene group (enantiotopic hydrogens) of ethanol generates CH_3CHDOH, i.e. two stereoisomers (a pair of enantiomers). And, to present a full picture, if one considers the methylenic hydrogens of citrate, adjacent to a chiral center (the central asymmetric carbon atom), replacement of H by D will lead to a pair of diastereoisomers: whereas the two groups $-CH_2COOH$ of citrate are enantiotopic, the two methylenic hydrogens in each of these are diastereotopic.

Entry of this concept into biochemistry, like those of many others discussed in this book, was tardy and not particularly easy! Somewhat as in the case of the proteic nature of enzyme whose recognition was blocked by Richard Willstätter, authority was on the side of obscurity (Krebs, 1941). Yet, there were some observations which could not be easily reconciled with the thinking prevalent at the time. Two groups had investigated in 1941 fixation of labeled carbon dioxide into oxaloacetate and the subsequent metabolism of oxaloacetate to labeled α-ketoglutarate, by liver preparations (Evans and Slotin, 1941; Wood et al., 1941). They found that the α-ketoglutarate product was labeled *solely* in the carboxyl group adjacent to the ketone carbonyl.

The explanation was provided by Ogston in a 1948 letter to *Nature* (Ogston, 1948). In his words,

"some time in 1948, when spending a spare hour looking through journals (mainly with a view to advising my pupils what to read) I came across a paper (Shemin, 1946) which seemed to show that aminomalonic acid could not be an intermediate in the conversion of serine to glycine. I read it with interest and, initially, with consent. Suddenly, something happened. One moment I thought: 'That's neat'; the next: 'But it's wrong'; the next: 'citrate!' It may have taken five seconds, perhaps less. A day or two later I sent off my letter to *Nature*" (Ogston, 1978).

The reader has guessed right: Ogston was trained as a chemist before he moved into biochemistry (Ogston, 1978).

"Ogston's letter stimulated lively discussion among biochemists who found it hard to accept that a symmetrical compound could behave in a non-symmetrical fashion" (Bentley, 1978).

Ogston's own vacillation when reading Shemin's paper was emulated by his readers' vacillation when reading his paper, as recalled by Vennesland:

"In a mood of desperation, I sat down with Ogston's paper and took some stick and ball models to assist my thinking. Ogston's paper didn't help, though I read it several times attentively. I thought he was implying a substrate might stay stuck to the enzyme, and that did not help one bit. While I was staring bleakly at the models, they suddenly told me the answer, but not in words, just in a picture. What happened in that moment of sudden insight seemed to be that I was suddenly graced with the ability to see the asymmetric carbon atom as Van 't Hoff had originally seen it. The feeling I experienced was a curious combination of exhilarating, sweet relief (because there was no contradiction between the two sets of experiments) and total dismay: 'Oh you idiot, why haven't you realized that before'. There were quite a few of us who had experiences similar to mine after the appearance of Ogston's paper. Even though I felt I had grasped the point quite well I found that when I tried to transmit my comprehensions to others it was like the tower of Babel. I distinctly remember hearing myself say 'of course I know it is symmetrical, but it isn't symmetrical' " (Vennesland, 1974).

Of importance in this first-hand account is the use of molecular models, which by 1948 had become of fairly general use in biochemistry (see Chapter 4).

As with many concepts, it is easy a posteriori to find precedents. In Ogston's case, it is important nevertheless to point out that D'Arcy Wentworth Thompson in his classic and influential book *On Growth and Form* had made this general statement, which, even though in a slightly different context, covers also the case illuminated by Ogston's 1948 letter to *Nature*:

"When we find ourselves investigating the forms assumed by chemical compounds under the peculiar circumstances of association with a living body, and when we find these forms to be characteristic or recognisable, and somehow different from those which the same substance is wont to assume under other circumstances, an analogy (. . .) presents itself to our minds between this subject of ours and certain synthetic problems of the organic chemist (. . .) molecular form is an important concept (. . .) Observe that it is only the first beginnings of chemical asymmetry that we need

discover; *for when asymmetry is once manifested, it is not disputed that it will con-*
tinue to 'beget assymmetry' " (D'Arcy, 1942; emphasis added).

Of course, authority reacted immediately after publication of
Ogston's letter to *Nature*: he received, just a few days after publication of his note, 'an excited letter from Sir Hans Krebs then at Shef-
field' (Ogston, 1978); till then Ogston had 'little notion of the size of
the iceberg beneath that tip' (Ogston, 1978)!

Following Ogston's letter (Ogston, 1948), new terminology was
introduced: a carbon atom of type Ca_2bd was labeled as prochiral
(Hanson, 1966) and the two enantiotopic a groups are referred to
nowadays as *pro-R* and *pro-S*, depending on whether elevation of
the group to be named over the other gives rise to a chiral center of R
or S configuration, in the Cahn–Ingold–Prelog sense (Cahn et al.,
1966).

A substantial body of work has derived from this notion of prochi-
rality, the stereospecificity of enzymatic reactions at such prochiral
centers is a valuable tool in understanding enzymatic mechanisms
and their fine tuning by evolutionary selection (Hanson, 1972; Rose,
1972; Hanson and Rose, 1975).

"One of the classic investigations in this area is the determination of the stereospecificity of a series of ketol-isomerase enzymes that was carried out by Rose and co-
workers (Rose and O'Connell, 1960; Rose et al., 1969). (. . .) Glucosephosphate isomerase removes the *pro-R* hydrogen from the C-1 of fructose-6-phosphate while
mannosephosphate isomerase removes the diastereotopic *pro-S* hydrogen:

Glucosephosphate isomerase produces an R center at C-2 of the resulting aldose while
the mannosephosphate isomerase produces an S center in the aldose. This latter
observation emphasizes that the carbonyl group of fructose is a center of prostereoisomerism. The addition of a hydrogen to one face of this trigonal center creates
glucose while addition of a hydrogen to the other face creates mannose. The stereoheterotopic faces of this trigonal center of prostereoisomerism are therefore diastereotopic faces. Following the suggestion of Hanson (1966) such stereoheterotopic
faces are designated as *re* (for *rectus*) and *si* (for *sinister*) based upon the priority

sequence of the three ligands attached to the trigonal center. In this example, glucose-phosphate isomerase adds hydrogen to the *re* face of the carbonyl while mannose-phosphate isomerase must add hydrogen to the *si* face. Rose's group found that the(se) relative stereochemical relationships exist in all ketol-isomerase catalyzed transformations. They also established that under appropriate conditions some intramolecular transfer from C-1 of the ketose to C-2 of the resulting aldose can be detected. These observations are most consistent with the evolution of ketol-isomerases where a single electrophilic group polarizes each of the two carbonyls and aids in the removal and transfer of hydrogen by a single base suprafacially via *cis*-enediol intermediates" (Alworth, 1979):

(B, basic group; A, electron-acceptor group, on the enzyme).

One last touch to bring the enzymatic receptors in the *static* picture to which I have limited myself here for brevity's and simplicity's sake, up to date: Koshland's 'induced fit'. Here it is, in his words:

"The long-standing explanation for enzyme specificity is the classic 'template' or 'lock and key' model of Fischer, Ehrlich et al. In this theory a part of the enzyme surface is considered to be a three-dimensional negative of at least a portion of the substrate. Substrate and enzyme have, prior to interaction, complementary topographies which fit together in the manner of a jigsaw puzzle.

Recently, however, it has been suggested (Koshland, 1958; 1959), that such a fitting is not sufficient to explain enzyme specificity, but rather some, and perhaps all, enzymes undergo a significant change in shape on interaction with their substrates. Not all of the enzyme in its native state would be a simple negative of the substrate and enzyme and substrate could not be paired from simple external contours. If a rough mechanical analogy for this new theory is needed, a 'glove' model might suffice. The glove is not a three-dimensional negative of a hand before the hand is introduced. It may be in any of a number of shapes, but only after introduction of the hand is the close three-dimensional fit obtained. In the enzyme case a substrate is presumed to induce a proper alignment of catalytic groups so that enzyme action ensues. Thus the fitting feature of the template hypothesis is retained, but an added requirement of a necessary induced change in the protein structure is added. With this added feature, a number of phenomena which are, at the least, puzzling and, at the most inexplicable by a template hypothesis can be understood" (Koshland, 1962).

Induced fit ought to be complemented with the concept of allosteric interactions (Monod et al., 1963; 1965; Koshland et al., 1966), but this is exceedingly familiar ground to the present day biochemists. Hence, I shall limit myself to this excellent prurient summary:

"Tickle one sensitive site on their surface - lip or tit - and another sensitive site - prick or cunt - opens up ready for action" (Waddington, 1974).

Allosteric interactions manifest themselves in enzymatic kinetics, the topic of the following Chapter.

REFERENCES

Alworth, W.L. (1979) in Asymmetry in Carbohydrates (Harmon, R.E., Ed.), Marcel Dekker, New York, p. 31.

Arrhenius, S. (1907) Immunochemistry, Macmillan, New York.

Arrhenius, S. (1915) Quantitative Laws in Biological Chemistry, Bells, London.

Bentley, R. (1978) Nature, 276, 673.

Byers, L.D. (1978) J. Theor. Biol., 74, 501.

Cahn, R.S., Ingold, C. and Prelog, IIV (1966) Angew. Chem., 78, 413; Angew. Chem. Int. Ed. Engl., 5, 385.

D'Arcy Wentworth Thompson (1942) On Growth and Form, Cambridge University Press, Cambridge, 2nd ed., pp. 649-651.

Debru, Cl. (1983) L'esprit des Protéines. Histoire et Philosophie Biochimiques, Hermann, Paris.

Ehrlich, P. (1897) Zur Kenntnis der Antitoxinwirkung, in Collected Papers, 1957, II, 29.

Ehrlich, P. (1900) Proc. Roy. Soc. London, B66, 420.

Ehrlich, P. (1904) Boston Med. Surg. J., 150, 443.

Ehrlich, P. (1909) Ber. Deutsch. Chem. Ges., 42, 17.

Ehrlich, P. (1913) Lancet, August 16, 445.

Ehrlich, P. (1956-1960) The Collected Papers of Paul Ehrlich (Himmelweit, P. et al., Eds.), 4 vols., London and New York, No. 7; quoted by Rubin (1980).

Evans Jr., E.A. and Slotin, L. (1941) J. Biol. Chem., 141, 439.

Fischer, E. (1894a) Ber. Chem. Ges. Frankfurt, 27, 2985.

Fischer, E. (1894b) Ber. Chem. Ges. Frankfurt, 27, 3479.

Fischer, E. (1895) Ber. Chem. Ges. Frankfurt, 28, 1429.

Fischer, E. (1898-1899) Hoppe-Seylers Z. Physiol. Chem., 26, 60.

Haldane, J.B.S. (1930) Enzymes, Longmans Green, London, p. 182.

Hanson, K.R. (1966) J. Am. Chem. Soc., 88, 2731.

Hanson, K.R. (1972) Annu. Rev. Plant Physiol., 23, 335.

Hanson, K.R. and Rose, I.A. (1975) Acc. Chem. Res., 8, 1.

Haurowitz, F. (1979) TIBS, N268.

Koshland Jr., D.E. (1958) Proc. Natl. Acad. Sci. USA, 44, 98.

Koshland Jr., D.E. (1959) J. Cellular Comp. Physiol. Suppl., 1, 54, 245.

Koshland Jr., D.E. (1962) in Horizons in Biochemistry (Kasha, M. and Pullmann, B., Eds.), Academic Press, New York, London, pp. 265-283.

Koshland Jr., D.E., Nemethy, G. and Filmer, D. (1966) Biochemistry, 5, 365.

Krebs, H.A. (1941) Nature, 147, 560.

Langley, J.N. (1878) J. Physiol. (London), 1, 339.

Langley, J.N. (1905) J. Physiol. (London), 33, 374.

Luchsinger, (1877) Pflüger's Arch., 15, 482.

Mislow, K. and Raban, M. (1967) in Topics in Stereochemistry (Allinger, N.L. and Eliel, E.L., Eds.), Vol. 1, p. 1.

Monod, J., Changeux, J.P. and Jacob, F. (1963) J. Mol. Biol., 6, 306.

Monod, J., Wyman, J. and Changeux, J.P. (1965) J. Mol. Biol., 12, 88.

Ogston, A.G. (1948) Nature, 162, 963.

Ogston, A.G. (1978) Nature, 276, 676.

Parascandola, J. (1980a) TIPS, 189.

Parascandola, J. (1980b) TIPS, 417.

Pauling, L. (1946) Chem. Eng. News, 24, 1375.

Pauling, L. (1948a) Am. Sci., 36, 51.

Pauling, L. (1948b) Nature, 161, 707.

Quastel, J.H. (1980) TIBS, 199.

Rose, I.A. (1972) Crit. Rev. Biochem., 1, 1.

Rose, I.A. and O'Connell, E.L. (1960) Biochim. Biophys. Acta, 42, 159.

Rose, I.A., O'Connell, E.L. and Mortlock, R.P. (1969) Biochim. Biophys. Acta, 178, 376.

Rubin, L.P. (1980) J. Hist. Med., 35, 397.

Temkin, C.L. (1950) Bull. Hist. Med., 24, 333, quoting Salomonsen, C.J. reminiscences.

Vennesland, B. (1974) Topics Curr. Chem., 48, 39.

Waddington, C.H. (1974) N.Y. Rev. Books, November 28, 4.

Weigert, C. (1896) Verh. Ges. Dt. Naturforsch. Ärzte, 1, 121.

Wolfenden, R. (1969) Nature, 223, 704.

Wolfenden, R. (1972) Acc. Chem. Res., 5, 10.

Wolfenden, R. (1976) Annu. Rev. Biophys. Bioeng., 5, 271.

Wood, H.G., Werkman, C.H., Hemingway, A. and Nier, A.O. (1941) J. Biol. Chem., 139, 483.

Chapter 36

Enzymatic Kinetics

The reader is well familiar with the plot of the velocity of an enzyme-catalyzed reaction as a function of substrate concentration: a fast initial increase curves gradually to a much slower rise leading asymptotically toward a limiting value (V_{max}). The history of enzymatic kinetics could be illustrated by a similar plot – in reverse! Slow progress during the period 1850–1900, then a sudden acceleration at the turn of the century and the regular reaping of important results and concepts with the appearance in succession of the papers of V. Henri, in 1901–1905, A.J. Brown, in 1902, Michaelis and Menten, in 1913, Hitchcock, in 1926, Borsook and Schott, in 1931, and Lineweaver and Burk, in 1934.

I shall rely for the 'prehistory' of enzymatic kinetics upon the account given by Debru (1983). The second half of the 19th century included scattered studies. These go back to the work of Ludwig Wilhelmy (1812–1864), who in 1850 measured the velocity of cane sugar inversion by acids, using polarimetry. He found a logarithmic rate law, as befits a first-order reaction.

Then Adolphe Würtz (1817–1884) in 1881 explained the action of soluble ferments, by their binding incessantly to the substrate, in ever-renewed but fleeting encounters. A brilliant intuition! Accumulation of reaction products explained the gradual slowdown, as the ferment appeared to lose its activity as the reaction proceeded (Mayer, 1882). And in 1890, O'Sullivan and Thompson repeated the Wilhelmy work, but using enzymes to catalyze cane sugar inversion. They found the same logarithmic rate law as he (O'Sullivan and Thompson, 1890).

Then came the transition, with the studies by Victor Henri and by A.J. Brown at the turn of the century. These experiments and the attendant theorizing have a dual basis and significance: they investigate enzymes as molecular objects subject to the same laws of chemistry, such as mass action or Guldberg and Waage, as ordinary chemical entities; this audacity is clearly the result of Emil Fischer's 'key and lock' work of 1894 and of Eduard Buchner's test-tube fermentation of 1897. And these studies place themselves in the framework of the new physical chemistry powerful at the time, due to the influential work of Van 't Hoff and Arrhenius, of Willard Gibbs and Wilhelm Ostwald.

The positive influence of Arrhenius has been praised by Northrop:

"Arrhenius, in his *Quantitative Laws in Biological Chemistry*, showed that all the various peculiarities of enzyme reactions could be found in inorganic reactions, and thus began the application of physical chemistry to vital processes (Arrhenius, 1915). He was violently attacked, especially for attempting to apply chemical theory to immunological reactions, but his viewpoint, of course, was correct, and led to the elucidation of enzyme kinetics by Michaelis and Menten" (Northrop, 1961).

The other name which stands out is that of Wilhelm Ostwald (1853–1932). He was responsible for re-emphasizing the catalytic nature of enzymes and for encouraging chemists to measure their kinetics (Fruton, 1982). He also studied carefully the earlier literature, starting with the Wilhelmy 1850 work which signaled for him the onset of rational attempts at understanding enzymatic catalysis. His influence on this field of enzymatic kinetics was quite beneficial.

Indeed, the modern reader finds the remarkable series of papers Victor Henri published between 1901 and 1905 congenial for their insight and their use of the formalism of elementary kinetics (Henri, 1901–1905). It was Henri who observed that, while the velocity V is proportional to the substrate concentration [S], when [S] is small, at high [S] V becomes nearly independent of [S]. In so doing, Henri sets the record straight: in contradistinction to what his predecessors (Emile Duclaux (1898) especially) had fancied, despite enzymes behaving differently from acids, their action is entirely consistent with the normal chemical laws of mass action and of Guldberg and Waage. Victor Henri thus reduces enzymes to a normal status, just as Emil Fischer and Eduard Buchner had done with their altogether different approaches. Victor Henri also assumes formation of a transient intermediate ES associating enzyme and substrate: in this manner, the corresponding kinetic equations account handsomely for the observations. This intermediate ES has pronounced resemblance with the complex which Emil Fischer had postulated during the previous decade: Victor Henri of course draws attention to this convergence. Victor Henri explains convincingly on this basis retardation by the reaction products: 'the substrate S and the product P compete for combining with the enzyme' (Debru, 1983). Victor Henri thus practices what he preaches. V. Henri was an original thinker. He was opposed to vitalistic speculation, because it was unnecessary and because it stifled research. He had been molded by physical chemistry. He had trained in Wilhelm Ostwald's laboratory in Leipzig, before learning physiology with Dastre. A contemporary of Victor Henri, Brown made related observations, but he did not have Henri's vision nor his formal rigor of conceptualization (Brown, 1902).

Leonor Michaelis (1875-1949) and his co-worker Maud L. Menten, in their classic 1913 paper, improve upon Victor Henri: not so much conceptually, but in the experimental details (constant pH; taking into account mutarotation in the reaction products) and in pushing forward the mathematical formalism. They developed what is henceforth known as the Michaelis–Menten equation for the initial reaction velocity:

$$V = V_{max} \frac{[S]}{[S] + K_m}$$

where $K_m = \dfrac{k_2 + k_3}{k_1}$,

k_1 is the forward and k_2 the reverse rate constant for formation of the ES intermediate, k_3 is the rate constant for decomposition to reaction products of this intermediate. Michaelis and Menten deserve credit for pointing out that the observation of a maximum velocity implies formation of a discrete ES complex (Michaelis and Menten, 1913).

Another 13 years, and Hitchcock pointed out the analogy between

Michaelis and Menten kinetics and the Langmuir adsorption iso-
therm (Hitchcock, 1926). Then came a small stumbling block. Victor Henri, with his usual clarity of mind, had recognized that a catalyst speeds up both the forward and the reverse reaction, the position of the equilibrium remaining unchanged. As he wrote (Henri,
1903) 'this elementary result is of great importance for the study of
diastases [enzymes]'. Could it be checked experimentally, this was
the question raised by Borsook and Schott in their 1931 paper (Borsook and Schott, 1931).

The alarm had been raised by J.B.S. Haldane in his gem of a book,
Enzymes; he discusses 'critically various cases occurring in the literature in which an enzyme appears to affect the equilibrium reached,
or not to accelerate the two opposing reactions to an equal extent'
(Slater, 1981). Haldane goes on, half seriously, half jokingly, to
say:

"If enzymes do not act as mere catalysts, a possibility exists of obtaining supplies of
free energy in defiance of the second law of thermodynamics.
    As we have seen all enzymatic reactions so far studied obey this law. Nevertheless,
if anything analogous to a Maxwell demon exists outside textbooks, it presumably has
about the dimensions of an enzyme molecule, and hence researches which show that
the second law holds in the case of enzyme action possess a very general interest"
(Haldane, 1930).

The following year, 'in the general interest', Henry Borsook and
Herman F. Schott embarked upon such a chase. They did not catch
Maxwell's demon! They measured

"electrometrically the oxidation-reduction potentials, after equilibrium had been
reached, of a mixture of succinate and fumarate (concentration ratios from 9 : 1 to
1 : 9) in the presence of methylene blue and succinate dehydrogenase prepared from
beef heart or beef diaphragm in phosphate buffer. (. . .) the mean value of $\bar{E}$ was
found to be - 0.437. (. . .) after applying various corrections (. . .) to the earlier published data, Borsook and Schott calculated values of $\bar{E}$ at 25°C of - 0.438 V from
Quastel and Whetham's experiments (enzyme from *Escherichia coli*) and - 0.436 V
from Thunberg's (quoted by Lehmann) and - 0.437 V from Lehmann's experiments
(enzyme from horse skeletal muscle).
    *This definitely excluded the possibility that the position of equilibrium is depen-
dent upon the source of the enzyme*" (Slater, 1981; emphasis added).

The only additional kinetic basis enzymology needed was a nice graphic linearization method, which Lineweaver and Burk provided with their double reciprocal plot of 1/V versus 1/[S] (Lineweaver and Burk, 1934).

REFERENCES

Borsook, H. and Schott, H.F. (1931) J. Biol. Chem., 92, 535.

Brown, A.J. (1902) J. Chem. Soc., 73.

Debru, Cl. (1983) L'esprit des Protéines.Histoire et Philosophie Biochimiques, Hermann, Paris.

Duclaux, E. (1898) Ann. Inst. Pasteur, 12, 96.

Fruton, J.S. (1982) J. Clin. Chem. Clin. Biochem., 20, 243.

Haldane, J.B.S. (1930) Enzymes, Longmans Green, London, p. 182.

Henri, V. (1901) C. R. Acad. Sci. Paris, 133, 891.

Henri, V. (1902) C. R. Acad. Sci. Paris, 134, 916.

Henri, V. (1903) Lois Générales de l'Action des Diastases, Hermann, Paris.

Henri, V. (1904) C. R. Soc. Biol., 57, (a) p. 173; (b) p. 385; (c) p. 467.

Henri, V. (1905) C. R. Soc. Biol., 58, 610.

Hitchcock, D.I. (1926) J. Am. Chem. Soc., 48, 2870.

Lineweaver, H. and Burk, D. (1934) J. Am. Chem. Soc., 56, 658.

Mayer, A. (1882) Die Lehre von den chemischen Fermenten oder Enzymologie, Heidelberg.

Michaelis, L. and Menten, M. (1913) Biochem. Z., 49, 333.

Northrop, J.H. (1961) Annu. Rev. Biochem., 30, 1.

O'Sullivan, C. and Thompson, F.N. (1890) J. Chem. Soc., 57, 834.

Slater, E.C. (1981) TIBS, 2, 280.

Chapter 37

X-Ray Studies on Enzymes

Back to our old friend pepsin, which has featured prominently throughout this history of the recognition of enzymes as molecules. We had left it in Northrop's laboratory at Rockefeller, purified and crystallized, shortly before 1930.

Just a few years afterwards, in the spring of 1934, John Philpot was purifying pepsin during a short visit to Uppsala, Sweden. He left his preparation in the refrigerator over the week-end when he went skiing. He was delighted by what he found upon return: handsome crystals, 2 mm long or more, in the shape of hexagonal pyramids. When Glen Millikan, a physiologist from California and Cambridge, visited the laboratory, Philpot could not resist showing him the pepsin crystals. Millikan's comment was: 'I know a man in Cambridge who would give his eyes for those crystals'. Philpot offered him some, and Millikan carried them back to Britain in his coat pocket. 'It was lucky for protein crystallography that Millikan took the crystals in the tube in which they were growing in their mother liquor. This enabled Desmond Bernal' to whom Millikan had brought back the crystals, 'to make his first critical observation: that the crystals lost birefringence when removed from their liquid of crystallization' (Crowfoot-Hodgkin and Riley, 1968). Bernal 'first took a crystal out of its mother liquor (. . .) mounted it and took an X-ray photograph, which showed nothing but vague blackening. He then thought of drawing crystals in their mother liquor into thin-walled capillary tubes of Lindemann glass – fortunately available in the laboratory since he and Helen Megaw had been growing ice crystals in them to study the expansion of ice' (Crowfoot-Hodgkin, 1979). These wet crystals gave individual, rather blurred, x-ray reflections:

"That night, Bernal, full of excitement, wandered about the streets of Cambridge, thinking of the future and how much it might be possible to know about the structure of proteins if the photographs he had just taken could be interpreted in every detail" (Crowfoot-Hodgkin and Riley, 1968).

As soon as Dorothy Crowfoot came back to the lab after a brief absence, she helped Bernal to take many more x-ray photographs on a 3-cm radius cylindrical camera in a series of 5° oscillations: these showed 'hundreds of X-ray reflections, rather large, corresponding with the size of the crystals, and noticeably even in intensity distribution' (Crowfoot-Hodgkin, 1979). The axial ratio Bernal had determined from the crystals under a petrographic microscope was $c/a = 2.3 \pm 0.1$. The cell dimensions were about 67 Å for the a axis, and a multiple of 154 Å for the c axis – a value derived from the axial ratio, since the reflections defining the long c axis were poorly resolved for a number of reasons. These values, published in a note in *Nature* (Bernal and Crowfoot, 1934) come actually very close to the accurate data obtained many years later by Max Perutz and Tom Blundell, with $a = 67.9$ Å and $c = 292$ Å (Crowfoot-Hodgkin, 1979). Unfortunately the original x-ray diffractograms of pepsin have disappeared, 'it seems almost certain that (they) were destroyed in the bombing of Birkbeck College during the war' (Crowfoot-Hodgkin, 1979).

Bernal had found an old paper (Schimper, 1881) on the swelling and shrinking of crystals of cell globulins, which was his basis for determination of the axial ratio by optical microscopy. The large lattice constants published in the letter to *Nature* agreed with the molecular weight of 40,000 Svedberg had determined for pepsin with the ultracentrifuge (see Chapters 1 and 11).

After the initial excitement, which was quite general, work on x-rays of protein crystals did not progress for a while: 'it is interesting that the observations on pepsin crystals were not immediately taken up in other countries although at the time there was considerable interest both in proteins and in x-ray diffraction. The research workers most immediately implicated all turned their efforts in other directions' (Crowfoot-Hodgkin, 1979), for various reasons.

It was only in 1938 that x-ray work was resumed on an enzyme, chymotrypsin (insulin and excelsin had had their x-ray pictures taken shortly before, in 1935 and 1936, respectively). Chymotrypsin was studied, jointly with hemoglobin, by Bernal, Fankuchen and Perutz (1938), in the context of the following all-important question:

"From the taking of the first pepsin photographs the problem before us was clear. How could the thousands of observable X-ray spectra be used in practice to give us a view of the electron density in protein crystals? And from the beginning this question was posed in the form, how could the appropriate phase constants be directly determined?" (Crowfoot-Hodgkin, 1979).

A hint of a solution became apparent from the comparison of the x-ray photographs for the wet and for the dry crystals:

"the dried crystals of chymotrypsin show not only alterations of spacing but also of relative intensities of reflections. If we assume that drying takes place by the removal of water from between protein molecules, studies of these changes provide an opportunity of separating the effects of inter- and intra-molecular scattering. This may make possible the direct Fourier analysis of the molecular structure, once complete sets of reflections are available in different states of hydration" (Bernal et al., 1938).

The difference in the x-ray photographs of protein crystals on their wet or dry state is a matter of considerable importance: it made all the difference, with the original Philpot pepsin crystals, that Bernal had this intuition of genius of keeping the crystals in contact with their mother liquor. To explain why I shall reproduce an exchange between Drs. Gordon and D.C. Hodgkin during the 1978 Symposium on the Origins of Modern Biochemistry, sponsored by the New York Academy of Sciences:

"Dr. Gordon: 'I got the impression that the pepsin crystal when it was wet was so very much better than when it was dry as tried first. Now is it a fortunate thing that Bernal happened to find such an enormous difference in the sharpness of the spots for that one? Or, if he had used another crystalline protein and got some sort of pattern from the dried crystal might he not have bothered to go to the wet one?'
    Dr. Hodgkin: 'I am always glad that Bernal took the first X-ray photographs of pepsin himself a little hurriedly, and saw almost no diffraction effects as others had observed before him, who all had tried taking X-ray photographs of dried crystals. We now know that it is possible to obtain extensive X-ray reflections from most wet

protein crystals and more limited data from most dried protein crystals, including pepsin, provided they are allowed to dry sufficiently slowly. In the wet crystals the molecules are in regular contact with one another, with water filling the spaces between them. If you let the water out, the molecules sag irregularly, order diminishes, and most of the X-ray reflections fade or become diffuse' (. . .)" (Crowfoot-Hodgkin, 1979).

These 1934 studies of Bernal and Crowfoot on pepsin were the pioneering experiments. It took a lot more work, and the advent of computers at the end of World War II, after these heroic beginnings, to make it possible *to solve* the structure of an enzyme. This was lysozyme, whose structure was determined using x-ray diffraction by David C. Phillips and his co-workers, and published during the period 1965–1967 (Blake et al., 1965; 1967a, b; Johnson and Phillips, 1965; Phillips, 1966; 1967; Johnson, 1967). I shall resort extensively to the excellent account published by Phillips in *Scientific American* (Phillips, 1966).

Alexander Fleming thrived on serendipity as Americans depend on vitamins: he discovered lysozyme in 1922, while suffering from a bad cold. For the heck of it, he had introduced a drop of his nasal mucus into one of his bacterial cultures. He named the active substance 'lysozyme' for its ability to lyse bacterial cells. Lysozyme consists of a single polypeptide chain with 129 residues, cross-linked with four disulfide bridges.

In 1962, about 400 diffraction maxima were used for a low-resolution picture. 'In 1965, after the development of more efficient methods of measurement and computation, an image was calculated on the basis of nearly 10 000 diffraction maxima, which resolved features separated by two angstroms' (Phillips, 1966). The interest in this structural determination was what the x-ray could reveal about the detailed mechanism of enzymatic activity for this molecule, lysozyme.

The normal substrate of lysozyme is a bacterial cell-wall polysaccharide made of alternating N-acetylglucosamine and N-acetylmuramic acid units. Lysozyme breaks the ether bridges between C-1 of the latter and C-4 of the former.

Phillips and his co-workers took advantage of an observation which Martin Wenzel in Berlin had made: N-acetylglucosamine is a competitive inhibitor of the enzyme's activity. Hence they studied the structure of the *stable* complex which tri-*N*-acetylglucosamine (tri-NAG) happens to form with lysozyme. Tri-NAG fills only half of the cleft, which suggests that 'more sugar residues, filling the remainder of the cleft, are required for the formation of a reactive enzyme-substrate complex' (Phillips, 1966). Detailed examination of the structure suggested the mechanism for bond cleavage which operates, in the functioning complex, when the equivalent of an hexasaccharide sits inside the cleft. The bond broken is that between the fourth and fifth sugars (labeled D and E), with sugar residue D

"somewhat distorted from its usual conformation. (. . .) (then glutamic acid) residue 35 transfers (. . .) a hydrogen ion (H^+) to the glycosidic oxygen, thus bringing about cleavage of the bond between that oxygen and carbon atom 1 of sugar residue D. This creates a positively charged carbonium ion (C^+) where the oxygen has been severed from carbon atom 1.

(. . .) this carbonium ion is stabilized by its interaction with the negatively charged aspartic side chain of residue 52 until it can combine with a hydroxide ion (OH^-) (. . .) from the surrounding water, thereby completing the reaction.

It is not clear from this description that the distortion of sugar residue D plays any part in the reaction, but in fact it probably does so for a very interesting reason. R.H. Lemieux and G. Huber of the National Research Council of Canada showed (Lemieux and Huber, 1955) that when a sugar molecule such as *N*-acetylglucosamine incorporates a carbonium ion at the carbon-1 position, it tends to take up the same conformation that is forced on ring D by its interaction with the enzyme molecule. This seems to be an example, therefore, of activation of the substrate by distortion the Haldane–Pauling concept, which has long been a favorite idea of enzymologists" (Phillips, 1966).

This is a fitting end to the story of molecular correlates of biological function: the physiological machinery of organisms could now be unraveled, the mechanism of enzymatic action could now be established with similar accuracy and detail that physical organic chemists had earlier shown in their studies of the reactivity of small molecules. Indeed organic chemists (Lemieux and Huber) had interjected precious information at a crucial stage in the elucidation of the lysozyme mechanism.

REFERENCES

Bernal, J.D. and Crowfoot, D. (1934) Nature, 133, 794.

Bernal, J.D., Fankuchen, I. and Perutz, M. (1938) Nature, 141, 523.

Blake, C.C.F., Koenig, D.F., Mair, G.A., North, A.C.T., Phillips, D.C. and Sarma, I.V.R. (1965) Nature, 206, 757.

Blake, C.C.F., Mair, G.A., North, A.C.T., Phillips, D.C. and Sarma, I.V.R. (1967a), Proc. Roy. Soc. (London), B 167, 365.

Blake, C.C.F., Johnson, L.N., Mair, G.A., North, A.C.T., Phillips, D.C. and Sarma, I.V.R. (1967b) Proc. Roy. Soc. (London), B 167, 378.

Crowfoot-Hodgkin, D. (1979) Ann. N.Y. Acad. Sci., 325, 121.

Crowfoot-Hodgkin, D. and Riley, D.P. (1968) in Structural Chemistry and Molecular Biology (Rich, A. and Davidson, N., Eds.), Freeman, San Francisco, p. 15.

Johnson, L.N. (1967) Proc. Roy. Soc. (London), B 167, 439.

Johnson, L.N. and Phillips, D.C. (1965) Nature, 206, 759.

Lemieux, R.H. and Huber, G. (1955) Can. J. Chem., 33, 128.

Phillips, D.C. (1966) Sci. Amer., November, 78.

Schimper, A.F.W. (1881) Z. Krist., 5, 131.

PART V

Hemoglobin and Enzymes

(Chapter 38)

Chapter 38

Tentative Sketch of a Comparative History

We come finally, and by way of a conclusion to this volume of Marcel Florkin's *History of Biochemistry*, to an interesting question: putting side by side the two case histories which have been presented here at some length, does a coherent pattern appear? I was somewhat surprised but delighted that the answer is affirmative.

The periodizations are similar; domination of the field by scientists from a given country occurred at certain times, in both of these lines of research. Technical breakthroughs speeded along both problems to a solution - mostly in a single area, microscopy taken in a general sense. Both histories benefited, from an early stage on, from the input of simple - not to say simplistic - chemical concepts. Furthermore, and this is no surprise, both histories show the deep ambivalence characteristic of all science history: the 'world-view' of a scientist colors his work to an enormous extent! In an historical perspective, all these various contributing paradigms, *themata*, myths, private obsessions, individual or collective research programs, ideologies, institutional factors, etc. appear as *unpredictable*: discovery thrives even on error. To this day, common names such as 'glycine' - or even 'protein' - preserve such rewarding errors from the past! However, the emphasis nowadays - no doubt because of mission-oriented research - is on identification of the epistemological obstacles which the historian, from his God-like stance, can point to. The histories of hemoglobin and of enzymes as molecules had a number of such very real stumbling blocks, which I shall briefly list.

Even if the very title of the present volume (which had been chosen by Marcel Florkin and which I had no good reason not to maintain) embodies an optimistic ideology of progress and thus a fairy-

tale view of history, with good guys and villains, I hold little sympathy with such a manichean point of view. As I shall make clear, I do
not believe that a convincing case can be made for either the positive
effect of what from hindsight appears to be an enlightened research
program (say Jacques Loeb's), or for the negative effect of what we
now hold as an aberrant systematization (say Wolfgang Ostwald's).

First, a brief look at the parallel histories of the emergence of
molecular concepts for respiratory pigments, and for enzymes (Table XIII). The comparison is rewarding; both the similarities and
the contrasts are striking.

TABLE XIII

Approximate chronology for some of the important steps in the acquisition of
molecular information for respiratory pigments and for enzymes

	Hemoglobin	Enzymes
Isolation	ca. 1830 (Berzelius, Chevreul, Lecanu, etc.)	ca. 1840 (pepsin, Th. Schwann) 1897 ('zymase', E. Buchner)
Crystallization	1852–1853 (Funke, Lehmann, Teichmann, etc.)	1929 (Northrop)
Naming	1864 (Felix Hoppe-Seyler)	1876 (Kühne)
Abiotic functioning	ca. 1890 (Christian Bohr)	1897 (Eduard Buchner)
X-ray structure	ca. 1960 (Max Perutz)	1965 (D.C. Phillips)

The first isolation of an enzyme (pepsin) and that of a respiratory
pigment (hemoglobin) occurred roughly during the same period, of
the 1830s–1840s, when many chemists vied for priority in isolation
and characterization of natural products from all possible sources.
This 'paradigm' antedates the rise of the modern chemical industry,
during the second half of the 19th century. It was centered on a quite
different endeavor, synthesis of artificial dyes.

In a number of important aspects, the story of enzymes has lagged
behind that of hemoglobin (Table XIII). It is tempting to single out
vitalism as the main drag force. Should it be held responsible for

abiotic functioning or crystallization of an enzyme, two related discoveries being made relatively late by comparison with corresponding events in the hemoglobin story? This is certainly what a polemical account, such as one of the participants', Northrop's for instance (Northrop, 1961) maintains. A more practical consideration explains why hemoglobin was crystallized so long before enzymes were: it is very abundant, and it will crystallize upon lysis in concentrated solution with the addition of a little acid (Edsall, 1985). But the crystallization of enzymes *was* unduly delayed:

"I do think that the rise of the colloidal school, around 1900-1905, did have an inhibitory effect on the attempt to purify enzymes; the colloidalists believed that proteins and other colloidal constituents of living cells were heterogeneous mixtures and aggregates" (Edsall, 1985).

What I find somewhat surprising is that the parallel events in the hemoglobin story did *not* strike scientists of the time as particularly earthshaking, despite all the archetypal myths and fantasies about blood! The explanation for reductionism being strongly resisted in the enzyme, but not in the hemoglobin story, stems probably from Pasteur's ambivalence about fermentations: he stayed on the fence, neither coming out firmly in favor of a vitalistic interpretation nor in favor of a materialistic position.

As for naming, this is a far from unimportant point: I see naming as the index for the historian to the start of a distinct field. One might claim that 1876 marks the start of enzymology as a separate discipline, and that likewise 1864 is the birth date of chemical hematology. Both these disciplines were initiated in Germany, as part of the rise of physiological chemistry, excellently documented in Kohler's book (Kohler, 1982).

This brings up another point, national strengths and weaknesses. A superficial look at the histories of respiratory pigments and enzymes suggests or gives the impression of French scientists leading the pack till about 1830, with German scientists then taking over till the end of the nineteenth century, when the contributions of British and American biochemists came to the fore. Furthermore, it is no accident if such apportioning of national pride to scientific successes became deliberate from the beginning of the nineteenth century. It ran hand in hand with the rise of European nationalisms and of economic rivalries between European states. It would make an interesting study in intellectual history to compare political nationalism and scientific chauvinism, as stemming both from a Romantic belief in originality and creativity.

Mary-Jo Nye has devoted a fine recent article to the debunking of such national comparisons (Nye, 1984). She has shown quite convincingly that chauvinistic bias by historians of science is responsible in many cases for ascribing scientific ascendence to one country at a certain time. She has shown that the various 'objective' measurements which have been proposed as proofs for such contentions are fraught with methodological faults. And she has gallantly risen to the defense of the quality of French science, in the last third of the nineteenth century. She feels it has been downgraded without good justification. She argues that this is due, to a far from negligible extent, to positivistic bias on the part of historians of science, who have perhaps been guilty of underestimating the usually more theoretical contributions from French scientists: an example at hand, which has been implicit in part III of this book, is the concept of allosteric interactions, as proposed by Monod, Wyman and Changeux (1965). Perutz, as an experimentalist, could build upon this theoretical base for the molecular explanation of hemoglobin cooperativity in dioxygen binding.

Both the hemoglobin (Part III) and the enzymes story (Part IV) demonstrate the impact of technological innovation. Again the two stories point to a common feature which thereby may have more general meaning: the new techniques, whose introduction gave renewed impetus to the search, made possible improved iconic representations. Improved optical microscopy allowed Cagniard-Latour in the 1830s to spy upon yeast cells happily budding in their sweet medium broth (Ch. 31). Much more recently, in the 1950s, improved computers made it possible to decipher x-ray diffraction patterns at near to atomic resolution, thus enabling crystallographers to build structural models for molecules such as the proteins myoglobin, hemoglobin, and lysozyme.

Chromatography and ultra-centrifugation were very important

technical developments: 'I would name chromatography, of various sorts, as one of the really revolutionary developments' (Edsall, 1985).

However important these technological advances proved to be, I do not consider their role to have been determining. I believe it to have been relatively secondary. The main role was that of concepts, and techniques always remained subsidiary to ideas, at least in the two stories examined here. A case at hand, already mentioned at the beginning of this volume (Ch. 2 and Ch. 3), is R.A. Zsigmondy's ultra-microscope: because he worked within the colloidal paradigm, his impact upon modern biochemistry has remained modest.

One technique has had enormous importance for the advent of modern biochemistry. This is x-ray crystallography. But I wonder if things would not have been considerably different, had it not been for the presence of personalities of genius, highly cultured scientists who have been genuine Renaissance men of our time; I am referring of course to Desmond Bernal and Linus Pauling. Some of their contemporaries, such as W.T. Astbury, became interested also quite early on, in the early 1930s in finding protein structure from the x-ray diffraction pattern. But Bernal and Pauling, by their dedication to the protein problem throughout the years, by their encouragement to their co-workers to tackle such problems (I am thinking here, for example, of Max Perutz and Fankuchen, two of Bernal's students), by their conviction that detailed knowledge of the amino acid component parts was a pre-requisite to elucidation of protein structure, had the main role. Bernal's chance discovery of the importance of wet crystals for production of nice photographs was seminal. When he made this observation on pepsin crystals, protein crystallography was started. This was immediately clear to Linus Pauling:

"I suppose if Bernal had not realised the necessity of keeping the crystals wet, in time someone else would have done so. But very good crystallographers had been trying to get protein crystals to give X-ray photographs for twenty years already without making the observation, taking the step he took. (. . .) Linus Pauling told Guy Dodson that when he saw the pepsin *Nature* letter [Bernal and Crowfoot, 1934] he tried the experiment on haemoglobin crystals - but regarded the problem presented by so many X-ray reflections as too difficult to pursue further at that time" (Crowfoot Hodgkin, 1984).

The key factor, in my opinion, was both Bernal's and Pauling's many-sided rich personalities, pregnant with all sorts of varied interests. This catholicity of interests influenced the whole field of protein structural determinations from the start. It provided the urge to go beyond mere data collection, and to provide explanations, at the molecular level, for the biological role of proteins. Bernal was probably set also in this direction by his work on simpler organic molecules, steroids especially, which put him in contact with some of the leading organic chemists in the 1930s and 1940s, i.e. at the time when the question of reaction mechanisms started being raised by the British school, by Christopher Ingold in particular. Thus, when D.C. Phillips used triNAG as lysozyme inhibitor and was able to elucidate in detail the mechanism of bond cleavage by lysozyme in an hexasaccharide, or when Max Perutz was able to pinpoint the atomic motions of hemoglobin sub-units relative to one another, and could see the iron atom diving into its porphyrin hole – they were the heirs to an intellectual tradition of transcending one's tool which went back to the early 1930s and to the beneficial, long-lasting influence of Desmond Bernal and Linus Pauling.

Most probably, such long-range vision enabled the attendant enormous investment in time and effort to be made:

"(. . .) the protein crystallographers have immersed themselves for years in some problem, seeing the hoped-for solution perhaps years ahead" (Kendrew, 1968).

I borrow also from the same interesting paper by John C. Kendrew his insider description of the mental make-up of the x-ray crystallographers who ushered in molecular biology:

"[the] cast of mind (. . .) needed [consists in]: the capacity for tedious experiments in the chemical laboratory, for collecting X-ray data over months and years, and for conducting lengthy computations, all without any immediate reward in the shape of new biological insights; and at the end of it all a complex and irregular structure about which no easy generalizations can be made – at least up to the present time" (Kendrew, 1968).

Another frame of mind was also important. Both of the histories I am pitting in parallel here benefited greatly from the injection of simple chemical concepts. Reductionist thinking about hemoglobin finds its root in the 1777 analogy by Lavoisier between oxygenation of blood and O_2 binding to heavy metals such as Hg and Pb. Likewise, it was a stroke of genius by Berzelius to recognize from the start enzymes as catalysts, in 1835-1836. Hemoglobin was once more ahead of enzymes in its acceptance as an ordinary molecule subject to the normal physicochemical laws, when Gustav Hüfner in 1883 determined its molecular weight on the basis of the elemental composition. For enzymes, the comparable advance came in the early 1900s, primarily under the influence of Svante Arrhenius, with the kinetic studies of Victor Henri between 1901 and 1905 and with the physicochemical measurements by Arrhenius himself, published in his 1907 book; the premises in this school of thought were that enzymes are molecular systems subject to normal mass action and Guldberg and Waage laws, just like simpler molecules. The parallel between the two histories, those of hemoglobin and of enzymes, as *conceptual* histories, continues with Lawrence J. Henderson, during the period 1920-1930, treating blood as a physicochemical system. This move was contemporary with the solubility measurements which Northrop was performing in the 1920s, and which stemmed from the earlier work by Jacques Loeb on the osmotic pressure and Donnan equilibrium for solutions of proteins.

If these two histories thrived upon the impact of very simple physicochemical ideas, they both were also subject to the influx of equally simple concepts, albeit of a more mythical nature, whose overall effect can only be adjudged as delaying rather than beneficial.

Such hurdles were hidden in the very language introduced to denote new scientific objects. Mulder's name, 'protein', by implying the existence of just one single and universal protein, stifled research somewhat and for quite a few years. In like manner, Mulder's great contemporary, Justus von Liebig, was undoubtedly influenced in his theorization of enzyme action as an *analogon* of the kinetic theory of gases (also being evolved during this same time of the mid-19th century), by the word 'ferment', with the underlying notion of violent turmoil at the microscopic level.

There were obstacles from mythical thought or beliefs. The myth of blood purity was rampant in the controversy of the 1890s between

Christian Bohr and Gustav Hüfner, when the former reported having isolated and crystallized several hemoglobins from the blood of a single animal species.

There were other oppositions, stemming from the institutional organization of science. Physiology being the province of physicians, their reluctance towards admitting chemists into their fold, except in relatively subservient positions, as documented in Kohler's book (Kohler, 1982), was a significant factor. We have seen it at work against Eduard Buchner during the early stages of his career; and we have noticed its influence again, a generation later, against Lawrence J. Henderson, whose intransigent physicochemical research program could not be swallowed by the Harvard Medical School. And

"Henderson's nomographic analysis simply baffled most of the clinicians. The mathematics, though in some senses quite simple, was novel and therefore bewildering. Also they could not see how to turn it directly to clinical use" (Edsall, 1985).

And there were conceptual obstacles, too. The colloidal paradigm, at the start of the 20th century, was one. Because of it, chemical physiologists were unsure whether the substrate was attached to the enzyme by non-specific physical forces of adsorption or by specific chemical bonds. The same question was asked about the bond between dioxygen and hemoglobin, till Coryell and Pauling gave it an unambiguous experimental answer in 1936.

Another conceptual obstacle has had a more recent impact. This is the widespread influence of mechanistic ideas. Debru thus takes Perutz to task for his springs-and-levers descriptions of the conversion of hemoglobin from the deoxy to the oxygenated form:

"(there is a) great temptation for the crystallographer to spontaneously adopt a mechanistic viewpoint, a temptation to which it turns out he is unable to resist. Thus the representation made by Perutz in 1970 [Perutz, 1970] is fraught with mechanism. Based upon the two different atomic structures for the oxygenated and deoxygenated conformations of hemoglobin, Perutz imagines and justifies an hypothetical mechanism, an ordered sequence of coordinated motions leading from one form to the other. Hence a number of simplistic metaphors, hence a whole mechanics of transmissions: the F helix, for instance, has become a lever. (. . .) the central problem in the Perutz representation is what occurs at the contact between the heme and the globin, and which he describes primarily as a triggering phenomenon. *An initially rudimentary form of mechanics arises there in fact from a partial and naive philosophy.* The mechanistic representation inevitably implies the search for the origin of motion. From an image of its transmission, of its communication by way of ordered deformations, it is natural to seek its source. In the 'natural' search for an hypothetical mechanism, Perutz introduces a somewhat metaphysical question: what is the prime motor? Clearly, the answer to this question is embodied in the question itself: the prime motor is oxygen and there cannot be any other. *The crystallographer has trapped himself inside his representation"* (Debru, 1983; my translation and emphasis).

Granted that all these obstacles had a delaying action upon the emergence of biochemistry as a molecular science, an heroic view of science history might point to the converse, to progressive research programs having had a positive influence. Two such research programs immediately come to mind. The colloidal school of Ostwald was such a research program. I am taking here the attitude that passing a negative value-judgment from hindsight is unwarranted. The Ostwald-directed school of colloidal chemistry was self-explicitly an attempt to explore a new state of matter, and to find its governing laws. By and large, seeking to identify its positive impact upon biochemistry, this research program has had a widespread, lasting influence. However, it is more difficult to identify the resulting successes, they have been much more modest than the ideological penetration. I can think of one such blatant success in our story, the phenomenological treatment of dioxygen binding to hemoglobin provided by A.V. Hill.

Rather surprisingly, if we turn a jaundiced eye upon the lasting achievements of the opposite school of thought, the research program of physicochemical studies associated with the names of Jacques Loeb, Henderson and Cohn, and Sørensen, despite its riding high in the collective psyche of modern biochemists, there also the real achievements are not on a par with the widespread and long-lasting influence of the ideas. This research program, I venture to submit, *was altogether not very successful,* despite some isolated breakthroughs: the main discoveries were made elsewhere, they came from other sectors, primarily from structural chemistry and from crystallography. In many ways, the intellectual battle between Ehrlich and Arrhenius, at the outset of the 20th century, set the poles and the lines of force for the intellectual history to follow. And, if the 'Loeb School' of biophysical chemistry has come, deservedly, to be seen as the magnificent preamble for an extremely important component of present-day biochemistry, and if it is clearly the offspring of the van 't Hoff–Arrhenius school of physical chemistry, still the present-day historian must admit that its achievements have been overshadowed by the successes of the structural organic chemistry research program, that was associated in the 1890s and early 1900s with the names of Emil Fischer and Paul Ehrlich.

Here is the testimony of an active participant:

"Through the work of Sanger, Kendrew and Perutz biochemistry and molecular biology entered into a new world, as far as proteins were concerned. They became definite chemical substances, folding into well defined, though complicated, three dimensional structures. We, in Cohn's laboratory, had always held to the faith that it would turn out this way: not that we envisaged the kind of three dimensional structure that actually existed. But it was Sanger, Kendrew and Perutz who really brought us into the promised land" (Edsall, 1985).

This conclusion is apparent also from consideration, in the parallel histories of hemoglobin and of enzymes, of the appearance of what turned out as the key notions: molecular complementarity (see Ch. 6); correlation between molecular structure and biological activity, where the *mechanism* of action is elucidated, as in D.C. Phillips's work on lysozyme or in M.F. Perutz's work on hemoglobin; the logically anterior notion of transferability of properties from the test-tube to the organism, as represented, inter alia by Lavoisier's equating the red color of blood with the red color of rust, or by Pauling's analogy between hemoglobin and Werner-type complexes of iron(II); and the powerful notion, which surprisingly had very limited success till Aaron Klug's work on icosahedral viruses, of the beautiful order, regularity, and symmetry which nearly everyone expected to find in biomolecules. Three of these four notions are of structural type, and were transferred successfully from organic chemistry to biochemistry; and the fourth notion, that of transferability of biological properties, is also since the time of Liebig and Wöhler, one of the central tenets of organic chemistry – and one which has naturally accreted around itself a structural imagery.

The writer is a physical organic chemist, which explains why this appears a satisfying conclusion to him. However, I cannot help wondering how differently, how much differently a biochemist would have presented the same story. The biochemist I have in mind is, of course, Marcel Florkin. He had already started gathering notes for the writing of this volume.

In June 1979, in Singapore I boarded a flight to Melbourne. I unfolded a copy of *Le Monde*, which I had just bought, to discover in it the announcement of Marcel Florkin's death. He was an old friend. I have thus accepted to write this volume even though I felt unequal to the task. My regret is not having been able to argue with Florkin over it. He surely would have disagreed with some of my statements.

He was strong-minded, he could not stand fools. The catholicity of his interests, his joyful erudition, his conversational talents made him precious. Thus it is with the same teasing affection he gave his friends that I dedicate this book to him.

REFERENCES

Bernal, J.D. and Crowfoot, D. (1934) Nature, 133, 794.

Crowfoot-Hodgkin, D. (1984) Letter of May 1st to Pierre Laszlo.

Debru, Cl. (1983) L'esprit des Protéines. Histoire et Philosophie Biochimiques, Hermann, Paris, p. 295.

Edsall, J.T. (1985) Letter of March 25 to Pierre Laszlo.

Kendrew, J.C. (1968) in Structural Chemistry and Molecular Biology (Rich, A. and Davidson, N., Eds.), Freeman, San Francisco, p. 187.

Kohler, R.E. (1982) From Medical Chemistry to Biochemistry. The Making of a Biomedical Discipline, Cambridge University Press, Cambridge, ch. 2.

Monod, J., Wyman, J. and Changeux, J.P. (1965) J. Mol. Biol., 12, 88.

Northrop, J.H. (1961) Annu. Rev. Biochem., 30, 1.

Nye, M.J. (1984) Isis, 75, 684.

Perutz, M.F. (1970) Nature, 228, 726.

Subject Index

Name Index